An Overview of Orbital Mechanics & Astrodynamics

" The Mathematics of Simulating & Maneuvering Objects in Orbit "

Edited by Paul F. Kisak

Contents

1 Orbital mechanics   **1**
   1.1   History 1
   1.2   Practical techniques 1
        1.2.1   Rules of thumb 1
   1.3   Laws of astrodynamics 3
        1.3.1   Escape velocity 4
        1.3.2   Formulae for free orbits 4
        1.3.3   Circular orbits 4
        1.3.4   Elliptical orbits 5
        1.3.5   Parabolic orbits 7
        1.3.6   Hyperbolic orbits 8
   1.4   Calculating trajectories 9
        1.4.1   Kepler's equation 9
        1.4.2   Conic orbits 10
        1.4.3   The patched conic approximation 10
        1.4.4   The universal variable formulation 10
        1.4.5   Perturbations 10
   1.5   Orbital maneuver 11
        1.5.1   Orbital transfer 11
        1.5.2   Gravity assist and the Oberth effect 12
        1.5.3   Interplanetary Transport Network and fuzzy orbits 13
   1.6   See also 13
   1.7   References 14
   1.8   External links 14
   1.9   Further reading 14

2 Orbit   **16**
   2.1   History 17
   2.2   Planetary orbits 19
        2.2.1   Understanding orbits 21

2.3	Newton's laws of motion		22
	2.3.1	Newton's law of gravitation and laws of motion for two-body problems	24
	2.3.2	Defining gravitational potential energy	24
	2.3.3	Orbital energies and orbit shapes	24
	2.3.4	Kepler's laws	24
	2.3.5	Limitations of Newton's law of gravitation	25
	2.3.6	Approaches to many-body problems	25
2.4	Newtonian analysis of orbital motion		25
2.5	Relativistic orbital motion		28
2.6	Orbital planes		28
2.7	Orbital period		28
2.8	Specifying orbits		28
2.9	Orbital perturbations		29
	2.9.1	Radial, prograde and transverse perturbations	29
	2.9.2	Orbital decay	29
	2.9.3	Oblateness	30
	2.9.4	Multiple gravitating bodies	30
	2.9.5	Light radiation and stellar wind	30
2.10	Astrodynamics		30
2.11	Earth orbits		30
2.12	Scaling in gravity		31
2.13	Patents		31
2.14	See also		31
2.15	References		32
2.16	Further reading		32
2.17	External links		33

3	Kepler orbit		34
3.1	Introduction		34
	3.1.1	Johannes Kepler	34
	3.1.2	Isaac Newton	36
3.2	Simplified two body problem		37
	3.2.1	Keplerian elements	38
3.3	Mathematical solution of the differential equation (1) above		40
	3.3.1	Alternate derivation	42
	3.3.2	Properties of trajectory equation	42
3.4	Some additional formulae		45
3.5	Determination of the Kepler orbit that corresponds to a given initial state		47
3.6	The osculating Kepler orbit		49

3.7	See also	50
3.8	Citations	50
3.9	References	50
3.10	External links	50

4 Orbital state vectors — 51
- 4.1 Frame of reference — 52
- 4.2 Position vector — 52
  - 4.2.1 Derivation — 52
- 4.3 See also — 53

5 Semi-major axis — 54
- 5.1 Ellipse — 55
- 5.2 Hyperbola — 55
- 5.3 Astronomy — 55
  - 5.3.1 Orbital period — 55
  - 5.3.2 Average distance — 56
  - 5.3.3 Energy; calculation of semi-major axis from state vectors — 57
- 5.4 References — 57
- 5.5 External links — 57

6 Orbital eccentricity — 58
- 6.1 Definition — 58
- 6.2 Etymology — 60
- 6.3 Calculation — 60
- 6.4 Examples — 61
- 6.5 Mean eccentricity — 61
- 6.6 Climatic effect — 62
- 6.7 See also — 62
- 6.8 References — 62
- 6.9 External links — 63

7 Orbital inclination — 64
- 7.1 Orbits — 65
  - 7.1.1 Natural and artificial satellites — 65
  - 7.1.2 Exoplanets and multiple star systems — 65
- 7.2 Other meanings — 66
- 7.3 Calculation — 66
- 7.4 See also — 66
- 7.5 References — 66

8 Argument of periapsis — 68
8.1 Calculation — 68
8.2 See also — 69
8.3 References — 70
8.4 External links — 70

9 Longitude of the ascending node — 71
9.1 Calculation from state vectors — 72
9.2 See also — 72
9.3 References — 73

10 Longitude of the periapsis — 74
10.1 Calculation from state vectors — 74
10.2 References — 74
10.3 External links — 74

11 True anomaly — 75
11.1 Formulas — 76
11.1.1 From state vectors — 76
11.1.2 From the eccentric anomaly — 77
11.1.3 Radius from true anomaly — 77
11.2 Notes — 77
11.3 See also — 77
11.4 References — 78

12 Mean anomaly — 79
12.1 Formulas — 79
12.2 See also — 80
12.3 References — 80

13 Eccentric anomaly — 81
13.1 Graphical Representation — 81
13.2 Formulas — 81
13.2.1 Radius and eccentric anomaly — 81
13.2.2 From the true anomaly — 83
13.2.3 From the mean anomaly — 83
13.3 In-line references and notes — 83
13.4 Background references — 84
13.5 See also — 84

14 Epoch (astronomy) — 85

- 14.1 Epoch versus equinox 85
  - 14.1.1 Date-references for coordinate systems 85
  - 14.1.2 Epochs and periods of validity 86
- 14.2 Changing the standard equinox and epoch 87
- 14.3 Specifying an epoch or equinox 87
- 14.4 Besselian years 88
- 14.5 Julian years and J2000 88
- 14.6 Epoch of the day 89
- 14.7 See also 89
- 14.8 References 89
- 14.9 External links 90

15 Orbital node 91
- 15.1 Planes of reference 92
- 15.2 Node distinction 92
- 15.3 Symbols and nomenclature 92
- 15.4 Earth orbit nodes 93
- 15.5 Lunar nodes 93
- 15.6 See also 93
- 15.7 References 93

16 Precession 95
- 16.1 Torque-free 95
- 16.2 Torque-induced 96
  - 16.2.1 Classical (Newtonian) 98
  - 16.2.2 Relativistic 100
- 16.3 Astronomy 100
  - 16.3.1 Axial precession (precession of the equinoxes) 100
  - 16.3.2 Perihelion precession 101
- 16.4 See also 103
- 16.5 References 103
- 16.6 External links 103

17 Orbital station-keeping 104
- 17.1 Station-keeping in low-earth orbit 104
- 17.2 Station-keeping for Earth observation spacecraft 105
- 17.3 Station-keeping in geostationary orbit 105
- 17.4 Station-keeping at libration points 106
- 17.5 See also 106

18 Ground track — 108

- 17.6 References . . . 106
- 17.7 External links . . . 107

18 Ground track — 108
- 18.1 Aircraft ground tracks . . . 109
- 18.2 Satellite ground tracks . . . 109
  - 18.2.1 Direct and retrograde motion . . . 109
  - 18.2.2 Orbital period and ground track . . . 109
  - 18.2.3 Inclination and ground track . . . 110
  - 18.2.4 Argument of perigee and ground track . . . 111
- 18.3 See also . . . 111
- 18.4 References . . . 111
- 18.5 External links . . . 112

19 Orbital maneuver — 113
- 19.1 General . . . 113
  - 19.1.1 Rocket equation . . . 113
  - 19.1.2 Delta-v . . . 113
  - 19.1.3 Delta-v budget . . . 114
  - 19.1.4 Impulsive maneuvers . . . 114
  - 19.1.5 Applying a low thrust over a longer period of time . . . 115
  - 19.1.6 Assists . . . 115
- 19.2 Transfer orbits . . . 116
  - 19.2.1 Hohmann transfer . . . 116
  - 19.2.2 Bi-elliptic transfer . . . 117
  - 19.2.3 Low energy transfer . . . 119
  - 19.2.4 Orbital inclination change . . . 119
  - 19.2.5 Constant Thrust Trajectory . . . 119
- 19.3 Rendezvous and docking . . . 119
  - 19.3.1 Orbit phasing . . . 120
  - 19.3.2 Space rendezvous and docking . . . 120
- 19.4 See also . . . 121
- 19.5 References . . . 121
- 19.6 External links . . . 121

20 Hohmann transfer orbit — 122
- 20.1 Explanation . . . 122
- 20.2 Calculation . . . 122
- 20.3 Example . . . 125

- 20.4 Worst case, maximum delta-v ... 125
- 20.5 Application to interplanetary travel ... 126
- 20.6 Hohmann transfer versus low thrust orbits ... 126
  - 20.6.1 Low-thrust transfer ... 126
  - 20.6.2 Interplanetary Transport Network ... 127
- 20.7 See also ... 127
- 20.8 References ... 127
- 20.9 External links ... 127

21 Delta-v — 128
- 21.1 Definition ... 128
- 21.2 Specific cases ... 128
- 21.3 Orbital maneuvers ... 129
- 21.4 Producing delta-v ... 131
- 21.5 Multiple maneuvers ... 131
- 21.6 Delta-v budgets ... 131
  - 21.6.1 Oberth effect ... 132
  - 21.6.2 Porkchop plot ... 132
  - 21.6.3 Delta-vs around the Solar System ... 132
- 21.7 See also ... 132
- 21.8 References ... 133

22 Bi-elliptic transfer — 135
- 22.1 Calculation ... 135
  - 22.1.1 Delta-v ... 135
  - 22.1.2 Transfer time ... 137
- 22.2 Example ... 137
- 22.3 See also ... 137
- 22.4 References ... 138

23 Gravity assist — 139
- 23.1 Explanation ... 140
- 23.2 Historical origins of the method ... 142
- 23.3 Why gravitational slingshots are used ... 143
- 23.4 Limits to slingshot use ... 144
- 23.5 Timeline of notable examples ... 145
  - 23.5.1 Mariner 10 – first use in an interplanetary trajectory ... 145
  - 23.5.2 Voyager 1 – farthest human-made object ... 145
  - 23.5.3 Galileo – a change of plan ... 146

- 23.5.4 The Ulysses probe changed the plane of its trajectory 146
- 23.5.5 MESSENGER 146
- 23.5.6 The Cassini probe – multiple gravity assists 146
- 23.5.7 Solar Probe+ 146
- 23.5.8 Rosetta – first spacecraft to match orbit with a comet 147
- 23.6 See also 148
- 23.7 References 148
- 23.8 External links 149

24 Lagrangian point 150
- 24.1 History 150
- 24.2 Lagrange points 150
- 24.3 Natural objects at Lagrangian points 153
- 24.4 Mathematical details 153
  - 24.4.1 L_1 154
  - 24.4.2 L_2 155
  - 24.4.3 L_3 156
  - 24.4.4 L_4 and L_5 156
- 24.5 Stability 156
- 24.6 Spaceflight applications 156
  - 24.6.1 Spacecraft at Sun–Earth L_1 157
  - 24.6.2 Spacecraft at Sun–Earth L_2 158
  - 24.6.3 List of missions to Lagrangian points 158
- 24.7 See also 158
- 24.8 Notes 159
- 24.9 References 159
- 24.10 External links 161

25 n-body problem 162
- 25.1 History 162
- 25.2 General formulation 163
- 25.3 Special cases 165
  - 25.3.1 Two-body problem 165
  - 25.3.2 Three-body problem 167
  - 25.3.3 Planetary problem 168
  - 25.3.4 Central configurations 169
  - 25.3.5 *n*-body choreography 170
- 25.4 Analytic approaches 170
  - 25.4.1 Power series solution 170

| 25.4.2   A generalized Sundman global solution 170
         25.4.3   Singularities of the n-body problem 171
   25.5 Simulation 171
         25.5.1   Few bodies 171
         25.5.2   Many bodies 171
         25.5.3   Strong gravitation 172
   25.6 Other n-body problems 172
   25.7 See also 172
   25.8 Notes 173
   25.9 References 175
   25.10 Further reading 177
   25.11 External links 177

26 Kepler's laws of planetary motion   179
   26.1 Comparison to Copernicus 179
   26.2 Nomenclature 181
   26.3 History 181
   26.4 Formulary 181
         26.4.1   First law 181
         26.4.2   Second law 184
         26.4.3   Third law 185
   26.5 Planetary acceleration 185
         26.5.1   Acceleration vector 186
         26.5.2   The inverse square law 187
         26.5.3   Newton's law of gravitation 189
   26.6 Position as a function of time 189
         26.6.1   Mean anomaly, M 190
         26.6.2   Eccentric anomaly, E 191
         26.6.3   True anomaly, θ 192
         26.6.4   Distance, r 193
   26.7 See also 193
   26.8 Notes 193
   26.9 References 193
   26.10 Bibliography 195
   26.11 External links 196

27 Tsiolkovsky rocket equation   197
   27.1 History 197
   27.2 Derivation 198

27.3 Terms of the equation 200
    27.3.1 Delta-v 200
    27.3.2 Mass fraction 201
    27.3.3 Effective exhaust velocity 201
27.4 Applicability 201
27.5 Examples 201
27.6 Stages 202
27.7 Common misconceptions 202
27.8 See also 203
27.9 References 203
27.10 External links 203

28 Vis-viva equation   204
28.1 Equation 204
28.2 Derivation for Elliptic Orbits ($0 \leq$ eccentricity < 1) 204
28.3 Practical applications 206
28.4 References 206

29 Payload fraction   207
29.1 Examples 207
29.2 References 207

30 Propellant mass fraction   208
30.1 Formulation 208
30.2 Significance 209
30.3 See also 209
30.4 References 209

31 Mass ratio   210
31.1 Derivation 210
31.2 See also 211
31.3 References 211

32 Apsis   212
32.1 Mathematical formulae 213
32.2 Terminology 214
    32.2.1 Terminology graph 215
32.3 Perihelion and aphelion of the Earth 215
32.4 Planetary perihelion and aphelion 215
32.5 See also 215

- 32.6 References 216
- 32.7 External links 216

33 Eccentricity vector 217
- 33.1 Calculation 217
- 33.2 See also 217
- 33.3 References 218

34 Non-inclined orbit 219
- 34.1 See also 219

35 Euler angles 220
- 35.1 Proper Euler angles 220
  - 35.1.1 Classic definition 220
  - 35.1.2 Alternative definition 222
  - 35.1.3 Conventions 222
  - 35.1.4 Signs and ranges 223
  - 35.1.5 Geometric derivation 223
- 35.2 Tait–Bryan angles 226
  - 35.2.1 Conventions 226
  - 35.2.2 Alternative names 226
- 35.3 Relationship with physical motions 227
  - 35.3.1 Intrinsic rotations 229
  - 35.3.2 Extrinsic rotations 230
  - 35.3.3 Conversion between intrinsic and extrinsic rotations 230
- 35.4 Gimbal motion relationship 231
  - 35.4.1 Gimbal analogy 231
  - 35.4.2 Intermediate frames 233
- 35.5 Relationship to other representations 233
  - 35.5.1 Rotation matrix 233
  - 35.5.2 Quaternions 234
  - 35.5.3 Geometric algebra 234
- 35.6 Properties 235
- 35.7 Higher dimensions 235
- 35.8 Applications 235
  - 35.8.1 Vehicles and moving frames 235
  - 35.8.2 Crystallographic texture 237
  - 35.8.3 Others 238
- 35.9 See also 239

36 Drag (physics) — 241

- 36.1 Examples of drag — 241
- 36.2 Types of drag — 241
- 36.3 Drag at high velocity — 243
  - 36.3.1 Power — 244
  - 36.3.2 Velocity of a falling object — 244
- 36.4 Very low Reynolds numbers: Stokes' drag — 245
- 36.5 Drag in aerodynamics — 247
  - 36.5.1 Lift-induced drag — 247
  - 36.5.2 Parasitic drag — 247
  - 36.5.3 Power curve in aviation — 247
  - 36.5.4 Wave drag in transonic and supersonic flow — 248
- 36.6 d'Alembert's paradox — 250
- 36.7 See also — 251
- 36.8 Notes — 251
- 36.9 References — 252
- 36.10 External links — 253

37 Theory of relativity — 254

- 37.1 Scope — 255
  - 37.1.1 Two-theory view — 255
- 37.2 On the theory of relativity — 255
- 37.3 Special relativity — 255
- 37.4 General relativity — 257
- 37.5 Experimental evidence — 257
  - 37.5.1 Tests of special relativity — 257
  - 37.5.2 Tests of general relativity — 259
- 37.6 History — 259
- 37.7 Everyday applications — 260
- 37.8 See also — 260
- 37.9 References — 260
- 37.10 Further reading — 261
- 37.11 External links — 262

38 Radiation pressure — 263

(35.10 References — 240; 35.11 Bibliography — 240; 35.12 External links — 240)

- 38.1 Discovery 263
- 38.2 Theory 263
  - 38.2.1 Radiation pressure by absorption (using classical electromagnetism: waves) 264
  - 38.2.2 Radiation pressure by reflection (using particle model: photons) 265
  - 38.2.3 Radiation pressure by emission 265
  - 38.2.4 Moderating factors 266
  - 38.2.5 Compression in a uniform radiation field 267
- 38.3 Solar radiation pressure 267
  - 38.3.1 Pressures of absorption and reflection 267
  - 38.3.2 Radiation pressure perturbations 268
  - 38.3.3 Solar sails 268
- 38.4 Cosmic effects of radiation pressure 269
  - 38.4.1 The early universe 269
  - 38.4.2 Galaxy formation and evolution 269
  - 38.4.3 Clouds of dust and gases 269
  - 38.4.4 Clusters of stars 269
  - 38.4.5 Star formation 269
  - 38.4.6 Stellar planetary systems 270
  - 38.4.7 Stellar interiors 270
  - 38.4.8 Comets 270
- 38.5 Laser applications of radiation pressure 271
- 38.6 See also 272
- 38.7 References 273
- 38.8 Further reading 273

39 Electromagnetism 274
- 39.1 History of the theory 275
- 39.2 Fundamental forces 277
- 39.3 Classical electrodynamics 277
- 39.4 Quantum mechanics 278
  - 39.4.1 Photoelectric effect 278
  - 39.4.2 Quantum electrodynamics 278
  - 39.4.3 Electroweak interaction 278
- 39.5 Quantities and units 279
- 39.6 See also 279
- 39.7 References 280
- 39.8 Further reading 280
  - 39.8.1 Web sources 280
  - 39.8.2 Lecture notes 280

	39.8.3 Textbooks	280
	39.8.4 General references	281
39.9	External links	282

40 Two-line element set — 287

- 40.1 History — 287
- 40.2 Format — 288
- 40.3 Applications — 289
- 40.4 References — 289

41 Proper orbital elements — 290

- 41.1 See also — 291
- 41.2 References — 291
- 41.3 External links — 291
- 41.4 Text and image sources, contributors, and licenses — 294
  - 41.4.1 Text — 294
  - 41.4.2 Images — 303
  - 41.4.3 Content license — 309

Chapter 1

Orbital mechanics

Orbital mechanics or **astrodynamics** is the application of ballistics and celestial mechanics to the practical problems concerning the motion of rockets and other spacecraft. The motion of these objects is usually calculated from Newton's laws of motion and Newton's law of universal gravitation. It is a core discipline within space mission design and control. Celestial mechanics treats more broadly the orbital dynamics of systems under the influence of gravity, including both spacecraft and natural astronomical bodies such as star systems, planets, moons, and comets. Orbital mechanics focuses on spacecraft trajectories, including orbital maneuvers, orbit plane changes, and interplanetary transfers, and is used by mission planners to predict the results of propulsive maneuvers. General relativity is a more exact theory than Newton's laws for calculating orbits, and is sometimes necessary for greater accuracy or in high-gravity situations (such as orbits close to the Sun).

1.1 History

Until the rise of space travel in the twentieth century, there was little distinction between orbital and celestial mechanics, and at the time of Sputnik, the field was called Space Dynamics (ref. the 1961 book by William Thompson of that name). The fundamental techniques, such as those used to solve the Keplerian problem (determining position as a function of time), are therefore the same in both fields. Furthermore, the history of the fields is almost entirely shared.

Johannes Kepler was the first to successfully model planetary orbits to a high degree of accuracy, publishing his laws in 1605. Isaac Newton published more general laws of celestial motion in his 1687 book, *Philosophiæ Naturalis Principia Mathematica*.

1.2 Practical techniques

Further information: List of orbits

1.2.1 Rules of thumb

The following rules of thumb are useful for situations approximated by classical mechanics under the standard assumptions of astrodynamics. The specific example discussed is of a satellite orbiting a planet, but the rules of thumb could also apply to other situations, such as orbits of small bodies around a star such as the Sun.

- Kepler's laws of planetary motion, which can be mathematically derived from Newton's laws, hold strictly only in describing the motion of two gravitating bodies, in the absence of non-gravitational forces, or approximately when the gravity of a single massive body like the Sun dominates other effects:

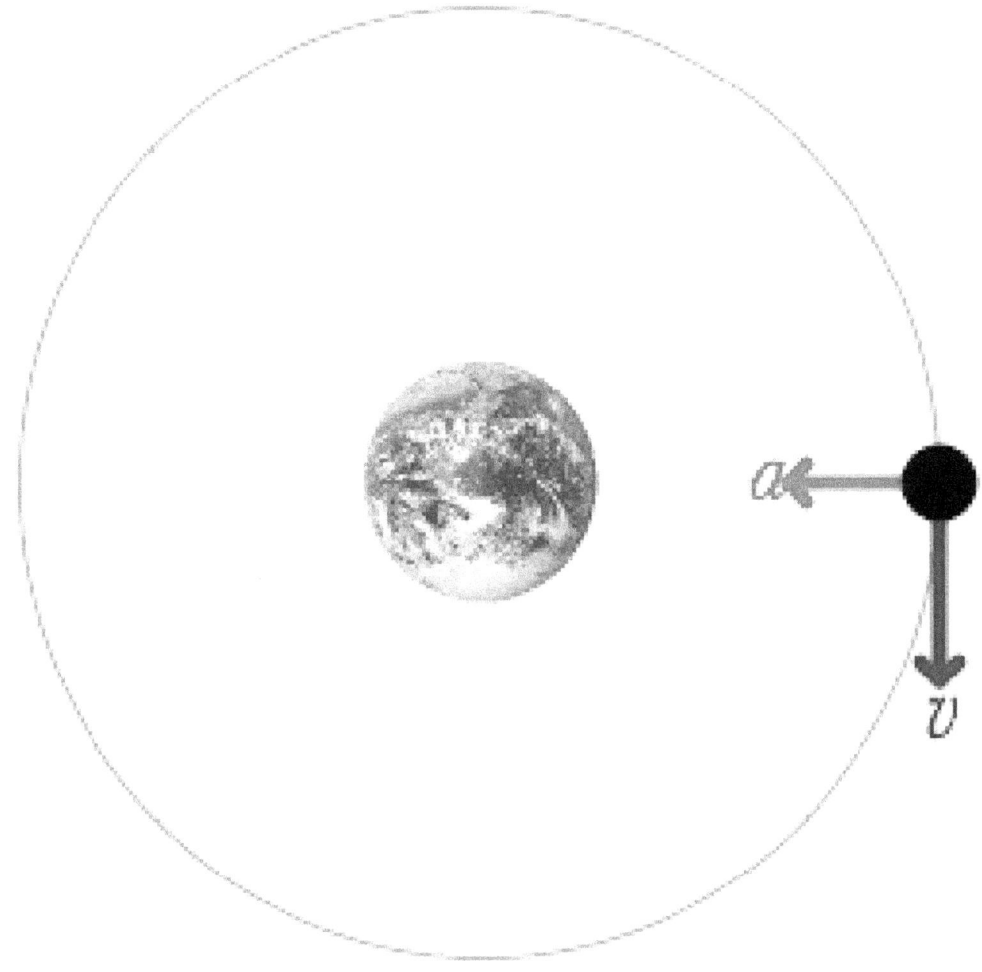

A satellite orbiting the earth has a tangential velocity and an inward acceleration.

- Orbits are elliptical, with the heavier body at one focus of the ellipse. Special cases of this are circular orbits (a circle being simply an ellipse of zero eccentricity) with the planet at the center, and parabolic orbits (which are ellipses with eccentricity of exactly 1, which is simply an infinitely long ellipse) with the planet at the focus.
- A line drawn from the planet to the satellite sweeps out *equal areas in equal times* no matter which portion of the orbit is measured.
- The square of a satellite's orbital period is proportional to the cube of its average distance from the planet.

- Without applying force (such as firing a rocket engine), the height and shape of the satellite's orbit won't change.

- A satellite in a low orbit (or low part of an elliptical orbit) moves more quickly with respect to the surface of the planet than a satellite in a higher orbit (or a high part of an elliptical orbit), due to the stronger gravitational attraction closer to the planet.

- If thrust is applied at only one point in the satellite's orbit, it will return to that same point on each subsequent orbit, though the rest of its path will change. Thus to move from one circular orbit to another, at least two brief applications of thrust are needed.

- From a circular orbit, thrust applied in a direction opposite to the satellite's motion creates an elliptical orbit with a lower periapse (lowest orbital point) at 180 degrees away from the firing point. Thrust applied in the direction of the satellite's motion creates an elliptical orbit with a higher apoapse 180 degrees away from the firing point.

The consequences of the rules of orbital mechanics are sometimes counter-intuitive. For example, if two spacecraft are in the same circular orbit and wish to dock, unless they are very close, the trailing craft cannot simply fire its engines to go faster. This will change the shape of its orbit, causing it to gain altitude and actually slow down relative to the leading craft, missing the target. One approach is to thrust retrograde, or opposite to the direction of motion, and then thrust again to re-circularize the orbit at a lower altitude. Because lower orbits are faster than higher orbits, the trailing craft will begin to catch up. A third firing at the right time will put the trailing craft in an elliptical orbit that intersects the path of the leading craft, approaching from below. Another approach is to thrust both tangentially and radially inward, so that the total centripetal acceleration is greater than gravity alone, allowing it to maintain altitude but gain speed; this is much more expensive in fuel terms however.

To the degree that the standard assumptions of astrodynamics do not hold, actual trajectories will vary from those calculated. For example, simple atmospheric drag is another complicating factor for objects in Earth orbit. These rules of thumb are decidedly inaccurate when describing two or more bodies of similar mass, such as a binary star system (see n-body problem). (Celestial mechanics uses more general rules applicable to a wider variety of situations.) The differences between classical mechanics and general relativity can also become important near large objects like stars.

1.3 Laws of astrodynamics

See also: Laplace–Runge–Lenz vector

The fundamental laws of astrodynamics are Newton's law of universal gravitation and Newton's laws of motion, while the fundamental mathematical tool is his differential calculus.

Every orbit and trajectory outside atmospheres is in principle reversible, i.e., in the space-time function the time is reversed. The velocities are reversed and the accelerations are the same, including those due to rocket bursts. Thus if a rocket burst is in the direction of the velocity, in the reversed case it is opposite to the velocity. Of course in the case of rocket bursts there is no full reversal of events, both ways the same delta-v is used and the same mass ratio applies.

Standard assumptions in astrodynamics include non-interference from outside bodies, negligible mass for one of the bodies, and negligible other forces (such as from the solar wind, atmospheric drag, etc.). More accurate calculations can be made without these simplifying assumptions, but they are more complicated. The increased accuracy often does not make enough of a difference in the calculation to be worthwhile.

Kepler's laws of planetary motion may be derived from Newton's laws, when it is assumed that the orbiting body is subject only to the gravitational force of the central attractor. When an engine thrust or propulsive force is present, Newton's laws still apply, but Kepler's laws are invalidated. When the thrust stops, the resulting orbit will be different but will once again be described by Kepler's laws. The three laws are:

1. The orbit of every planet is an ellipse with the sun at one of the foci.

2. A line joining a planet and the sun sweeps out equal areas during equal intervals of time.

3. The squares of the orbital periods of planets are directly proportional to the cubes of the semi-major axis of the orbits.

1.3.1 Escape velocity

Main article: Escape velocity

The formula for escape velocity is easily derived as follows. The specific energy (energy per unit mass) of any space vehicle is composed of two components, the specific potential energy and the specific kinetic energy. The specific potential energy associated with a planet of mass M is given by

$$-\frac{GM}{r}$$

while the specific kinetic energy of an object is given by

$$\frac{v^2}{2}$$

Since energy is conserved, the total specific orbital energy

$$\frac{v^2}{2} - \frac{GM}{r}$$

does not depend on the distance, r, from the center of the central body to the space vehicle in question. Therefore, the object can reach infinite r only if this quantity is nonnegative, which implies

$$v \geq \sqrt{\frac{2GM}{r}}$$

The escape velocity from the Earth's surface is about 11 km/s, but that is insufficient to send the body an infinite distance because of the gravitational pull of the Sun. To escape the Solar System from a location at a distance from the Sun equal to the distance Sun–Earth, but not close to the Earth, requires around 42 km/s velocity, but there will be "part credit" for the Earth's orbital velocity for spacecraft launched from Earth, if their further acceleration (due to the propulsion system) carries them in the same direction as Earth travels in its orbit.

1.3.2 Formulae for free orbits

Orbits are conic sections, so, naturally, the formula for the distance of a body for a given angle corresponds to the formula for that curve in polar coordinates, which is:

$$r = \frac{p}{1 + e \cos \theta}$$
$$\mu = G(m_1 + m_2)$$
$$p = h^2/\mu$$

where μ is called the gravitational parameter which is G * M, where M is Mass, G is the gravitational constant, m_1 and m_2 are the masses of objects 1 and 2, and h is the specific angular momentum of object 2 with respect to object 1. The parameter θ is known as the true anomaly, p is the semi-latus rectum, while e is the orbital eccentricity, all obtainable from the various forms of the six independent orbital elements.

1.3.3 Circular orbits

All bounded orbits where the gravity of a central body dominates are elliptical in nature. A special case of this is the circular orbit, which is an ellipse of zero eccentricity. The formula for the velocity of a body in a circular orbit at distance r from the center of gravity of mass M is

1.3. LAWS OF ASTRODYNAMICS

$$v = \sqrt{\frac{GM}{r}}$$

where G is the gravitational constant, equal to

6.673 84 × 10^{-11} m^3/(kg·s^2)

To properly use this formula, the units must be consistent; for example, M must be in kilograms, and r must be in meters. The answer will be in meters per second.

The quantity GM is often termed the standard gravitational parameter, which has a different value for every planet or moon in the Solar System.

Once the circular orbital velocity is known, the escape velocity is easily found by multiplying by the square root of 2:

$$v = \sqrt{2}\sqrt{\frac{GM}{r}} = \sqrt{\frac{2GM}{r}}.$$

1.3.4 Elliptical orbits

If 0<e<1, then the denominator of the equation of free orbits varies with the true anomaly θ, but remains positive, never becoming zero. Therefore, the relative position vector remains bounded, having its smallest magnitude at periapsis r_p, which is given by:

$$r_p = \frac{p}{1+e}$$

The maximum value r is reached when θ = 180. This point is called the apoapsis, and its radial coordinate, denoted r_a, is

$$r_a = \frac{p}{1-e}$$

Let 2a be the distance measured along the apse line from periapsis P to apoapsis A, as illustrated in the equation below:

$$2a = r_p + r_a$$

Substituting the equations above, we get:

$$a = \frac{p}{1-e^2}$$

a is the semimajor axis of the ellipse. Solving for p, and substituting the result in the conic section curve formula above, we get:

$$r = \frac{a(1-e^2)}{1+e\cos\theta}$$

Orbital period

Under standard assumptions the orbital period (T) of a body traveling along an elliptic orbit can be computed as:

$$T = 2\pi \sqrt{\frac{a^3}{\mu}}$$

where:

- μ is standard gravitational parameter,
- a is length of semi-major axis.

Conclusions:

- The orbital period is equal to that for a circular orbit with the orbit radius equal to the semi-major axis (a).
- For a given semi-major axis the orbital period does not depend on the eccentricity (See also: Kepler's third law).

Velocity

Under standard assumptions the orbital speed (v) of a body traveling along an **elliptic orbit** can be computed from the Vis-viva equation as:

$$v = \sqrt{\mu \left(\frac{2}{r} - \frac{1}{a}\right)}$$

where:

- μ is the standard gravitational parameter,
- r is the distance between the orbiting bodies.
- a is the length of the semi-major axis.

The velocity equation for a hyperbolic trajectory has either $+\frac{1}{a}$, or it is the same with the convention that in that case a is negative.

Energy

Under standard assumptions, specific orbital energy (ϵ) of elliptic orbit is negative and the orbital energy conservation equation (the Vis-viva equation) for this orbit can take the form:

$$\frac{v^2}{2} - \frac{\mu}{r} = -\frac{\mu}{2a} = \epsilon < 0$$

where:

- v is the speed of the orbiting body,
- r is the distance of the orbiting body from the center of mass of the central body,

1.3. LAWS OF ASTRODYNAMICS

- a is the semi-major axis,
- μ is the standard gravitational parameter.

Conclusions:

- For a given semi-major axis the specific orbital energy is independent of the eccentricity.

Using the virial theorem we find:

- the time-average of the specific potential energy is equal to 2ε
  - the time-average of r^{-1} is a^{-1}
- the time-average of the specific kinetic energy is equal to $-\varepsilon$

1.3.5 Parabolic orbits

If the eccentricity equals 1, then the orbit equation becomes:

$$r = \frac{h^2}{\mu} \frac{1}{1 + \cos\theta}$$

where:

- r is the radial distance of the orbiting body from the mass center of the central body,
- h is specific angular momentum of the orbiting body,
- θ is the true anomaly of the orbiting body,
- μ is the standard gravitational parameter.

As the true anomaly θ approaches 180°, the denominator approaches zero, so that r tends towards infinity. Hence, the energy of the trajectory for which $e=1$ is zero, and is given by:

$$\epsilon = \frac{v^2}{2} - \frac{\mu}{r} = 0$$

where:

- v is the speed of the orbiting body.

In other words, the speed anywhere on a parabolic path is:

$$v = \sqrt{\frac{2\mu}{r}}$$

1.3.6 Hyperbolic orbits

If $e>1$, the orbit formula,

$$r = \frac{h^2}{\mu} \frac{1}{1 + e\cos\theta}$$

describes the geometry of the hyperbolic orbit. The system consists of two symmetric curves. the orbiting body occupies one of them. The other one is its empty mathematical image. Clearly, the denominator of the equation above goes to zero when $\cos\theta = -1/e$, we denote this value of true anomaly

$$\theta_\infty = \cos^{-1}(-1/e)$$

since the radial distance approaches infinity as the true anomaly approaches θ_∞. θ_∞ is known as the *true anomaly of the asymptote*. Observe that θ_∞ lies between 90° and 180°. From the trig identity $\sin^2\theta + \cos^2\theta = 1$ it follows that:

$$\sin\theta_\infty = (e^2-1)^{1/2}/e$$

Energy

Under standard assumptions, specific orbital energy (ϵ) of a hyperbolic trajectory is greater than zero and the orbital energy conservation equation for this kind of trajectory takes form:

$$\epsilon = \frac{v^2}{2} - \frac{\mu}{r} = \frac{\mu}{-2a}$$

where:

- v is the orbital velocity of orbiting body,
- r is the radial distance of orbiting body from central body,
- a is the negative semi-major axis,
- μ is standard gravitational parameter.

Hyperbolic excess velocity

See also: Characteristic energy

Under standard assumptions the body traveling along hyperbolic trajectory will attain in infinity an orbital velocity called hyperbolic excess velocity (v_∞) that can be computed as:

$$v_\infty = \sqrt{\frac{\mu}{-a}}$$

where:

- μ is standard gravitational parameter,

- a is the negative semi-major axis of orbit's hyperbola.

The hyperbolic excess velocity is related to the specific orbital energy or characteristic energy by

$$2\epsilon = C_3 = v_\infty^2$$

1.4 Calculating trajectories

1.4.1 Kepler's equation

One approach to calculating orbits (mainly used historically) is to use Kepler's equation:

$$M = E - \epsilon \cdot \sin E$$

where M is the mean anomaly, E is the eccentric anomaly, and ϵ is the eccentricity.

With Kepler's formula, finding the time-of-flight to reach an angle (true anomaly) of θ from periapsis is broken into two steps:

1. Compute the eccentric anomaly E from true anomaly θ

2. Compute the time-of-flight t from the eccentric anomaly E

Finding the eccentric anomaly at a given time (the inverse problem) is more difficult. Kepler's equation is transcendental in E, meaning it cannot be solved for E algebraically. Kepler's equation can be solved for E analytically by inversion.

A solution of Kepler's equation, valid for all real values of ϵ is:

$$E = \begin{cases} \sum_{n=1}^{\infty} \frac{M^{\frac{n}{3}}}{n!} \lim_{\theta \to 0} \left(\frac{d^{n-1}}{d\theta^{n-1}} \left(\frac{\theta}{\sqrt[3]{\theta - \sin(\theta)}} \right)^n \right), & \epsilon = 1 \\ \sum_{n=1}^{\infty} \frac{M^n}{n!} \lim_{\theta \to 0} \left(\frac{d^{n-1}}{d\theta^{n-1}} \left(\frac{\theta}{\theta - \epsilon \cdot \sin(\theta)} \right)^n \right), & \epsilon \neq 1 \end{cases}$$

Evaluating this yields:

$$E = \begin{cases} x + \frac{1}{60}x^3 + \frac{1}{1400}x^5 + \frac{1}{25200}x^7 + \frac{43}{17248000}x^9 + \frac{1213}{7207200000}x^{11} + \frac{151439}{12713500800000}x^{13} \cdots \mid x = (6M)^{\frac{1}{3}}, & \epsilon = 1 \\ \frac{1}{1-\epsilon}M - \frac{\epsilon}{(1-\epsilon)^4}\frac{M^3}{3!} + \frac{(9\epsilon^2 + \epsilon)}{(1-\epsilon)^7}\frac{M^5}{5!} - \frac{(225\epsilon^3 + 54\epsilon^2 + \epsilon)}{(1-\epsilon)^{10}}\frac{M^7}{7!} + \frac{(11025\epsilon^4 + 4131\epsilon^3 + 243\epsilon^2 + \epsilon)}{(1-\epsilon)^{13}}\frac{M^9}{9!} & \epsilon \neq 1 \end{cases}$$

Alternatively, Kepler's Equation can be solved numerically. First one must guess a value of E and solve for time-of-flight; then adjust E as necessary to bring the computed time-of-flight closer to the desired value until the required precision is achieved. Usually, Newton's method is used to achieve relatively fast convergence.

The main difficulty with this approach is that it can take prohibitively long to converge for the extreme elliptical orbits. For near-parabolic orbits, eccentricity ϵ is nearly 1, and plugging $\epsilon = 1$ into the formula for mean anomaly, $E - \sin E$, we find ourselves subtracting two nearly-equal values, and accuracy suffers. For near-circular orbits, it is hard to find the periapsis in the first place (and truly circular orbits have no periapsis at all). Furthermore, the equation was derived on the assumption of an elliptical orbit, and so it does not hold for parabolic or hyperbolic orbits. These difficulties are what led to the development of the universal variable formulation, described below.

1.4.2 Conic orbits

For simple procedures, such as computing the delta-v for coplanar transfer ellipses, traditional approaches are fairly effective. Others, such as time-of-flight are far more complicated, especially for near-circular and hyperbolic orbits.

1.4.3 The patched conic approximation

Main article: Patched conic approximation

The Hohmann transfer orbit alone is a poor approximation for interplanetary trajectories because it neglects the planets' own gravity. Planetary gravity dominates the behaviour of the spacecraft in the vicinity of a planet and in most cases Hohmann severely overestimates delta-v, and produces highly inaccurate prescriptions for burn timings.

A relatively simple way to get a first-order approximation of delta-v is based on the 'Patched Conic Approximation' technique. One must choose the one dominant gravitating body in each region of space through which the trajectory will pass, and to model only that body's effects in that region. For instance, on a trajectory from the Earth to Mars, one would begin by considering only the Earth's gravity until the trajectory reaches a distance where the Earth's gravity no longer dominates that of the Sun. The spacecraft would be given escape velocity to send it on its way to interplanetary space. Next, one would consider only the Sun's gravity until the trajectory reaches the neighbourhood of Mars. During this stage, the transfer orbit model is appropriate. Finally, only Mars's gravity is considered during the final portion of the trajectory where Mars's gravity dominates the spacecraft's behaviour. The spacecraft would approach Mars on a hyperbolic orbit, and a final retrograde burn would slow the spacecraft enough to be captured by Mars.

The size of the "neighborhoods" (or spheres of influence) vary with radius r_{SOI} :

$$r_{SOI} = a_p \left(\frac{m_p}{m_s} \right)^{2/5}$$

where a_p is the semimajor axis of the planet's orbit relative to the Sun; m_p and m_s are the masses of the planet and Sun, respectively.

This simplification is sufficient to compute rough estimates of fuel requirements, and rough time-of-flight estimates, but it is not generally accurate enough to guide a spacecraft to its destination. For that, numerical methods are required.

1.4.4 The universal variable formulation

To address computational shortcomings of traditional approaches for solving the 2-body problem, the universal variable formulation was developed. It works equally well for the circular, elliptical, parabolic, and hyperbolic cases, the differential equations converging well when integrated for any orbit. It also generalizes well to problems incorporating perturbation theory.

1.4.5 Perturbations

The universal variable formulation works well with the variation of parameters technique, except now, instead of the six Keplerian orbital elements, we use a different set of orbital elements: namely, the satellite's initial position and velocity vectors x_0 and v_0 at a given epoch $t = 0$. In a two-body simulation, these elements are sufficient to compute the satellite's position and velocity at any time in the future, using the universal variable formulation. Conversely, at any moment in the satellite's orbit, we can measure its position and velocity, and then use the universal variable approach to determine what its initial position and velocity *would have been* at the epoch. In perfect two-body motion, these orbital elements would be invariant (just like the Keplerian elements would be).

However, perturbations cause the orbital elements to change over time. Hence, we write the position element as $x_0(t)$ and the velocity element as $v_0(t)$, indicating that they vary with time. The technique to compute the effect of perturbations becomes one of finding expressions, either exact or approximate, for the functions $x_0(t)$ and $v_0(t)$.

The following are some effects which make real orbits differ from the simple models based on a spherical earth. Most of them can be handled on short timescales (perhaps less than a few thousand orbits) by perturbation theory because they are small relative to the corresponding two-body effects.

- Equatorial bulges cause precession of the node and the perigee
- Tesseral harmonics[1] of the gravity field introduce additional perturbations
- Lunar and solar gravity perturbations alter the orbits
- Atmospheric drag reduces the semi-major axis unless make-up thrust is used

Over very long timescales (perhaps millions of orbits), even small perturbations can dominate, and the behaviour can become chaotic. On the other hand, the various perturbations can be orchestrated by clever astrodynamicists to assist with orbit maintenance tasks, such as station-keeping, ground track maintenance or adjustment, or phasing of perigee to cover selected targets at low altitude.

1.5 Orbital maneuver

Main article: Orbital maneuver

In spaceflight, an **orbital maneuver** is the use of propulsion systems to change the orbit of a spacecraft. For spacecraft far from Earth—for example those in orbits around the Sun—an orbital maneuver is called a *deep-space maneuver (DSM)*.

1.5.1 Orbital transfer

Transfer orbits are usually elliptical orbits that allow spacecraft to move from one (usually substantially circular) orbit to another. Usually they require a burn at the start, a burn at the end, and sometimes one or more burns in the middle.

- The Hohmann transfer orbit requires a minimal delta-v.
- A bi-elliptic transfer can require less energy than the Hohmann transfer, if the ratio of orbits is 11.94 or greater,[2] but comes at the cost of increased trip time over the Hohmann transfer.
- Faster transfers may use any orbit that intersects both the original and destination orbits, at the cost of higher delta-v.
- Transfer orbit using Electrical Propulsion or Low Thrust engines requires the initial orbit to be supersynchronous to the final circular orbit and by thrusting continuously in the direction of the velocity at Apogee, the transfer orbit transforms to a circular one. This method however takes much longer to achieve due to the low thrust injected into the orbit [3]

For the case of orbital transfer between non-coplanar orbits, the change-of-plane thrust must be made at the point where the orbital planes intersect (the "node").

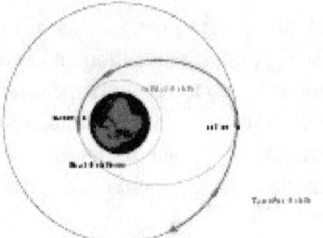

A Hohmann transfer from a low circular orbit to a higher circular orbit

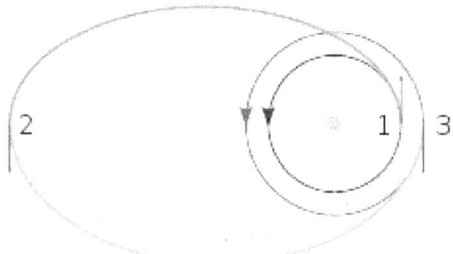

A bi-elliptic transfer from a low circular starting orbit (dark blue), to a higher circular orbit (red)

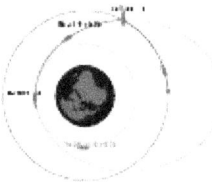

Generic two-impulse elliptical transfer between two circular orbits

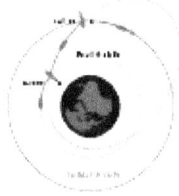

A general transfer from a low circular orbit to a higher circular orbit

An optimal sequence for transferring a satellite from a supersynchronous to a geosynchronous orbit using electric propulsion

1.5.2 Gravity assist and the Oberth effect

In a gravity assist, a spacecraft swings by a planet and leaves in a different direction, at a different speed. This is useful to speed or slow a spacecraft instead of carrying more fuel.

This maneuver can be approximated by an elastic collision at large distances, though the flyby does not involve any physical contact. Due to Newton's Third Law (equal and opposite reaction), any momentum gained by a spacecraft must be lost by the planet, or vice versa. However, because the planet is much, much more massive than the spacecraft, the effect on the planet's orbit is negligible.

The Oberth effect can be employed, particularly during a gravity assist operation. This effect is that use of a propulsion system works better at high speeds, and hence course changes are best done when close to a gravitating body; this can multiply the effective delta-v.

1.5.3 Interplanetary Transport Network and fuzzy orbits

Main article: Interplanetary Transport Network
See also: Low energy transfers

It is now possible to use computers to search for routes using the nonlinearities in the gravity of the planets and moons of the Solar System. For example, it is possible to plot an orbit from high earth orbit to Mars, passing close to one of the Earth's Trojan points. Collectively referred to as the Interplanetary Transport Network, these highly perturbative, even chaotic, orbital trajectories in principle need no fuel beyond that needed to reach the Lagrange point (in practice keeping to the trajectory requires some course corrections). The biggest problem with them is they can be exceedingly slow, taking many years to arrive. In addition launch windows can be very far apart.

They have, however, been employed on projects such as Genesis. This spacecraft visited the Earth-Sun Lagrange L_1 point and returned using very little propellant.

1.6 See also

- Kepler orbit
- Spacecraft propulsion
- Tsiolkovsky rocket equation
- Aerodynamics
- Astrophysics
- Celestial mechanics
- Universal variable formulation
- Chaos theory
- Lagrangian point
- N-body problem
- Orbit
- Orders of magnitude (speed)
- Roche limit
- Canonical units
- Aerospace Engineering
- Mechanical Engineering

1.7 References

[1] "Tesseral Harmonic".

[2] Vallado, David Anthony (2001). *Fundamentals of Astrodynamics and Applications*. Springer. p. 317. ISBN 0-7923-6903-3.

[3] Spitzer, Arnon (1997). *Optimal Transfer Orbit Trajectory using Electric Propulsion*. USPTO.

- Curtis, Howard D., (2009). *Orbital Mechanics for Engineering Students, 2e*. New York: Elsevier. ISBN 978-0-12-374778-5.

- Bate, Roger R.; Mueller, Donald D.; White, Jerry E. (1971). *Fundamentals of Astrodynamics*. New York: Dover Publications. ISBN 0-486-60061-0.

- Sellers, Jerry J.; Astore, William J.; Giffen, Robert B.; Larson, Wiley J. (2004). Kirkpatrick, Douglas H., ed. *Understanding Space: An Introduction to Astronautics* (2 ed.). McGraw Hill. p. 228. ISBN 0-07-242468-0.

- "Air University Space Primer, Chapter 8 - Orbital Mechanics" (PDF). USAF.

1.8 External links

- ORBITAL MECHANICS (Rocket and Space Technology)
- Java Astrodynamics Toolkit

1.9 Further reading

Many of the options, procedures, and supporting theory are covered in standard works such as:

- Bate, R.R.; Mueller, D.D.; White, J.E., (1971). *Fundamentals of Astrodynamics*. Dover Publications, New York. ISBN 978-0-486-60061-1.

- Vallado, D. A. (2001). *Fundamentals of Astrodynamics and Applications* (2nd ed.). Springer. ISBN 978-0-7923-6903-5.

- Battin, R.H. (1999). *An Introduction to the Mathematics and Methods of Astrodynamics*. American Institute of Aeronautics & Ast, Washington, D.C. ISBN 978-1-56347-342-5.

- Chobotov, V.A., ed. (2002). *Orbital Mechanics* (3rd ed.). American Institute of Aeronautics & Ast, Washington, D.C. ISBN 978-1-56347-537-5.

- Herrick, S. (1971). *Astrodynamics: Orbit Determination, Space Navigation, Celestial Mechanics, Volume 1*. Van Nostrand Reinhold, London. ISBN 978-0-442-03370-5.

- Herrick, S. (1972). *Astrodynamics: Orbit Correction, Perturbation Theory, Integration, Volume 2*. Van Nostrand Reinhold, London. ISBN 978-0-442-03371-2.

- Kaplan, M.H. (1976). *Modern Spacecraft Dynamics and Controls*. Wiley, New York. ISBN 978-0-471-45703-9.

- Tom Logsdon (1997). *Orbital Mechanics*. Wiley-Interscience, New York. ISBN 978-0-471-14636-0.

- John E. Prussing & Bruce A. Conway (1993). *Orbital Mechanics*. Oxford University Press, New York. ISBN 978-0-19-507834-3.

- M.J. Sidi (2000). *Spacecraft Dynamics and Control*. Cambridge University Press, New York. ISBN 978-0-521-78780-2.

- W.E. Wiesel (1996). *Spaceflight Dynamics* (2nd ed.). McGraw-Hill, New York. ISBN 978-0-07-070110-6.

- J.P. Vinti (1998). *Orbital and Celestial Mechanics*. American Institute of Aeronautics & Ast, Reston, Virginia. ISBN 978-1-56347-256-5.

- P. Gurfil (2006). *Modern Astrodynamics*. Butterworth-Heinemann. ISBN 978-0-12-373562-1.

Chapter 2

Orbit

This article is about orbits in celestial mechanics, due to gravity. For other uses, see Orbit (disambiguation).

In physics, an **orbit** is the gravitationally curved path of an object around a point in space, for example the orbit of

The International Space Station orbits above Earth.

a planet around the center of a star system, such as the Solar System.[1][2] Orbits of planets are typically elliptical, but unlike the ellipse followed by a pendulum or an object attached to a spring, the central object is at a focal point of the ellipse and not at its center.

Current understanding of the mechanics of orbital motion is based on Albert Einstein's general theory of relativity, which accounts for gravity as due to curvature of space-time, with orbits following geodesics. For ease of calculation, relativity is commonly approximated by the force-based theory of universal gravitation based on Kepler's laws of planetary motion.[3]

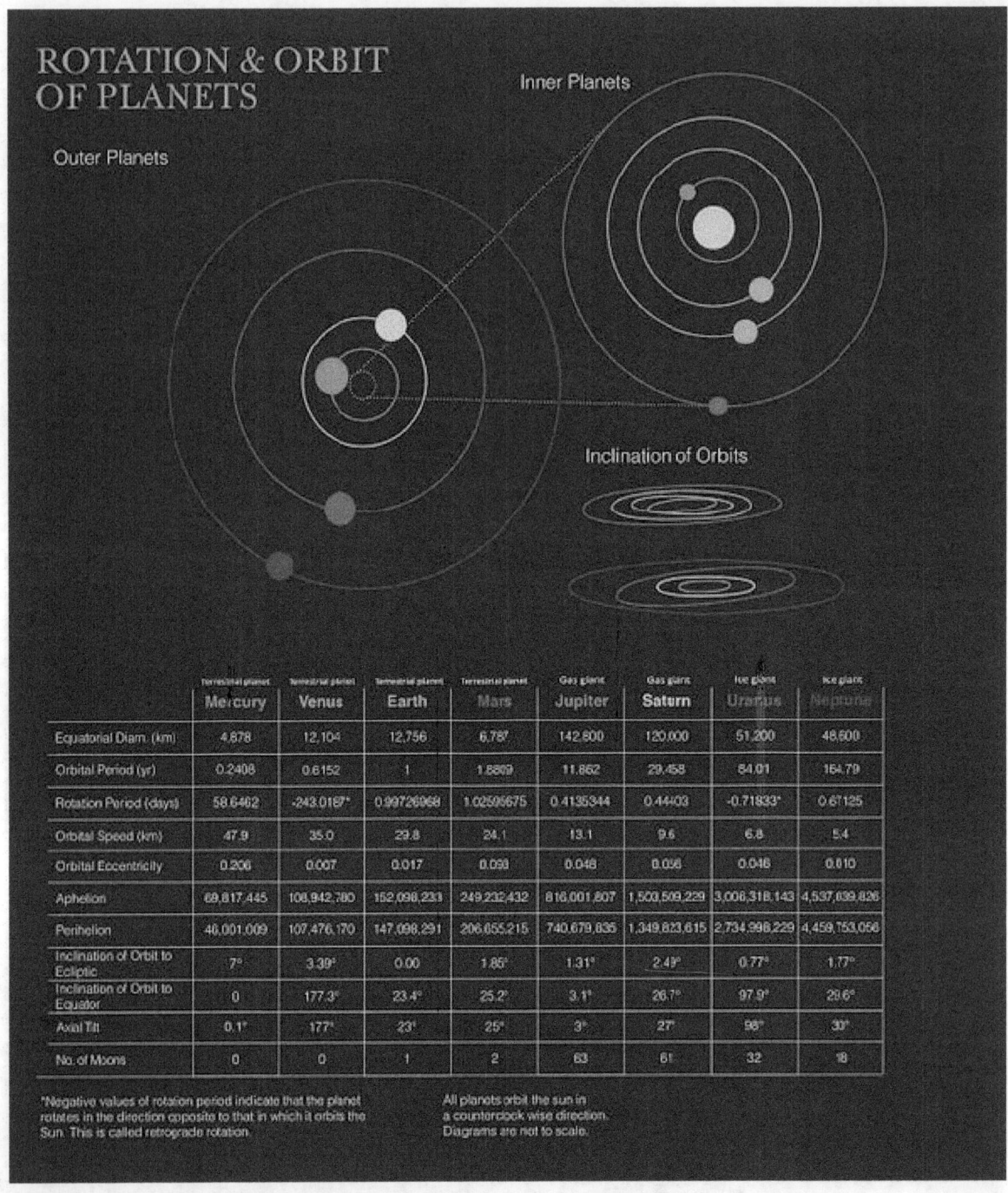

Planetary orbits

2.1 History

Historically, the apparent motions of the planets were first understood geometrically (and without regard to gravity) in terms of epicycles, which are the sums of numerous circular motions.[4] Theories of this kind predicted paths of the planets moderately well, until Johannes Kepler was able to show that the motions of planets were in fact (at least approximately) elliptical motions.[5]

In the geocentric model of the solar system, the celestial spheres model was originally used to explain the apparent motion

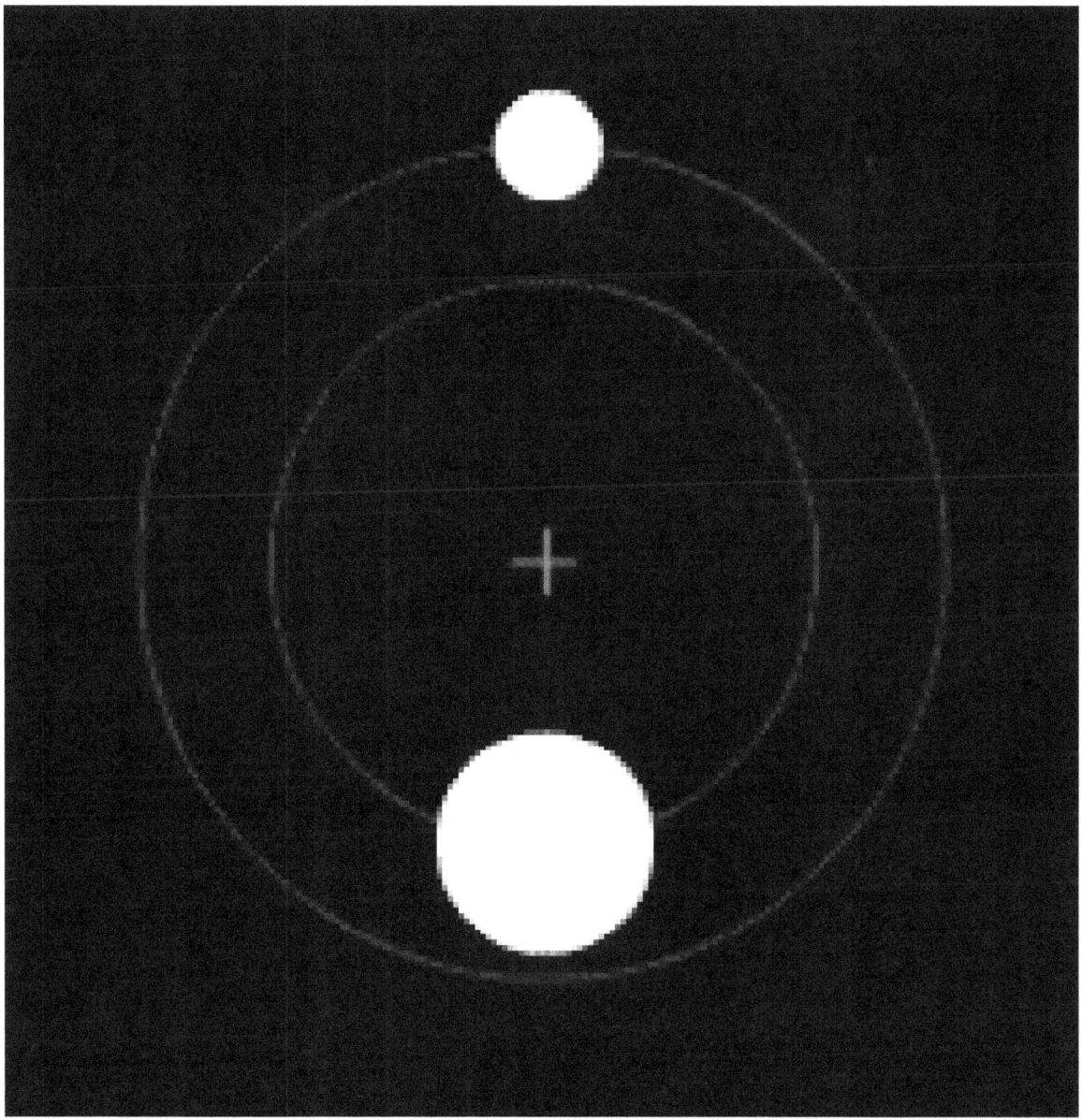

Two bodies of different mass orbiting a common barycenter. The relative sizes and type of orbit are similar to the Pluto–Charon system.

of the planets in the sky in terms of perfect spheres or rings, but after the planets' motions were more accurately measured, theoretical mechanisms such as deferent and epicycles were added. Although it was capable of accurately predicting the planets' position in the sky, more and more epicycles were required over time, and the model became more and more unwieldy.

The basis for the modern understanding of orbits was first formulated by Johannes Kepler whose results are summarised in his three laws of planetary motion. First, he found that the orbits of the planets in our solar system are elliptical, not circular (or epicyclic), as had previously been believed, and that the Sun is not located at the center of the orbits, but rather at one focus.[6] Second, he found that the orbital speed of each planet is not constant, as had previously been thought, but rather that the speed depends on the planet's distance from the Sun. Third, Kepler found a universal relationship between the orbital properties of all the planets orbiting the Sun. For the planets, the cubes of their distances from the Sun are proportional to the squares of their orbital periods. Jupiter and Venus, for example, are respectively about 5.2 and 0.723 AU distant from the Sun, their orbital periods respectively about 11.86 and 0.615 years. The proportionality is seen by the fact that the ratio for Jupiter, $5.2^3/11.86^2$, is practically equal to that for Venus, $0.723^3/0.615^2$, in accord with the relationship.

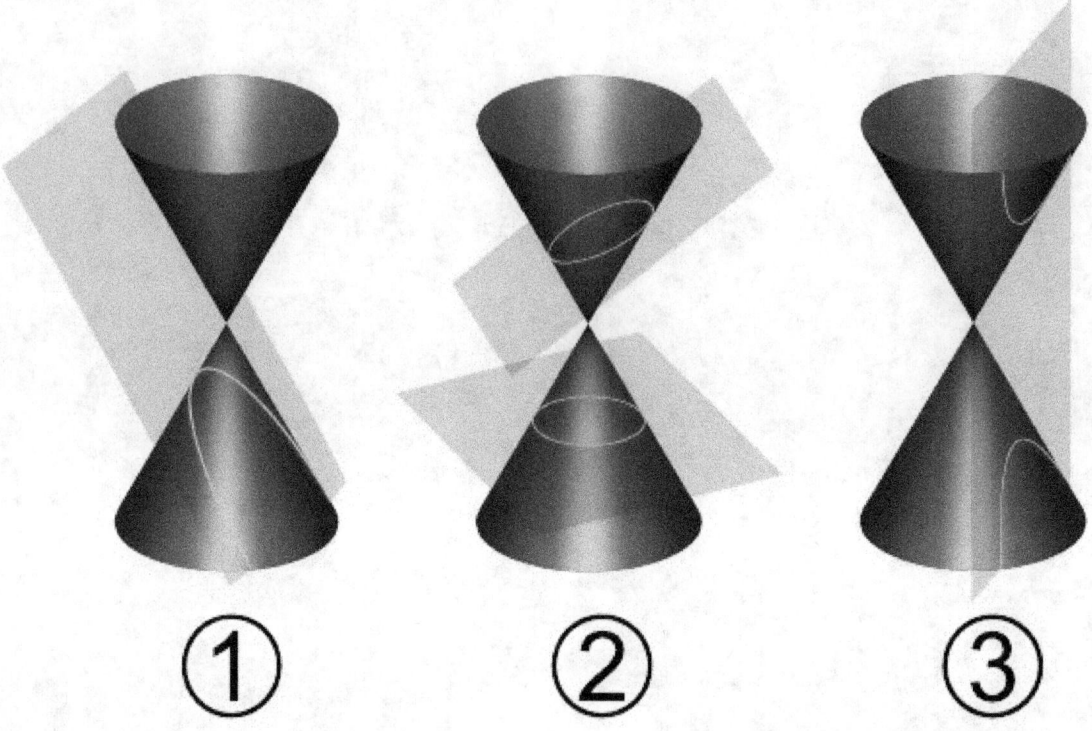

The lines traced out by orbits dominated by the gravity of a central source are conic sections: the shapes of the curves of intersection between a plane and a cone. Parabolic (1) and hyperbolic (3) orbits are escape orbits, whereas elliptical and circular orbits (2) are captive.

Isaac Newton demonstrated that Kepler's laws were derivable from his theory of gravitation and that, in general, the orbits of bodies subject to gravity were conic sections, if the force of gravity propagated instantaneously. Newton showed that, for a pair of bodies, the orbits' sizes are in inverse proportion to their masses, and that the bodies revolve about their common center of mass. Where one body is much more massive than the other, it is a convenient approximation to take the center of mass as coinciding with the center of the more massive body.

Albert Einstein was able to show that gravity was due to curvature of space-time, and thus he was able to remove Newton's assumption that changes propagate instantaneously. In relativity theory, orbits follow geodesic trajectories which approximate very well to the Newtonian predictions. However, there are differences that can be used to determine which theory describes reality more accurately. Essentially all experimental evidence that can distinguish between the theories agrees with relativity theory to within experimental measurement accuracy, but the differences from Newtonian mechanics are usually very small (except where there are very strong gravity fields and very high speeds). The first calculation of the relativistic distortion came from the speed of Mercury's orbit and the strength of the solar gravity field because these are enough to cause Mercury's orbital elements to change.

However, Newton's solution is still used for most short term purposes since it is significantly easier to use.

2.2 Planetary orbits

Within a planetary system, planets, dwarf planets, asteroids and other minor planets, comets, and space debris orbit the barycenter in elliptical orbits. A comet in a parabolic or hyperbolic orbit about a barycenter is not gravitationally bound to the star and therefore is not considered part of the star's planetary system. Bodies which are gravitationally bound to one

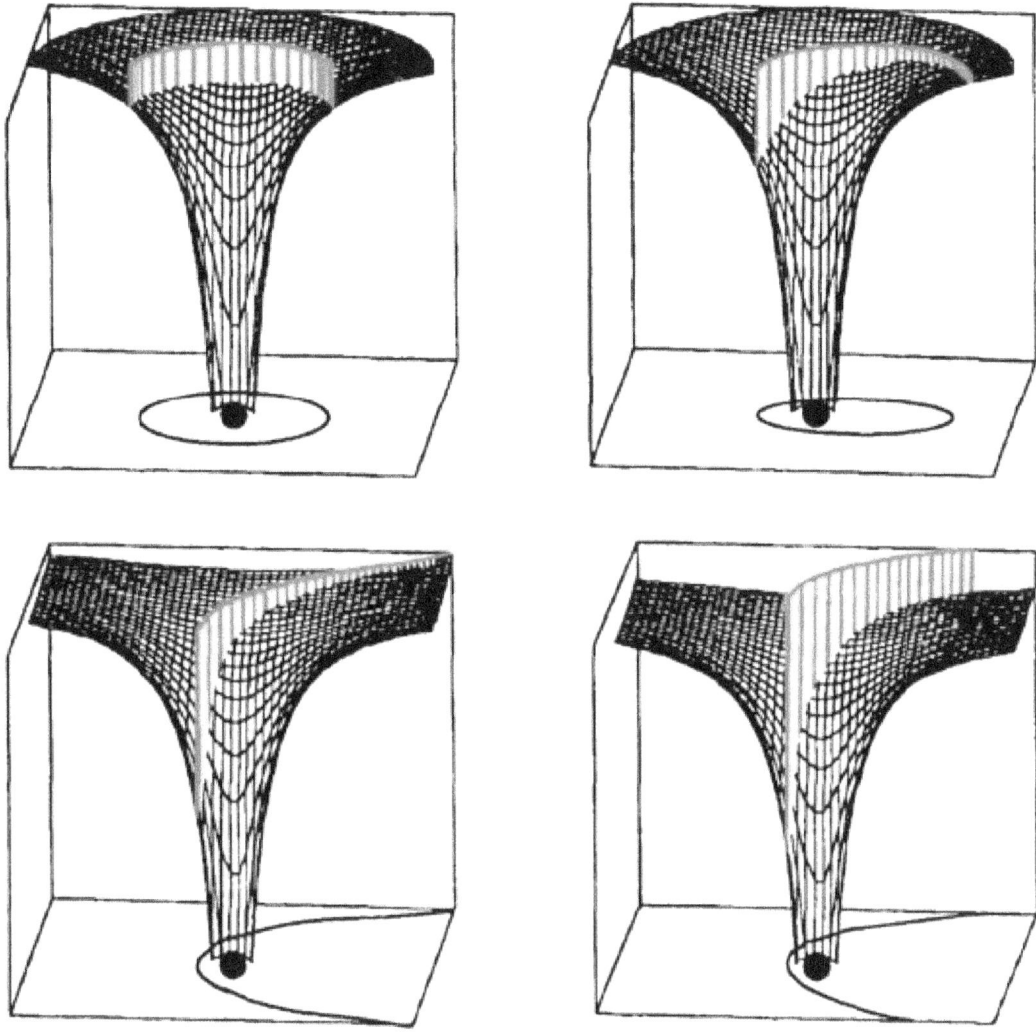

This image shows the four trajectory categories with the gravitational potential well of the central mass's field of potential energy shown in black and the height of the kinetic energy of the moving body shown in red extending above that, correlating to changes in speed as distance changes according to Kepler's laws.

of the planets in a planetary system, either natural or artificial satellites, follow orbits about a barycenter near or within that planet.

Owing to mutual gravitational perturbations, the eccentricities of the planetary orbits vary over time. Mercury, the smallest planet in the Solar System, has the most eccentric orbit. At the present epoch, Mars has the next largest eccentricity while the smallest orbital eccentricities are seen in Venus and Neptune.

As two objects orbit each other, the periapsis is that point at which the two objects are closest to each other and the apoapsis is that point at which they are the farthest from each other. (More specific terms are used for specific bodies. For example, *perigee* and *apogee* are the lowest and highest parts of an orbit around Earth, while *perihelion* and *aphelion* are the closest and farthest points of an orbit around the Sun.)

In the elliptical orbit, the center of mass of the orbiting-orbited system is at one focus of both orbits, with nothing present at the other focus. As a planet approaches periapsis, the planet will increase in speed, or velocity. As a planet approaches apoapsis, its velocity will decrease.

2.2.1 Understanding orbits

There are a few common ways of understanding orbits:

- As the object moves sideways, it falls toward the central body. However, it moves so quickly that the central body will curve away beneath it.

- A force, such as gravity, pulls the object into a curved path as it attempts to fly off in a straight line.

- As the object moves sideways (tangentially), it falls toward the central body. However, it has enough tangential velocity to miss the orbited object, and will continue falling indefinitely. This understanding is particularly useful for mathematical analysis, because the object's motion can be described as the sum of the three one-dimensional coordinates oscillating around a gravitational center.

As an illustration of an orbit around a planet, the Newton's cannonball model may prove useful (see image below). This is a 'thought experiment', in which a cannon on top of a tall mountain is able to fire a cannonball horizontally at any chosen muzzle velocity. The effects of air friction on the cannonball are ignored (or perhaps the mountain is high enough that the cannon will be above the Earth's atmosphere, which comes to the same thing).[7]

If the cannon fires its ball with a low initial velocity, the trajectory of the ball curves downward and hits the ground (A). As the firing velocity is increased, the cannonball hits the ground farther (B) away from the cannon, because while the ball is still falling towards the ground, the ground is increasingly curving away from it (see first point, above). All these motions are actually "orbits" in a technical sense – they are describing a portion of an elliptical path around the center of gravity – but the orbits are interrupted by striking the Earth.

If the cannonball is fired with sufficient velocity, the ground curves away from the ball at least as much as the ball falls – so the ball never strikes the ground. It is now in what could be called a non-interrupted, or circumnavigating, orbit. For any specific combination of height above the center of gravity and mass of the planet, there is one specific firing velocity (unaffected by the mass of the ball, which is assumed to be very small relative to the Earth's mass) that produces a circular orbit, as shown in (C).

As the firing velocity is increased beyond this, elliptic orbits are produced; one is shown in (D). If the initial firing is above the surface of the Earth as shown, there will also be elliptical orbits at slower velocities; these will come closest to the Earth at the point half an orbit beyond, and directly opposite, the firing point.

At a specific velocity called escape velocity, again dependent on the firing height and mass of the planet, an open orbit such as (E) results – a parabolic trajectory. At even faster velocities the object will follow a range of hyperbolic trajectories. In a practical sense, both of these trajectory types mean the object is "breaking free" of the planet's gravity, and "going off into space".

The velocity relationship of two moving objects with mass can thus be considered in four practical classes, with subtypes:

1. **No orbit**

2. **Suborbital trajectories**
    - Range of interrupted elliptical paths

3. **Orbital trajectories (or simply "orbits")**
    - Range of elliptical paths with closest point opposite firing point
    - Circular path
    - Range of elliptical paths with closest point at firing point

4. **Open (or escape) trajectories**

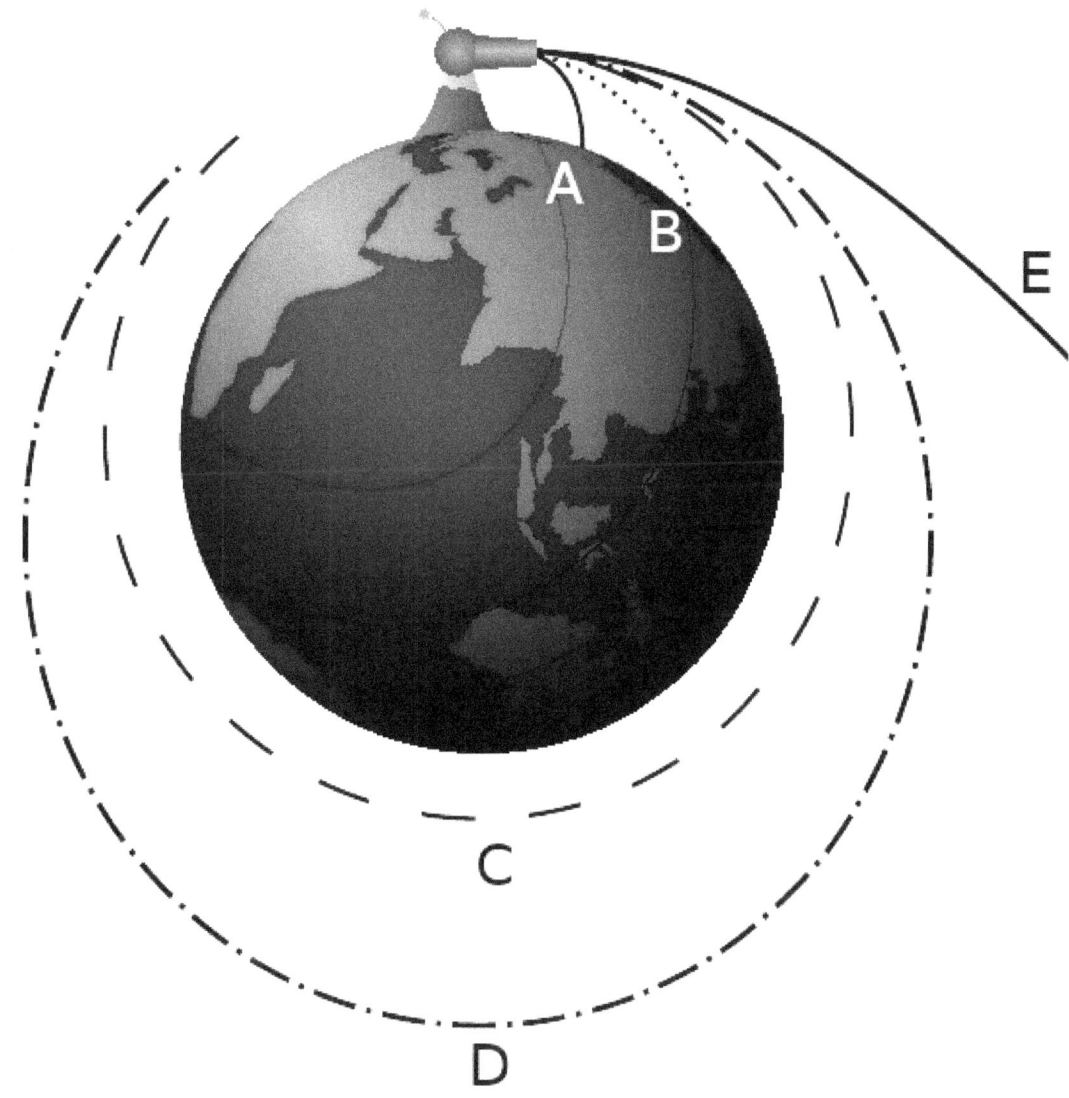

Newton's cannonball, an illustration of how objects can "fall" in a curve

- Parabolic paths
- Hyperbolic paths

It is worth noting that actual rockets launched from earth go vertically at first to get through the air (which causes frictional drag) then slowly pitch over, and end firing the rocket engine parallel to the atmosphere to achieve orbit.

Then, their orbits keep them above the atmosphere. If e.g., an elliptical orbit dips into dense air, the object will lose speed and re-enter (i.e. fall). Occasionally a space craft will intentionally intercept the atmosphere, in an act commonly referred to as an aerobraking maneuver

2.3 Newton's laws of motion

2.3. NEWTON'S LAWS OF MOTION

Conic sections describe the possible orbits (yellow) of small objects around the earth. A projection of these orbits onto the gravitational potential (blue) of the earth makes it possible to determine the orbital energy at each point in space.

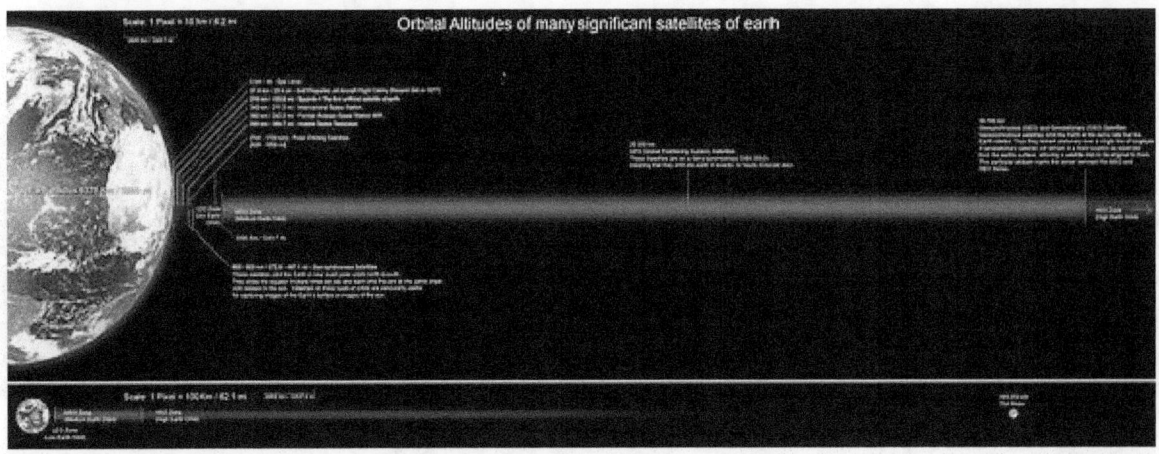

2.3.1 Newton's law of gravitation and laws of motion for two-body problems

In many situations relativistic effects can be neglected, and Newton's laws give a highly accurate description of the motion. The acceleration of each body is equal to the sum of the gravitational forces on it, divided by its mass, and the gravitational force between each pair of bodies is proportional to the product of their masses and decreases inversely with the square of the distance between them. To this Newtonian approximation, for a system of two point masses or spherical bodies, only influenced by their mutual gravitation (the two-body problem), the orbits can be exactly calculated. If the heavier body is much more massive than the smaller, as for a satellite or small moon orbiting a planet or for the Earth orbiting the Sun, it is accurate and convenient to describe the motion in a coordinate system that is centered on the heavier body, and we say that the lighter body is in orbit around the heavier. For the case where the masses of two bodies are comparable, an exact Newtonian solution is still available, and qualitatively similar to the case of dissimilar masses, by centering the coordinate system on the center of mass of the two.

2.3.2 Defining gravitational potential energy

Energy is associated with gravitational fields. A stationary body far from another can do external work if it is pulled towards it, and therefore has gravitational *potential energy*. Since work is required to separate two bodies against the pull of gravity, their gravitational potential energy increases as they are separated, and decreases as they approach one another. For point masses the gravitational energy decreases without limit as they approach zero separation, and it is convenient and conventional to take the potential energy as zero when they are an infinite distance apart, and then negative (since it decreases from zero) for smaller finite distances.

2.3.3 Orbital energies and orbit shapes

With two bodies, an orbit is a conic section. The orbit can be open (so the object never returns) or closed (returning), depending on the total energy (kinetic + potential energy) of the system. In the case of an open orbit, the speed at any position of the orbit is at least the escape velocity for that position; in the case of a closed orbit, always less. Since the kinetic energy is never negative, if the common convention is adopted of taking the potential energy as zero at infinite separation, the bound orbits have negative total energy, parabolic trajectories have zero total energy, and hyperbolic orbits have positive total energy.

An open orbit has the shape of a hyperbola (when the velocity is greater than the escape velocity), or a parabola (when the velocity is exactly the escape velocity). The bodies approach each other for a while, curve around each other around the time of their closest approach, and then separate again forever. This may be the case with some comets if they come from outside the solar system.

A closed orbit has the shape of an ellipse. In the special case that the orbiting body is always the same distance from the center, it is also the shape of a circle. Otherwise, the point where the orbiting body is closest to Earth is the perigee, called periapsis (less properly, "perifocus" or "pericentron") when the orbit is around a body other than Earth. The point where the satellite is farthest from Earth is called apogee, apoapsis, or sometimes apifocus or apocentron. A line drawn from periapsis to apoapsis is the line-of-apsides. This is the major axis of the ellipse, the line through its longest part.

2.3.4 Kepler's laws

Orbiting bodies in closed orbits repeat their paths after a constant period of time. This motion is described by the empirical laws of Kepler, which can be mathematically derived from Newton's laws. These can be formulated as follows:

1. The orbit of a planet around the Sun is an ellipse, with the Sun in one of the focal points of the ellipse. [This focal point is actually the barycenter of the Sun-planet system; for simplicity this explanation assumes the Sun's mass is infinitely larger than that planet's.] The orbit lies in a plane, called the orbital plane. The point on the orbit closest to the attracting body is the periapsis. The point farthest from the attracting body is called the apoapsis. There are also specific terms for orbits around particular bodies; things orbiting the Sun have a perihelion and aphelion,

things orbiting the Earth have a perigee and apogee, and things orbiting the Moon have a perilune and apolune (or periselene and aposelene respectively). An orbit around any star, not just the Sun, has a periastron and an apastron.

2. As the planet moves around its orbit during a fixed amount of time, the line from the Sun to planet sweeps a constant area of the orbital plane, regardless of which part of its orbit the planet traces during that period of time. This means that the planet moves faster near its perihelion than near its aphelion, because at the smaller distance it needs to trace a greater arc to cover the same area. This law is usually stated as "equal areas in equal time."

3. For a given orbit, the ratio of the cube of its semi-major axis to the square of its period is constant.

2.3.5 Limitations of Newton's law of gravitation

Note that while bound orbits around a point mass or around a spherical body with a Newtonian gravitational field are closed ellipses, which repeat the same path exactly and indefinitely, any non-spherical or non-Newtonian effects (as caused, for example, by the slight oblateness of the Earth, or by relativistic effects, changing the gravitational field's behavior with distance) will cause the orbit's shape to depart from the closed ellipses characteristic of Newtonian two-body motion. The two-body solutions were published by Newton in Principia in 1687. In 1912, Karl Fritiof Sundman developed a converging infinite series that solves the three-body problem; however, it converges too slowly to be of much use. Except for special cases like the Lagrangian points, no method is known to solve the equations of motion for a system with four or more bodies.

2.3.6 Approaches to many-body problems

Instead, orbits with many bodies can be approximated with arbitrarily high accuracy. These approximations take two forms:

> One form takes the pure elliptic motion as a basis, and adds perturbation terms to account for the gravitational influence of multiple bodies. This is convenient for calculating the positions of astronomical bodies. The equations of motion of the moons, planets and other bodies are known with great accuracy, and are used to generate tables for celestial navigation. Still, there are secular phenomena that have to be dealt with by post-Newtonian methods.
>
> The differential equation form is used for scientific or mission-planning purposes. According to Newton's laws, the sum of all the forces will equal the mass times its acceleration ($F = ma$). Therefore accelerations can be expressed in terms of positions. The perturbation terms are much easier to describe in this form. Predicting subsequent positions and velocities from initial values corresponds to solving an initial value problem. Numerical methods calculate the positions and velocities of the objects a short time in the future, then repeat the calculation. However, tiny arithmetic errors from the limited accuracy of a computer's math are cumulative, which limits the accuracy of this approach.

Differential simulations with large numbers of objects perform the calculations in a hierarchical pairwise fashion between centers of mass. Using this scheme, galaxies, star clusters and other large objects have been simulated.

2.4 Newtonian analysis of orbital motion

(See also Kepler orbit, orbit equation and Kepler's first law.)

The earth follows an ellipse round the sun. But unlike the ellipse followed by a pendulum or an object attached to a spring, the sun is at a focal point of the ellipse and not at its centre.

The following derives mathematically this orbit as Newton would have done. We start only with the Newtonian law that the gravitational acceleration towards the central body is related to the inverse of the square of the distance between them.

$$A = \frac{F}{m_2} = -\frac{1}{m_2}\frac{Gm_1 m_2}{r^2} = -\frac{\mu}{r^2}$$

where μ is the standard gravitational parameter, in this case Gm_1. We assume that the central body is heavy enough that it can be considered to be stationary and we ignore the more subtle effects of general relativity.

When a pendulum or an object attached to a spring swings in an ellipse, the inward acceleration/force is proportionate to the distance $A = F/m = -kr$. Due to the way vectors add, the component of the force in the $\hat{x}$ or in the $\hat{y}$ directions are also proportionate to the respective components of the distances, $r''_x = A_x = -kr_x$. Hence, the entire analysis can be done separately in these dimensions. This results in the harmonic parabolic equations $x = A\cos(t)$ and $y = B\sin(t)$ of the ellipse. But with the decreasing relationship $A = \mu/r^2$, the dimensions cannot be separated.

The reason that the sun is at the focal point of the orbit's ellipse and not at the centre is because if the object is moving really fast then, as it moves out, the gravitational pull decreases allowing the object to almost escape. If it does not quite escape, then its orbit cycles around first as a parabola. But then when quite far away, it slowly circles back. It picks up speed as it falls, only to boomerang back out into space again.

The location of the orbiting object at the current time t is located in the plane using Vector calculus in polar coordinates both with the standard Euclidean basis and with the polar basis with the origin coinciding with the center of force. Let r be the distance between the object and the center and θ be the angle it has rotated. Let $\hat{x}$ and $\hat{y}$ be the standard Euclidean bases and let $\hat{r} = \cos(\theta)\hat{x} + \sin(\theta)\hat{y}$ and $\hat{\theta} = -\sin(\theta)\hat{x} + \cos(\theta)\hat{y}$ be the radial and transverse polar basis with the first being the unit vector pointing from the central body to the current location of the orbiting object and the second being the orthogonal unit vector pointing in the direction that the orbiting object would travel if orbiting in a counter clockwise circle. Then the vector to the orbiting object is

$$\hat{O} = r\cos(\theta)\hat{x} + r\sin(\theta)\hat{y} = r\hat{r}$$

We use $\dot{r}$ and $\dot{\theta}$ to denote the standard derivatives of how this distance and angle change over time. But we also take the derivative of a vector to see how it changes over time by subtracting its location at time t from that at time $t + \delta t$ and dividing by δt. The result is also a vector. Because our basis vector $\hat{r}$ moves as the object orbits, we start by differentiating it. From time t to $t + \delta t$, the vector $\hat{r}$ keeps its beginning at the origin and rotates from angle θ to $\theta + \dot{\theta}\delta t$ which moves its head a distance $\dot{\theta}\delta t$ in the perpendicular direction $\hat{\theta}$ giving a derivative of $\dot{\theta}\hat{\theta}$.

$$\hat{r} = \cos(\theta)\hat{x} + \sin(\theta)\hat{y}$$

$$\frac{\delta \hat{r}}{\delta t} = \dot{\hat{r}} = -\sin(\theta)\dot{\theta}\hat{x} + \cos(\theta)\dot{\theta}\hat{y} = \dot{\theta}\hat{\theta}$$

$$\hat{\theta} = -\sin(\theta)\hat{x} + \cos(\theta)\hat{y}$$

$$\frac{\delta \hat{\theta}}{\delta t} = \dot{\hat{\theta}} = -\cos(\theta)\dot{\theta}\hat{x} - \sin(\theta)\dot{\theta}\hat{y} = -\dot{\theta}\hat{r}$$

We can now find the velocity and acceleration of our orbiting object.

$$\hat{O} = r\hat{r}$$
$$\dot{\hat{O}} = \frac{\delta r}{\delta t}\hat{r} + r\frac{\delta \hat{r}}{\delta t} = \dot{r}\hat{r} + r[\dot{\theta}\hat{\theta}]$$
$$\ddot{\hat{O}} = [\ddot{r}\hat{r} + \dot{r}\dot{\theta}\hat{\theta}] + [\dot{r}\dot{\theta}\hat{\theta} + r\ddot{\theta}\hat{\theta} - r\dot{\theta}^2\hat{r}]$$
$$= [\ddot{r} - r\dot{\theta}^2]\hat{r} + [r\ddot{\theta} + 2\dot{r}\dot{\theta}]\hat{\theta}$$

The coefficients of $\hat{r}$ and $\hat{\theta}$ are the radial and transverse components of the acceleration. As said, Newton gives that this first due to gravity is $-\mu/r^2$ and the second is zero.

2.4. NEWTONIAN ANALYSIS OF ORBITAL MOTION

Equation (2) can be rearranged using integration by parts.

$$r\ddot{\theta} + 2\dot{r}\dot{\theta} = \frac{1}{r}\frac{d}{dt}\left(r^2\dot{\theta}\right) = 0$$

We can divide through by r because it is not zero unless the orbiting object crashes. Then having the derivative be zero gives that the function is a constant.

which is actually the theoretical proof of Kepler's second law (A line joining a planet and the Sun sweeps out equal areas during equal intervals of time). The constant of integration, h, is the angular momentum per unit mass.

In order to get an equation for the orbit from equation (1), we need to eliminate time.[8] (See also Binet equation.) In polar coordinates, this would express the distance r of the orbiting object from the center as a function of its angle θ. However, it is easier to introduced the auxiliary variable $u = 1/r$ and to express u as a function of θ. Derivatives of r with respect to time may be rewritten as derivatives of u with respect to angle.

$$u = \frac{1}{r}$$

$$\dot{\theta} = \frac{h}{r^2} = hu^2$$

$$\frac{\delta u}{\delta \theta} = \frac{\delta}{\delta t}\left(\frac{1}{r}\right)\frac{\delta t}{\delta \theta} = -\frac{\dot{r}}{r^2\dot{\theta}} = -\frac{\dot{r}}{h}$$

$$\frac{\delta^2 u}{\delta \theta^2} = -\frac{1}{h}\frac{\delta \dot{r}}{\delta t}\frac{\delta t}{\delta \theta} = -\frac{\ddot{r}}{h\dot{\theta}} = -\frac{\ddot{r}}{h^2 u^2}$$

Plugging these into (1) gives

$$\ddot{r} - r\dot{\theta}^2 = -\frac{\mu}{r^2}$$

$$-h^2 u^2 \frac{\delta^2 u}{\delta \theta^2} - \frac{1}{u}(hu^2)^2 = -\mu u^2$$

$$\frac{\delta^2 u}{\delta \theta^2} + u = \frac{\mu}{h^2}$$

So for the gravitational force – or, more generally, for *any* inverse square force law – the right hand side of the equation becomes a constant and the equation is seen to be the harmonic equation (up to a shift of origin of the dependent variable). The solution is:

$$u(\theta) = \frac{\mu}{h^2} - A\cos(\theta - \theta_0)$$

where A and θ_0 are arbitrary constants. This resulting equation of the orbit of the object is that of an ellipse in Polar form relative to one of the focal points. This is put into a more standard form by letting $e \equiv h^2 A/\mu$ be the eccentricity, letting $a \equiv h^2/(\mu(1-e^2))$ be the semi-major axis. Finally, letting $\theta_0 \equiv 0$ so the long axis of the elipce is along the positive x coordinate.

$$r(\theta) = \frac{a(1-e^2)}{1 - e\cos\theta}$$

2.5 Relativistic orbital motion

The above classical (Newtonian) analysis of orbital mechanics assumes that the more subtle effects of general relativity, such as frame dragging and gravitational time dilation are negligible. Relativistic effects cease to be negligible when near very massive bodies (as with the precession of Mercury's orbit about the Sun), or when extreme precision is needed (as with calculations of the orbital elements and time signal references for GPS satellites.[9])

2.6 Orbital planes

Main article: Orbital plane (astronomy)

The analysis so far has been two dimensional; it turns out that an unperturbed orbit is two-dimensional in a plane fixed in space, and thus the extension to three dimensions requires simply rotating the two-dimensional plane into the required angle relative to the poles of the planetary body involved.

The rotation to do this in three dimensions requires three numbers to uniquely determine; traditionally these are expressed as three angles.

2.7 Orbital period

Main article: Orbital period

The orbital period is simply how long an orbiting body takes to complete one orbit.

2.8 Specifying orbits

Main article: Orbital trajectory
See also: Keplerian elements

Six parameters are required to specify a Keplerian orbit about a body. For example, the three numbers that specify the body's initial position, and the three values that specify its velocity will define a unique orbit that can be calculated forwards (or backwards). However, traditionally the parameters used are slightly different.

The traditionally used set of orbital elements is called the set of Keplerian elements, after Johannes Kepler and his laws. The Keplerian elements are six:

- Inclination (i)
- Longitude of the ascending node (Ω)
- Argument of periapsis (ω)
- Eccentricity (e)
- Semimajor axis (a)
- Mean anomaly at epoch (M_0).

In principle once the orbital elements are known for a body, its position can be calculated forward and backwards indefinitely in time. However, in practice, orbits are affected or perturbed, by other forces than simple gravity from an assumed point source (see the next section), and thus the orbital elements change over time.

2.9 Orbital perturbations

An orbital perturbation is when a force or impulse which is much smaller than the overall force or average impulse of the main gravitating body and which is external to the two orbiting bodies causes an acceleration, which changes the parameters of the orbit over time.

2.9.1 Radial, prograde and transverse perturbations

A small radial impulse given to a body in orbit changes the eccentricity, but not the orbital period (to first order). A prograde or retrograde impulse (i.e. an impulse applied along the orbital motion) changes both the eccentricity and the orbital period. Notably, a prograde impulse at periapsis raises the altitude at apoapsis, and vice versa, and a retrograde impulse does the opposite. A transverse impulse (out of the orbital plane) causes rotation of the orbital plane without changing the period or eccentricity. In all instances, a closed orbit will still intersect the perturbation point.

2.9.2 Orbital decay

Main article: Orbital decay

If an orbit is about a planetary body with significant atmosphere, its orbit can decay because of drag. Particularly at each periapsis, the object experiences atmospheric drag, losing energy. Each time, the orbit grows less eccentric (more circular) because the object loses kinetic energy precisely when that energy is at its maximum. This is similar to the effect of slowing a pendulum at its lowest point; the highest point of the pendulum's swing becomes lower. With each successive slowing more of the orbit's path is affected by the atmosphere and the effect becomes more pronounced. Eventually, the effect becomes so great that the maximum kinetic energy is not enough to return the orbit above the limits of the atmospheric drag effect. When this happens the body will rapidly spiral down and intersect the central body.

The bounds of an atmosphere vary wildly. During a solar maximum, the Earth's atmosphere causes drag up to a hundred kilometres higher than during a solar minimum.

Some satellites with long conductive tethers can also experience orbital decay because of electromagnetic drag from the Earth's magnetic field. As the wire cuts the magnetic field it acts as a generator, moving electrons from one end to the other. The orbital energy is converted to heat in the wire.

Orbits can be artificially influenced through the use of rocket engines which change the kinetic energy of the body at some point in its path. This is the conversion of chemical or electrical energy to kinetic energy. In this way changes in the orbit shape or orientation can be facilitated.

Another method of artificially influencing an orbit is through the use of solar sails or magnetic sails. These forms of propulsion require no propellant or energy input other than that of the Sun, and so can be used indefinitely. See statite for one such proposed use.

Orbital decay can occur due to tidal forces for objects below the synchronous orbit for the body they're orbiting. The gravity of the orbiting object raises tidal bulges in the primary, and since below the synchronous orbit the orbiting object is moving faster than the body's surface the bulges lag a short angle behind it. The gravity of the bulges is slightly off of the primary-satellite axis and thus has a component along the satellite's motion. The near bulge slows the object more than the far bulge speeds it up, and as a result the orbit decays. Conversely, the gravity of the satellite on the bulges applies torque on the primary and speeds up its rotation. Artificial satellites are too small to have an appreciable tidal effect on the planets they orbit, but several moons in the solar system are undergoing orbital decay by this mechanism. Mars' innermost moon Phobos is a prime example, and is expected to either impact Mars' surface or break up into a ring within 50 million years.

Orbits can decay via the emission of gravitational waves. This mechanism is extremely weak for most stellar objects, only becoming significant in cases where there is a combination of extreme mass and extreme acceleration, such as with black holes or neutron stars that are orbiting each other closely.

2.9.3 Oblateness

The standard analysis of orbiting bodies assumes that all bodies consist of uniform spheres, or more generally, concentric shells each of uniform density. It can be shown that such bodies are gravitationally equivalent to point sources.

However, in the real world, many bodies rotate, and this introduces oblateness and distorts the gravity field, and gives a quadrupole moment to the gravitational field which is significant at distances comparable to the radius of the body.

2.9.4 Multiple gravitating bodies

Main article: n-body problem

The effects of other gravitating bodies can be significant. For example, the orbit of the Moon cannot be accurately described without allowing for the action of the Sun's gravity as well as the Earth's. One approximate result is that bodies will usually have reasonably stable orbits around a heavier planet or moon, in spite of these perturbations, provided they are orbiting well within the heavier body's Hill sphere.

When there are more than two gravitating bodies it is referred to as an n-body problem. Most n-body problems have no closed form solution, although some special cases have been formulated.

2.9.5 Light radiation and stellar wind

For smaller bodies particularly, light and stellar wind can cause significant perturbations to the attitude and direction of motion of the body, and over time can be significant. Of the planetary bodies, the motion of asteroids is particularly affected over large periods when the asteroids are rotating relative to the Sun.

2.10 Astrodynamics

Main article: Orbital mechanics

Orbital mechanics or **astrodynamics** is the application of ballistics and celestial mechanics to the practical problems concerning the motion of rockets and other spacecraft. The motion of these objects is usually calculated from Newton's laws of motion and Newton's law of universal gravitation. It is a core discipline within space mission design and control. Celestial mechanics treats more broadly the orbital dynamics of systems under the influence of gravity, including spacecraft and natural astronomical bodies such as star systems, planets, moons, and comets. Orbital mechanics focuses on spacecraft trajectories, including orbital maneuvers, orbit plane changes, and interplanetary transfers, and is used by mission planners to predict the results of propulsive maneuvers. General relativity is a more exact theory than Newton's laws for calculating orbits, and is sometimes necessary for greater accuracy or in high-gravity situations (such as orbits close to the Sun).

2.11 Earth orbits

Main article: List of orbits

- Low Earth orbit (LEO): Geocentric orbits with altitudes up to 2,000 km (0–1,240 miles).[10]
- Medium Earth orbit (MEO): Geocentric orbits ranging in altitude from 2,000 km (1,240 miles) to just below geosynchronous orbit at 35,786 kilometers (22,236 mi). Also known as an intermediate circular orbit. These are "most commonly at 20,200 kilometers (12,600 mi), or 20,650 kilometers (12,830 mi), with an orbital period of 12 hours."[11]

- Both Geosynchronous orbit (GSO) and Geostationary orbit (GEO) are orbits around Earth matching Earth's sidereal rotation period. All geosynchronous and geostationary orbits have a semi-major axis of 42,164 km (26,199 mi).[12] All geostationary orbits are also geosynchronous, but not all geosynchronous orbits are geostationary. A geostationary orbit stays exactly above the equator, whereas a geosynchronous orbit may swing north and south to cover more of the Earth's surface. Both complete one full orbit of Earth per sidereal day (relative to the stars, not the Sun).

- High Earth orbit: Geocentric orbits above the altitude of geosynchronous orbit 35,786 km (22,240 miles).[11]

2.12 Scaling in gravity

The gravitational constant G has been calculated as:

- $(6.6742 \pm 0.001) \times 10^{-11}$ $(kg/m^3)^{-1} s^{-2}$.

Thus the constant has dimension density^{-1} time^{-2}. This corresponds to the following properties.

Scaling of distances (including sizes of bodies, while keeping the densities the same) gives similar orbits without scaling the time: if for example distances are halved, masses are divided by 8, gravitational forces by 16 and gravitational accelerations by 2. Hence velocities are halved and orbital periods remain the same. Similarly, when an object is dropped from a tower, the time it takes to fall to the ground remains the same with a scale model of the tower on a scale model of the Earth.

Scaling of distances while keeping the masses the same (in the case of point masses, or by reducing the densities) gives similar orbits; if distances are multiplied by 4, gravitational forces and accelerations are divided by 16, velocities are halved and orbital periods are multiplied by 8.

When all densities are multiplied by 4, orbits are the same; gravitational forces are multiplied by 16 and accelerations by 4, velocities are doubled and orbital periods are halved.

When all densities are multiplied by 4, and all sizes are halved, orbits are similar; masses are divided by 2, gravitational forces are the same, gravitational accelerations are doubled. Hence velocities are the same and orbital periods are halved.

In all these cases of scaling, if densities are multiplied by 4, times are halved; if velocities are doubled, forces are multiplied by 16.

These properties are illustrated in the formula (derived from the formula for the orbital period)

$$GT^2\sigma = 3\pi \left(\frac{a}{r}\right)^3,$$

for an elliptical orbit with semi-major axis a, of a small body around a spherical body with radius r and average density σ, where T is the orbital period. See also Kepler's Third Law.

2.13 Patents

The application of certain orbits or orbital maneuvers to specific useful purposes have been the subject of patents.[13]

2.14 See also

- Klemperer rosette
- List of orbits
- Molniya orbit

- Orbital spaceflight
- Perifocal coordinate system
- Polar Orbits
- Radial trajectory
- Rosetta (orbit)
- VSOP (planets)

2.15 References

[1] The Space Place :: What's a Barycenter

[2] orbit (astronomy) – Britannica Online Encyclopedia

[3] Kuhn, *The Copernican Revolution*, pp. 238, 246–252

[4] *Encyclopaedia Britannica*, 1968, vol. 2, p. 645

[5] M Caspar, *Kepler* (1959, Abelard-Schuman), at pp.131–140; A Koyré, *The Astronomical Revolution: Copernicus, Kepler, Borelli* (1973, Methuen), pp. 277–279

[6] Jones, Andrew. "Kepler's Laws of Planetary Motion". about.com. Retrieved 2008-06-01.

[7] See pages 6 to 8 in Newton's "Treatise of the System of the World" (written 1685, translated into English 1728, see Newton's 'Principia' - A preliminary version), for the original version of this 'cannonball' thought-experiment.

[8] Fitzpatrick, Richard (2006-02-02). "Planetary orbits". *Classical Mechanics – an introductory course*. The University of Texas at Austin. Archived from the original on 2006-05-23. Retrieved 2009-01-14.

[9] Pogge, Richard W.; "Real-World Relativity: The GPS Navigation System". Retrieved 25 January 2008.

[10] "NASA Safety Standard 1740.14, Guidelines and Assessment Procedures for Limiting Orbital Debris" (PDF). Office of Safety and Mission Assurance. 1 August 1995., pages 37-38 (6-1,6-2); figure 6-1.

[11] "Orbit: Definition". *Ancillary Description Writer's Guide, 2013*. National Aeronautics and Space Administration (NASA) Global Change Master Directory. Retrieved 2013-04-29.

[12] Vallado, David A. (2007). *Fundamentals of Astrodynamics and Applications*. Hawthorne, CA: Microcosm Press. p. 31.

[13] http://motherboard.vice.com/read/how-satellite-companies-patent-their-orbits

2.16 Further reading

- Abell; Morrison & Wolff (1987). *Exploration of the Universe* (fifth ed.). Saunders College Publishing.
- Linton, Christopher (2004). *From Eudoxus to Einstein*. Cambridge: University Press. ISBN 0-521-82750-7
- Swetz, Frank; et al. (1997). *Learn from the Masters!*. Mathematical Association of America. ISBN 0-88385-703-0
- Andrea Milani and Giovanni F. Gronchi. *Theory of Orbit Determination* (Cambridge University Press; 378 pages; 2010). Discusses new algorithms for determining the orbits of both natural and artificial celestial bodies.

2.17 External links

- CalcTool: Orbital period of a planet calculator. Has wide choice of units. Requires JavaScript.
- Java simulation on orbital motion. Requires Java.
- NOAA page on Climate Forcing Data includes (calculated) data on Earth orbit variations over the last 50 million years and for the coming 20 million years
- On-line orbit plotter. Requires JavaScript.
- Orbital Mechanics (Rocket and Space Technology)
- Orbital simulations by Varadi, Ghil and Runnegar (2003) provide another, slightly different series for Earth orbit eccentricity, and also a series for orbital inclination. Orbits for the other planets were also calculated, by F. Varadi; B. Runnegar; M. Ghil (2003). "Successive Refinements in Long-Term Integrations of Planetary Orbits". *The Astrophysical Journal* **592**: 620–630. Bibcode:2003ApJ...592..620V. doi:10.1086/375560., but only the eccentricity data for Earth and Mercury are available online.
- Understand orbits using direct manipulation. Requires JavaScript and Macromedia
- Merrifield, Michael. "Orbits (including the first manned orbit)". *Sixty Symbols*. Brady Haran for the University of Nottingham.
- Planetary orbit Simulator Astronoo

Chapter 3

Kepler orbit

In celestial mechanics, a **Kepler orbit** (or **Keplerian orbit**) describes the motion of an orbiting body as an ellipse, parabola, or hyperbola, which forms a two-dimensional orbital plane in three-dimensional space. (A Kepler orbit can also form a straight line.) It considers only the point-like gravitational attraction of two bodies, neglecting perturbations due to gravitational interactions with other objects, atmospheric drag, solar radiation pressure, a non-spherical central body, and so on. It is thus said to be a solution of a special case of the two-body problem, known as the Kepler problem. As a theory in classical mechanics, it also does not take into account the effects of general relativity. Keplerian orbits can be parametrized into six orbital elements in various ways.

In most applications, there is a large central body, the center of mass of which is assumed to be the center of mass of the entire system. By decomposition, the orbits of two objects of similar mass can be described as Kepler orbits around their common center of mass, their barycenter.

3.1 Introduction

From ancient times until the 16th and 17th centuries, the motions of the planets were believed to follow perfectly circular geocentric paths as taught by the ancient Greek philosophers Aristotle and Ptolemy. Variations in the motions of the planets were explained by smaller circular paths overlaid on the larger path (see epicycle). As measurements of the planets became increasingly accurate, revisions to the theory were proposed. In 1543, Nicolaus Copernicus published a heliocentric model of the solar system, although he still believed that the planets traveled in perfectly circular paths centered on the sun.

3.1.1 Johannes Kepler

In 1601, Johannes Kepler acquired the extensive, meticulous observations of the planets made by Tycho Brahe. Kepler would spend the next five years trying to fit the observations of the planet Mars to various curves. In 1609, Kepler published the first two of his three laws of planetary motion. The first law states:

> "The orbit of every planet is an ellipse with the sun at a focus."

More generally, the path of an object undergoing Keplerian motion may also follow a parabola or a hyperbola, which, along with ellipses, belong to a group of curves known as conic sections. Mathematically, the distance between a central body and an orbiting body can be expressed as:

$$r(\nu) = \frac{a(1-e^2)}{1 + e\cos(\nu)}$$

3.1. INTRODUCTION

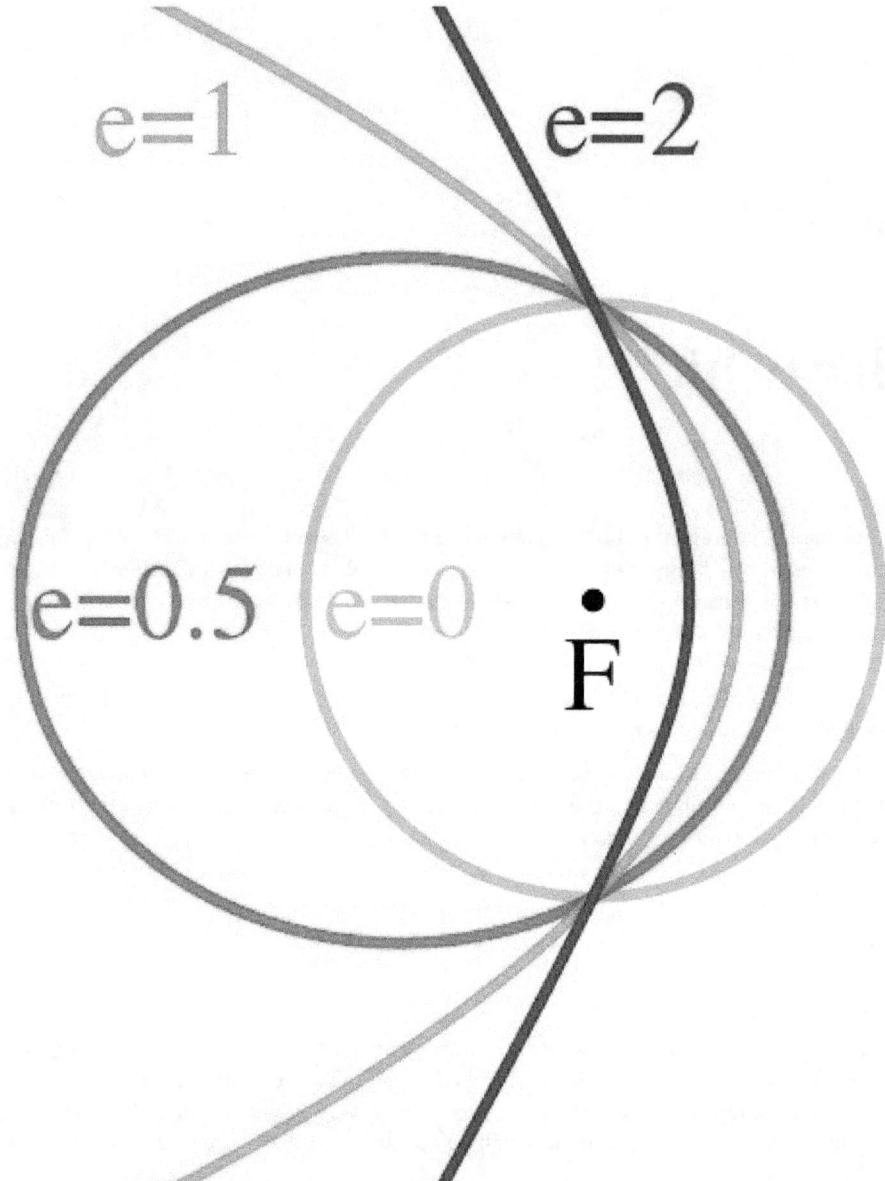

*A diagram of the various forms of the **Kepler Orbit** and their eccentricities. Blue is a hyperbolic trajectory (e > 1). Green is a parabolic trajectory (e = 1). Red is an elliptical orbit (0 < e < 1). Grey is a circular orbit (e = 0).*

where:

- r is the distance
- a is the semi-major axis, which defines the size of the orbit
- e is the eccentricity, which defines the shape of the orbit
- ν is the true anomaly, which is the angle between the current position of the orbiting object and the location in the orbit at which it is closest to the central body (called the periapsis)

Alternately, the equation can be expressed as:

$$r(\nu) = \frac{p}{1 + e\cos(\nu)}$$

Where p is called the semi-latus rectum of the curve. This form of the equation is particularly useful when dealing with parabolic trajectories, for which the semi-major axis is infinite.

Despite developing these laws from observations, Kepler was never able to develop a theory to explain these motions.[1]

3.1.2 Isaac Newton

Between 1665 to 1666, Isaac Newton developed several concepts related to motion, gravitation and differential calculus. However, these concepts were not published until 1687 in the Principia, in which he outlined his laws of motion and his law of universal gravitation. His second of his three laws of motion states:

> The acceleration a of a body is parallel and directly proportional to the net force acting on the body, is in the direction of the net force, and is inversely proportional to the mass of the body:

$$\mathbf{F} = m\mathbf{a} = m\frac{d^2\mathbf{r}}{dt^2}$$

Where:

- $\mathbf{F}$ is the force vector
- m is the mass of the body on which the force is acting
- $\mathbf{a}$ is the acceleration vector, the second time derivative of the position vector $\mathbf{r}$

Strictly speaking, this form of the equation only applies to an object of constant mass, which holds true based on the simplifying assumptions made below.

Newton's law of gravitation states:

> Every point mass attracts every other point mass by a force pointing along the line intersecting both points. The force is proportional to the product of the two masses and inversely proportional to the square of the distance between the point masses:

$$F = G\frac{m_1 m_2}{r^2}$$

where:

- F is the magnitude of the gravitational force between the two point masses
- G is the gravitational constant
- m_1 is the mass of the first point mass
- m_2 is the mass of the second point mass
- r is the distance between the two point masses

From the laws of motion and the law of universal gravitation, Newton was able to derive Kepler's laws, demonstrating consistency between observation and theory. The laws of Kepler and Newton formed the basis of modern celestial mechanics until Albert Einstein introduced the concepts of special and general relativity in the early 20th century. For most applications, Keplerian motion approximates the motions of planets and satellites to relatively high degrees of accuracy and is used extensively in astronomy and astrodynamics.

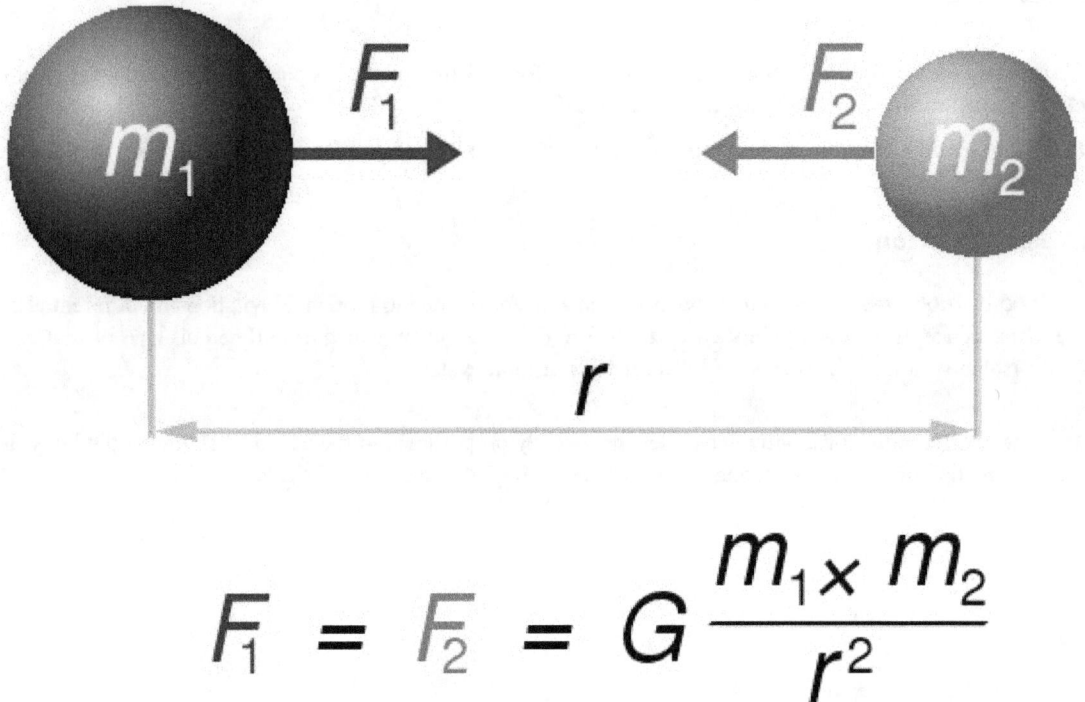

The mechanisms of Newton's law of universal gravitation; a point mass m_1 attracts another point mass m_2 by a force F_2 which is proportional to the product of the two masses and inversely proportional to the square of the distance (r) between them. Regardless of masses or distance, the magnitudes of $|F_1|$ and $|F_2|$ will always be equal. G is the gravitational constant.

3.2 Simplified two body problem

(See also Orbit Analysis)

To solve for the motion of an object in a two body system, two simplifying assumptions can be made:

1. The bodies are spherically symmetric and can be treated as point masses.
2. There are no external or internal forces acting upon the bodies other than their mutual gravitation.

The shapes of large celestial bodies are close to spheres. By symmetry, the net gravitational force attracting a mass point towards a homogeneous sphere must be directed towards its centre. The shell theorem (also proven by Isaac Newton) states that the magnitude of this force is the same as if all mass was concentrated in the middle of the sphere, even if the density of the sphere varies with depth (as it does for most celestial bodies). From this immediately follows that the attraction between two homogeneous spheres is as if both had its mass concentrated to its center.

Smaller objects, like asteroids or spacecraft often have a shape strongly deviating from a sphere. But the gravitational forces produced by these irregularities are generally small compared to the gravity of the central body. The difference between an irregular shape and a perfect sphere also diminishes with distances, and most orbital distances are very large when compared with the diameter of a small orbiting body. Thus for some applications, shape irregularity can be neglected without significant impact on accuracy.

Planets rotate at varying rates and thus may take a slightly oblate shape because of the centrifugal force. With such an oblate shape, the gravitational attraction will deviate somewhat from that of a homogeneous sphere. This phenomenon is quite noticeable for artificial Earth satellites, especially those in low orbits. At larger distances the effect of this oblateness becomes negligible. Planetary motions in the Solar System can be computed with sufficient precision if they are treated as point masses.

Two point mass objects with masses m_1 and m_2 and position vectors $\mathbf{r}_1$ and $\mathbf{r}_2$ relative to some inertial reference frame experience gravitational forces:

$$m_1 \ddot{\mathbf{r}}_1 = \frac{-G m_1 m_2}{r^2} \hat{\mathbf{r}}$$

$$m_2 \ddot{\mathbf{r}}_2 = \frac{G m_1 m_2}{r^2} \hat{\mathbf{r}}$$

where $\mathbf{r}$ is the relative position vector of mass 1 with respect to mass 2, expressed as:

$$\mathbf{r} = \mathbf{r}_1 - \mathbf{r}_2$$

and $\hat{\mathbf{r}}$ is the unit vector in that direction and r is the length of that vector.

Dividing by their respective masses and subtracting the second equation from the first yields the equation of motion for the acceleration of the first object with respect to the second:

$$\ddot{\mathbf{r}} = -\frac{\mu}{r^2} \hat{\mathbf{r}}$$

where μ is the gravitational parameter and is equal to

$$\mu = G(m_1 + m_2)$$

In many applications, a third simplifying assumption can be made:

> 3. When compared to the central body, the mass of the orbiting body is insignificant. Mathematically, $m_1 \gg m_2$, so $\mu = G(m_1 + m_2) \approx G m_1$.

This assumption is not necessary to solve the simplified two body problem, but it simplifies calculations, particularly with Earth-orbiting satellites and planets orbiting the sun. Even Jupiter's mass is less than the sun's by a factor of 1047,[2] which would constitute an error of 0.096% in the value of μ. Notable exceptions include the Earth-moon system (mass ratio of 81.3), the Pluto-Charon system (mass ratio of 8.9) and binary star systems.

Under these assumptions the differential equation for the two body case can be completely solved mathematically and the resulting orbit which follows Kepler's laws of planetary motion is called a "Kepler orbit". The orbits of all planets are to high accuracy Kepler orbits around the Sun. The small deviations are due to the much weaker gravitational attractions between the planets, and in the case of Mercury, due to general relativity. The orbits of the artificial satellites around the Earth are, with a fair approximation, Kepler orbits with small perturbations due to the gravitational attraction of the sun, the moon and the oblateness of the Earth. In high accuracy applications for which the equation of motion must be integrated numerically with all gravitational and non-gravitational forces (such as solar radiation pressure and atmospheric drag) being taken into account, the Kepler orbit concepts are of paramount importance and heavily used.

3.2.1 Keplerian elements

Main article: Keplerian elements

It is worth mentioning that any Keplerian trajectory can be defined by six parameters. The motion of an object moving in three-dimensional space is characterized by a position vector and a velocity vector. Each vector has three components,

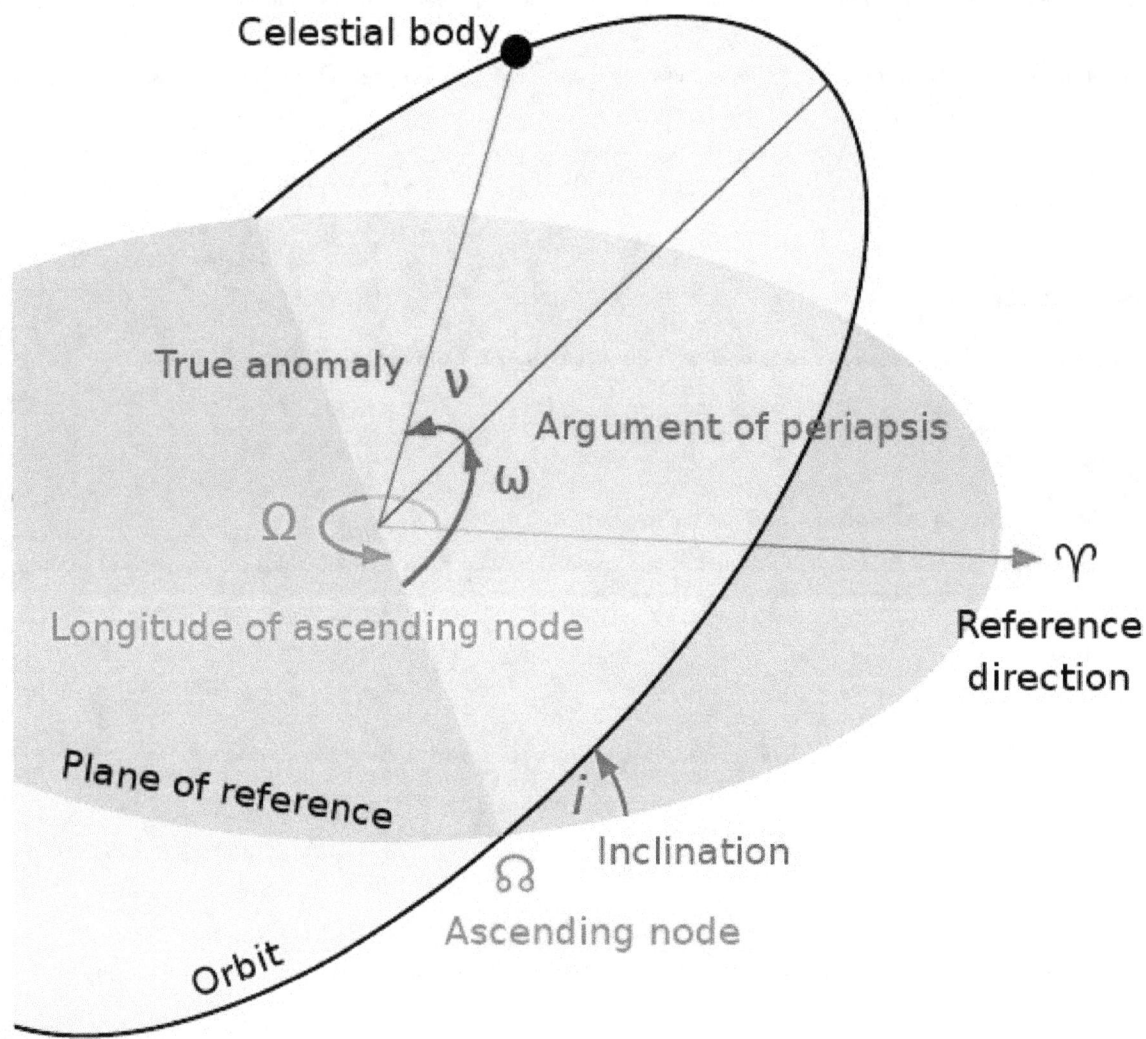

Keplerian orbital elements.

so the total number of values needed to define a trajectory through space is six. An orbit is generally defined by six elements (known as *Keplerian elements*) that can be computed from position and velocity, three of which have already been discussed. These elements are convenient in that of the six, five are unchanging for an unperturbed orbit (a stark contrast to two constantly changing vectors). The future location of an object within its orbit can be predicted and its new position and velocity can be easily obtained from the orbital elements.

Two define the size and shape of the trajectory:

- Semimajor axis (a)
- Eccentricity (e)

Three define the orientation of the orbital plane:

- Inclination (i) defines the angle between the orbital plane and the reference plane.
- Longitude of the ascending node (Ω) defines the angle between the reference direction and the upward crossing of the orbit on the reference plane (the ascending node).

- Argument of periapsis (ω) defines the angle between the ascending node and the periapsis.

And finally:

- True anomaly (ν) defines the position of the orbiting body along the trajectory, measured from periapsis. Several alternate values can be used instead of true anomaly, the most common being M the mean anomaly and T, the time since periapsis.

Because i, Ω and ω are simply angular measurements defining the orientation of the trajectory in the reference frame, they are not strictly necessary when discussing the motion of the object within the orbital plane. They have been mentioned here for completeness, but are not required for the proofs below.

3.3 Mathematical solution of the differential equation (1) above

For movement under any central force, i.e. a force parallel to $\mathbf{r}$, the specific relative angular momentum $\mathbf{H} = \mathbf{r} \times \dot{\mathbf{r}}$ stays constant:
$$\dot{\mathbf{H}} = \frac{d}{dt}(\mathbf{r} \times \dot{\mathbf{r}}) = \dot{\mathbf{r}} \times \dot{\mathbf{r}} + \mathbf{r} \times \ddot{\mathbf{r}} = 0 + 0 = 0$$

Since the cross product of the position vector and its velocity stays constant, they must lie in the same plane, orthogonal to $\mathbf{H}$. This implies the vector function is a plane curve.

Because the equation has symmetry around its origin, it is easier to solve in polar coordinates. However, it is important to note that equation (1) refers to linear acceleration ($\ddot{\mathbf{r}}$), as opposed to angular $\left(\ddot{\theta}\right)$ or radial ($\ddot{r}$) acceleration. Therefore, one must be cautious when transforming the equation. Introducing a cartesian coordinate system ($\hat{\mathbf{x}}$, $\hat{\mathbf{y}}$) and polar unit vectors ($\hat{\mathbf{r}}$, $\hat{\boldsymbol{\theta}}$) in the plane orthogonal to $\mathbf{H}$:

$$\hat{\mathbf{r}} = \cos(\theta)\hat{\mathbf{x}} + \sin(\theta)\hat{\mathbf{y}}$$
$$\hat{\boldsymbol{\theta}} = -\sin(\theta)\hat{\mathbf{x}} + \cos(\theta)\hat{\mathbf{y}}$$

We can now rewrite the vector function $\mathbf{r}$ and its derivatives as:

$$\mathbf{r} = r(\cos\theta\hat{\mathbf{x}} + \sin\theta\hat{\mathbf{y}}) = r\hat{\mathbf{r}}$$

$$\dot{\mathbf{r}} = \dot{r}\hat{\mathbf{r}} + r\dot{\theta}\hat{\boldsymbol{\theta}}$$

$$\ddot{\mathbf{r}} = (\ddot{r} - r\dot{\theta}^2)\hat{\mathbf{r}} + (r\ddot{\theta} + 2\dot{r}\dot{\theta})\hat{\boldsymbol{\theta}}$$

(see "Vector calculus"). Substituting these into (1), we find:

$$(\ddot{r} - r\dot{\theta}^2)\hat{\mathbf{r}} + (r\ddot{\theta} + 2\dot{r}\dot{\theta})\hat{\boldsymbol{\theta}} = \left(-\frac{\mu}{r^2}\right)\hat{\mathbf{r}} + (0)\hat{\boldsymbol{\theta}}$$

This gives the non-ordinary polar differential equation:

$$\ddot{r} - r\dot{\theta}^2 = -\frac{\mu}{r^2}$$

In order to solve this equation, we must first eliminate all time derivatives. We find that:
$$H = |\mathbf{r} \times \dot{\mathbf{r}}| = |(r\cos(\theta), r\sin(\theta), 0) \times (\dot{r}\cos(\theta) - r\sin(\theta)\dot{\theta}, \dot{r}\sin(\theta) + r\cos(\theta)\dot{\theta}, 0)| = |(0, 0, r^2\dot{\theta})| = r^2\dot{\theta}$$

3.3. MATHEMATICAL SOLUTION OF THE DIFFERENTIAL EQUATION (1) ABOVE

$$\dot{\theta} = \frac{H}{r^2}$$

Taking the time derivative of (3), we get

$$\ddot{\theta} = -\frac{2 \cdot H \cdot \dot{r}}{r^3}$$

Equations (3) and (4) allow us to eliminate the time derivatives of θ. In order to eliminate the time derivatives of r, we must use the chain rule to find appropriate substitutions:

$$\dot{r} = \frac{dr}{d\theta} \cdot \dot{\theta}$$

$$\ddot{r} = \frac{d^2 r}{d\theta^2} \cdot \dot{\theta}^2 + \frac{dr}{d\theta} \cdot \ddot{\theta}$$

Using these four substitutions, all time derivatives in (2) can be eliminated, yielding an ordinary differential equation for r as function of θ.

$$\ddot{r} - r\dot{\theta}^2 = -\frac{\mu}{r^2}$$

$$\frac{d^2 r}{d\theta^2} \cdot \dot{\theta}^2 + \frac{dr}{d\theta} \cdot \ddot{\theta} - r\dot{\theta}^2 = -\frac{\mu}{r^2}$$

$$\frac{d^2 r}{d\theta^2} \cdot \left(\frac{H}{r^2}\right)^2 + \frac{dr}{d\theta} \cdot \left(-\frac{2 \cdot H \cdot \dot{r}}{r^3}\right) - r \left(\frac{H}{r^2}\right)^2 = -\frac{\mu}{r^2}$$

$$\frac{H^2}{r^4} \cdot \left(\frac{d^2 r}{d\theta^2} - 2 \cdot \frac{\left(\frac{dr}{d\theta}\right)^2}{r} - r\right) = -\frac{\mu}{r^2}$$

The differential equation (7) can be solved analytically by the variable substitution

$$r = \frac{1}{s}$$

Using the chain rule for differentiation one gets:

$$\frac{dr}{d\theta} = -\frac{1}{s^2} \cdot \frac{ds}{d\theta}$$

$$\frac{d^2 r}{d\theta^2} = \frac{2}{s^3} \cdot \left(\frac{ds}{d\theta}\right)^2 - \frac{1}{s^2} \cdot \frac{d^2 s}{d\theta^2}$$

Using the expressions (10) and (9) for $\frac{d^2 r}{d\theta^2}$ and $\frac{dr}{d\theta}$ one gets

$$H^2 \cdot \left(\frac{d^2 s}{d\theta^2} + s\right) = \mu$$

with the general solution

$$s = \frac{\mu}{H^2} \cdot (1 + e \cdot \cos(\theta - \theta_0))$$

where e and θ_0 are constants of integration depending on the initial values for s and $\frac{ds}{d\theta}$.

Instead of using the constant of integration θ_0 explicitly one introduces the convention that the unit vectors $\hat{x}$, $\hat{y}$ defining the coordinate system in the orbital plane are selected such that θ_0 takes the value zero and e is positive. This then means that θ is zero at the point where s is maximal and therefore $r = \frac{1}{s}$ is minimal. Defining the parameter p as $\frac{H^2}{\mu}$ one has that

$$r = \frac{1}{s} = \frac{p}{1 + e \cdot \cos\theta}$$

3.3.1 Alternate derivation

Another way to solve this equation without the use of polar differential equations is as follows:
Define a unit vector $\mathbf{u}$ such that $\mathbf{r} = r\mathbf{u}$ and $\ddot{\mathbf{r}} = -\frac{\mu}{r^2}\mathbf{u}$. It follows that

$$\mathbf{H} = \mathbf{r} \times \dot{\mathbf{r}} = r\mathbf{u} \times \tfrac{d}{dt}(r\mathbf{u}) = r\mathbf{u} \times (r\dot{\mathbf{u}} + \dot{r}\mathbf{u}) = r^2(\mathbf{u} \times \dot{\mathbf{u}}) + r\dot{r}(\mathbf{u} \times \mathbf{u}) = r^2 \mathbf{u} \times \dot{\mathbf{u}}$$

Now consider

$$\ddot{\mathbf{r}} \times \mathbf{H} = -\tfrac{\mu}{r^2}\mathbf{u} \times (r^2 \mathbf{u} \times \dot{\mathbf{u}}) = -\mu \mathbf{u} \times (\mathbf{u} \times \dot{\mathbf{u}}) = -\mu[(\mathbf{u} \cdot \dot{\mathbf{u}})\mathbf{u} - (\mathbf{u} \cdot \mathbf{u})\dot{\mathbf{u}}]$$

(see Vector triple product). Notice that

$$\mathbf{u} \cdot \mathbf{u} = |\mathbf{u}|^2 = 1$$

$$\mathbf{u} \cdot \dot{\mathbf{u}} = \tfrac{1}{2}(\mathbf{u} \cdot \dot{\mathbf{u}} + \dot{\mathbf{u}} \cdot \mathbf{u}) = \tfrac{1}{2}\tfrac{d}{dt}(\mathbf{u} \cdot \mathbf{u}) = 0$$

Substituting these values into the previous equation, one gets:

$$\ddot{\mathbf{r}} \times \mathbf{H} = \mu \dot{\mathbf{u}}$$

Integrating both sides:

$$\dot{\mathbf{r}} \times \mathbf{H} = \mu \mathbf{u} + \mathbf{c}$$

Where $\mathbf{c}$ is a constant vector. Dotting this with $\mathbf{r}$ yields an interesting result:

$$\mathbf{r} \cdot (\dot{\mathbf{r}} \times \mathbf{H}) = \mathbf{r} \cdot (\mu \mathbf{u} + \mathbf{c}) = \mu \mathbf{r} \cdot \mathbf{u} + \mathbf{r} \cdot \mathbf{c} = \mu r(\mathbf{u} \cdot \mathbf{u}) + rc\cos(\theta) = r(\mu + c\cos(\theta))$$

Where θ is the angle between $\bar{r}$ and $\bar{c}$. Solving for r:

$$r = \tfrac{\mathbf{r} \cdot (\dot{\mathbf{r}} \times \mathbf{H})}{\mu + c\cos(\theta)} = \tfrac{(\mathbf{r} \times \dot{\mathbf{r}}) \cdot \mathbf{H}}{\mu + c\cos(\theta)} = \tfrac{|\mathbf{H}|^2}{\mu + c\cos(\theta)}$$

Notice that (r, θ) are effectively the polar coordinates of the vector function. Making the substitutions $p = \frac{|\mathbf{H}|^2}{\mu}$ and $e = \frac{c}{\mu}$, we again arrive at the equation

$$r = \frac{p}{1 + e \cdot \cos\theta}$$

This is the equation in polar coordinates for a conic section with origin in a focal point. The argument θ is called "true anomaly".

3.3.2 Properties of trajectory equation

For $e = 0$ this is a circle with radius p.

For $0 < e < 1$ this is an ellipse with

3.3. MATHEMATICAL SOLUTION OF THE DIFFERENTIAL EQUATION (1) ABOVE

$$a = \frac{p}{1 - e^2}$$

$$b = \frac{p}{\sqrt{1 - e^2}} = a \cdot \sqrt{1 - e^2}$$

For $e = 1$ this is a parabola with focal length $\frac{p}{2}$

For $e > 1$ this is a hyperbola with

$$a = \frac{p}{e^2 - 1}$$

$$b = \frac{p}{\sqrt{e^2 - 1}} = a \cdot \sqrt{e^2 - 1}$$

The following image illustrates an ellipse (red), a parabola (green) and a hyperbola (blue)

The point on the horizontal line going out to the right from the focal point is the point with $\theta = 0$ for which the distance to the focus takes the minimal value $\frac{p}{1+e}$, the pericentre. For the ellipse there is also an apocentre for which the distance to the focus takes the maximal value $\frac{p}{1-e}$. For the hyperbola the range for θ is

$$\left[-\cos^{-1}\left(-\frac{1}{e}\right) < \theta < \cos^{-1}\left(-\frac{1}{e}\right) \right]$$

and for a parabola the range is

$$[-\pi < \theta < \pi]$$

Using the chain rule for differentiation (5), the equation (2) and the definition of p as $\frac{H^2}{\mu}$ one gets that the radial velocity component is

$$V_r = \dot{r} = \frac{H}{p} \cdot e \cdot \sin\theta = \sqrt{\frac{\mu}{p}} \cdot e \cdot \sin\theta$$

and that the tangential component (velocity component perpendicular to V_r) is

$$V_t = r \cdot \dot{\theta} = \frac{H}{r} = \sqrt{\frac{\mu}{p}} \cdot (1 + e \cdot \cos\theta)$$

The connection between the polar argument θ and time t is slightly different for elliptic and hyperbolic orbits.

For an elliptic orbit one switches to the "eccentric anomaly" E for which

$$x = a \cdot (\cos E - e)$$
$$y = b \cdot \sin E$$

and consequently

$$\dot{x} = -a \cdot \sin E \cdot \dot{E}$$
$$\dot{y} = b \cdot \cos E \cdot \dot{E}$$

and the angular momentum H is

$$H = x \cdot \dot{y} - y \cdot \dot{x} = a \cdot b \cdot (1 - e \cdot \cos E) \cdot \dot{E}$$

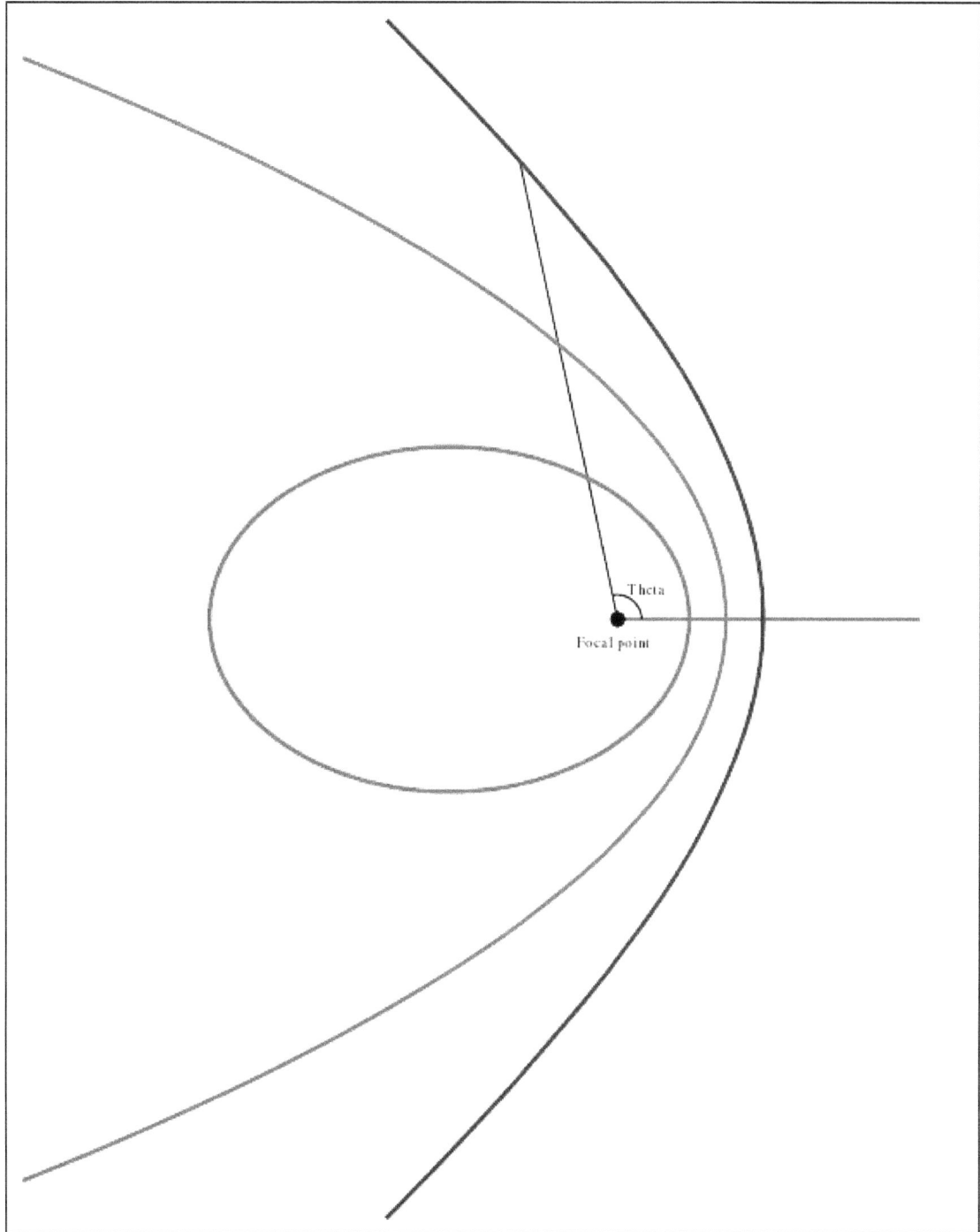

An elliptic Kepler orbit with an eccentricity of 0.7, a parabolic Kepler orbit and a hyperbolic Kepler orbit with an eccentricity of 1.3. The distance to the focal point is a function of the polar angle relative to the horizontal line as given by the equation (13)

Integrating with respect to time t one gets

$$H \cdot t = a \cdot b \cdot (E - e \cdot \sin E)$$

under the assumption that time $t = 0$ is selected such that the integration constant is zero.

As by definition of p one has

$$H = \sqrt{\mu \cdot p}$$

this can be written

$$t = a \cdot \sqrt{\frac{a}{\mu}} (E - e \cdot \sin E)$$

For a hyperbolic orbit one uses the hyperbolic functions for the parameterisation

$$x = a \cdot (e - \cosh E)$$
$$y = b \cdot \sinh E$$

for which one has

$$\dot{x} = -a \cdot \sinh E \cdot \dot{E}$$
$$\dot{y} = b \cdot \cosh E \cdot \dot{E}$$

and the angular momentum H is

$$H = x \cdot \dot{y} - y \cdot \dot{x} = a \cdot b \cdot (e \cdot \cosh E - 1) \cdot \dot{E}$$

Integrating with respect to time t one gets

$$H \cdot t = a \cdot b \cdot (e \cdot \sinh E - E)$$

i.e.

$$t = a \cdot \sqrt{\frac{a}{\mu}} (e \cdot \sinh E - E)$$

To find what time t that corresponds to a certain true anomaly θ one computes corresponding parameter E connected to time with relation (27) for an elliptic and with relation (34) for a hyperbolic orbit.

Note that the relations (27) and (34) define a mapping between the ranges

$$[-\infty < t < \infty] \longleftrightarrow [-\infty < E < \infty]$$

3.4 Some additional formulae

For an *elliptic orbit* one gets from (20) and (21) that

$$r = a \cdot (1 - e \cdot \cos E)$$

and therefore that

$$\cos \theta = \frac{x}{r} = \frac{\cos E - e}{1 - e \cdot \cos E}$$

From (36) then follows that

$$\tan^2\frac{\theta}{2} = \frac{1-\cos\theta}{1+\cos\theta} = \frac{1-\frac{\cos E - e}{1-e\cdot\cos E}}{1+\frac{\cos E - e}{1-e\cdot\cos E}} = \frac{1-e\cdot\cos E - \cos E + e}{1-e\cdot\cos E + \cos E - e} = \frac{1+e}{1-e}\cdot\frac{1-\cos E}{1+\cos E} = \frac{1+e}{1-e}\cdot\tan^2\frac{E}{2}$$

From the geometrical construction defining the eccentric anomaly it is clear that the vectors ($\cos E$, $\sin E$) and ($\cos\theta$, $\sin\theta$) are on the same side of the x-axis. From this then follows that the vectors $\left(\cos\frac{E}{2}, \sin\frac{E}{2}\right)$ and $\left(\cos\frac{\theta}{2}, \sin\frac{\theta}{2}\right)$ are in the same quadrant. One therefore has that

$$\tan\frac{\theta}{2} = \sqrt{\frac{1+e}{1-e}}\cdot\tan\frac{E}{2}$$

and that

$$\theta = 2\cdot\arg\left(\sqrt{1-e}\cdot\cos\frac{E}{2}, \sqrt{1+e}\cdot\sin\frac{E}{2}\right) + n\cdot 2\pi$$

$$E = 2\cdot\arg\left(\sqrt{1+e}\cdot\cos\frac{\theta}{2}, \sqrt{1-e}\cdot\sin\frac{\theta}{2}\right) + n\cdot 2\pi$$

where "$\arg(x, y)$" is the polar argument of the vector (x, y) and n is selected such that $|E - \theta| < \pi$

For the numerical computation of $\arg(x, y)$ the standard function ATAN2(y,x) (or in double precision DATAN2(y,x)) available in for example the programming language FORTRAN can be used.

Note that this is a mapping between the ranges

$$[-\infty < \theta < \infty] \longleftrightarrow [-\infty < E < \infty]$$

For a *hyperbolic orbit* one gets from (28) and (29) that

$$r = a\cdot(e\cdot\cosh E - 1)$$

and therefore that

$$\cos\theta = \frac{x}{r} = \frac{e - \cosh E}{e\cdot\cosh E - 1}$$

As

$$\tan^2\frac{\theta}{2} = \frac{1-\cos\theta}{1+\cos\theta} = \frac{1-\frac{e-\cosh E}{e\cdot\cosh E - 1}}{1+\frac{e-\cosh E}{e\cdot\cosh E - 1}} = \frac{e\cdot\cosh E - e + \cosh E}{e\cdot\cosh E + e - \cosh E} = \frac{e+1}{e-1}\cdot\frac{\cosh E - 1}{\cosh E + 1} = \frac{e+1}{e-1}\cdot\tanh^2\frac{E}{2}$$

and as $\tan\frac{\theta}{2}$ and $\tanh\frac{E}{2}$ have the same sign it follows that

$$\tan\frac{\theta}{2} = \sqrt{\frac{e+1}{e-1}}\cdot\tanh\frac{E}{2}$$

This relation is convenient for passing between "true anomaly" and the parameter E, the latter being connected to time through relation (34). Note that this is a mapping between the ranges

$$\left[-\cos^{-1}\left(-\frac{1}{e}\right) < \theta < \cos^{-1}\left(-\frac{1}{e}\right)\right] \longleftrightarrow [-\infty < E < \infty]$$

and that $\frac{E}{2}$ can be computed using the relation

$$\tanh^{-1} x = \frac{1}{2} \ln\left(\frac{1+x}{1-x}\right)$$

From relation (27) follows that the orbital period P for an elliptic orbit is

$$P = 2\pi \cdot a \cdot \sqrt{\frac{a}{\mu}}$$

As the potential energy corresponding to the force field of relation (1) is

$$-\frac{\mu}{r}$$

it follows from (13), (14), (18) and (19) that the sum of the kinetic and the potential energy

$$\frac{V_r^2 + V_t^2}{2} - \frac{\mu}{r}$$

for an elliptic orbit is

$$-\frac{\mu}{2 \cdot a}$$

and from (13), (16), (18) and (19) that the sum of the kinetic and the potential energy for a hyperbolic orbit is

$$\frac{\mu}{2 \cdot a}$$

Relative the inertial coordinate system

$$\hat{x}, \hat{y}$$

in the orbital plane with $\hat{x}$ towards pericentre one gets from (18) and (19) that the velocity components are

$$V_x = \cos\theta \cdot V_r - \sin\theta \cdot V_t = -\sqrt{\frac{\mu}{p}} \cdot \sin\theta$$
$$V_y = \sin\theta \cdot V_r + \cos\theta \cdot V_t = \sqrt{\frac{\mu}{p}} \cdot (e + \cos\theta)$$

See also Equation of the center – Analytical expansions
The Equation of the center relates mean anomaly to true anomaly for elliptical orbits, for small numerical eccentricity.

3.5 Determination of the Kepler orbit that corresponds to a given initial state

This is the "initial value problem" for the differential equation (1) which is a first order equation for the 6-dimensional "state vector" $(\vec{r}, \vec{v})$ when written as

$$\dot{\vec{v}} = -\mu \cdot \frac{\vec{r}}{r^2}$$
$$\dot{\vec{r}} = \vec{v}$$

For any values for the initial "state vector" ($\vec{r}_0$, $\vec{v}_0$) the Kepler orbit corresponding to the solution of this initial value problem can be found with the following algorithm:

Define the orthogonal unit vectors ($\hat{r}$, $\hat{t}$) through

$$\vec{r}_0 = r \cdot \hat{r}$$
$$\vec{v}_0 = V_r \cdot \hat{r} + V_t \cdot \hat{t}$$

with $r > 0$ and $V_t > 0$

From (13), (18) and (19) follows that by setting

$$p = \frac{(r \cdot V_t)^2}{\mu}$$

and by defining $e \geq 0$ and θ such that

$$e \cdot \cos\theta = \frac{V_t}{V_0} - 1$$
$$e \cdot \sin\theta = \frac{V_r}{V_0}$$

where

$$V_0 = \sqrt{\frac{\mu}{p}}$$

one gets a Kepler orbit that for true anomaly θ has the same r, V_r and V_t values as those defined by (50) and (51).

If this Kepler orbit then also has the same ($\hat{r}$, $\hat{t}$) vectors for this true anomaly θ as the ones defined by (50) and (51) the state vector ($\vec{r}$, $\vec{v}$) of the Kepler orbit takes the desired values ($\vec{r}_0$, $\vec{v}_0$) for true anomaly θ.

The standard inertially fixed coordinate system ($\hat{x}$, $\hat{y}$) in the orbital plane (with $\hat{x}$ directed from the centre of the homogeneous sphere to the pericentre) defining the orientation of the conical section (ellipse, parabola or hyperbola) can then be determined with the relation

$$\hat{x} = \cos\theta \cdot \hat{r} - \sin\theta \cdot \hat{t}$$
$$\hat{y} = \sin\theta \cdot \hat{r} + \cos\theta \cdot \hat{t}$$

Note that the relations (53) and (54) has a singularity when $V_r = 0$ and

$$V_t = V_0 = \sqrt{\frac{\mu}{p}} = \sqrt{\frac{\mu}{\frac{(r \cdot V_t)^2}{\mu}}}$$

i.e.

$$V_t = \sqrt{\frac{\mu}{r}}$$

which is the case that it is a circular orbit that is fitting the initial state ($\vec{r}_0$, $\vec{v}_0$)

3.6 The osculating Kepler orbit

Main article: Osculating orbit

For any state vector $(\bar{r}, \bar{v})$ the Kepler orbit corresponding to this state can be computed with the algorithm defined above. First the parameters p, e, θ are determined from r, V_r, V_t and then the orthogonal unit vectors in the orbital plane $\hat{x}, \hat{y}$ using the relations (56) and (57).

If now the equation of motion is

$$\ddot{\bar{r}} = \bar{F}(\bar{r}, \dot{\bar{r}}, t)$$

where

$$\bar{F}(\bar{r}, \dot{\bar{r}}, t)$$

is a function other than

$$-\mu \cdot \frac{\hat{r}}{r^2}$$

the resulting parameters

$p, e, \theta, \hat{x}, \hat{y}$

defined by $\bar{r}, \dot{\bar{r}}$ will all vary with time as opposed to the case of a Kepler orbit for which only the parameter θ will vary

The Kepler orbit computed in this way having the same "state vector" as the solution to the "equation of motion" (59) at time t is said to be "osculating" at this time.

This concept is for example useful in case

$$\bar{F}(\bar{r}, \dot{\bar{r}}, t) = -\mu \cdot \frac{\hat{r}}{r^2} + \bar{f}(\bar{r}, \dot{\bar{r}}, t)$$

where

$$\bar{f}(\bar{r}, \dot{\bar{r}}, t)$$

is a small "perturbing force" due to for example a faint gravitational pull from other celestial bodies. The parameters of the osculating Kepler orbit will then only slowly change and the osculating Kepler orbit is a good approximation to the real orbit for a considerable time period before and after the time of osculation.

This concept can also be useful for a rocket during powered flight as it then tells which Kepler orbit the rocket would continue in in case the thrust is switched off.

For a "close to circular" orbit the concept "eccentricity vector" defined as $\bar{e} = e \cdot \hat{x}$ is useful. From (53), (54) and (56) follows that

$$\bar{e} = \frac{(V_t - V_0) \cdot \hat{r} - V_r \cdot \hat{t}}{V_0}$$

i.e. $\bar{e}$ is a smooth differentiable function of the state vector $(\bar{r}, \bar{v})$ also if this state corresponds to a circular orbit.

3.7 See also

- Two-body problem
- Gravitational two-body problem
- Kepler problem
- Kepler's laws of planetary motion
- Elliptic orbit
- Hyperbolic trajectory
- Parabolic trajectory
- Radial trajectory
- Orbit modeling

3.8 Citations

[1] Bate, Mueller, White. pp 177–181

[2] http://ssd.jpl.nasa.gov

3.9 References

- El'Yasberg "Theory of flight of artificial earth satellites", **Israel program for Scientific Translations (1967)**
- Bate, Roger; Mueller, Donald; White, Jerry (1971). *Fundamentals of Astrodynamics.* Dover Publications, Inc., New York. ISBN 0-486-60061-0.

3.10 External links

- JAVA applet animating the orbit of a satellite in an elliptic Kepler orbit around the Earth with any value for semi-major axis and eccentricity.

Chapter 4

Orbital state vectors

In astrodynamics and celestial dynamics, the **orbital state vectors** (sometimes **state vectors**) of an orbit are cartesian vectors of position (**r**) and velocity (**v**) that together with their time (epoch) (t) uniquely determine the trajectory of the orbiting body in space.

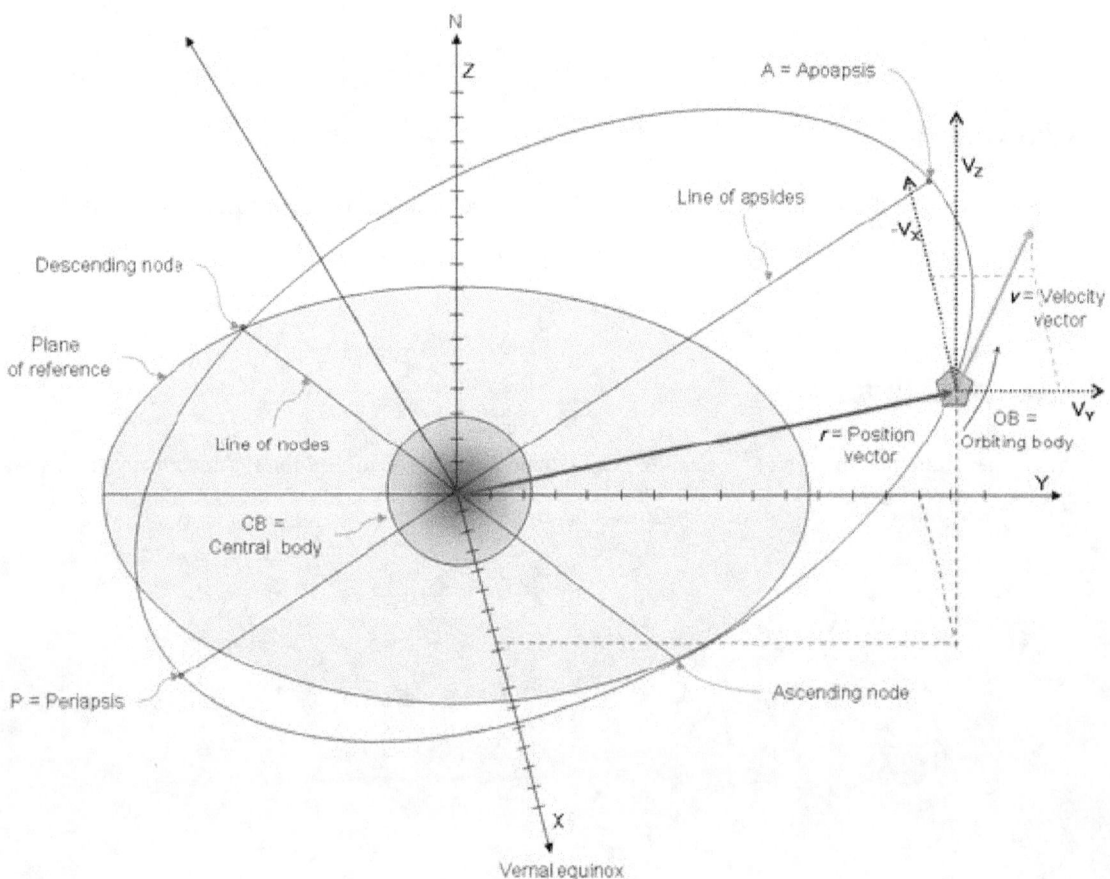

Orbital position vector, orbital velocity vector, other orbital elements

4.1 Frame of reference

State vectors are defined with respect to some frame of reference, usually but not always an inertial reference frame. One of the more popular reference frames for the state vectors of bodies moving near Earth is the Earth-centered equatorial system defined as follows:

- The origin is Earth's center of mass;
- The Z axis is coincident with Earth's rotational axis, positive northward;
- The X/Y plane coincides with Earth's equatorial plane, with the +X axis pointing toward the vernal equinox and the Y axis completing a right-handed set.

This reference frame is not truly inertial because of the slow, 26,000 year precession of Earth's axis, so the reference frames defined by Earth's orientation at a standard astronomical epoch such as B1950 or J2000 are also commonly used.

Many other reference frames can be used to meet various application requirements, including those centered on the Sun or on other planets or moons, the one defined by the barycenter and total angular momentum of the solar system, or even a spacecraft's own orbital plane and angular momentum.

4.2 Position vector

The **position vector r** describes the position of the body in the chosen frame of reference, while the **velocity vector v** describes its velocity in the same frame at the same time. Together, these two vectors and the time at which they are valid uniquely describe the body's trajectory.

The body does not actually have to be in orbit for its state vector to determine its trajectory; it only has to move ballistically, i.e., solely under the effects of its own inertia and gravity. For example, it could be a spacecraft or missile in a suborbital trajectory. If other forces such as drag or thrust are significant, they must be added vectorially to those of gravity when performing the integration to determine future position and velocity.

For any object moving through space, the velocity vector is tangent to the trajectory. If $\hat{u}_t$ is the unit vector tangent to the trajectory, then

$$\mathbf{v} = v\hat{\mathbf{u}}_t$$

4.2.1 Derivation

The **velocity vector v** can be derived from **position vector r** by differentiation with respect to time:

$$\mathbf{v} = \frac{d\mathbf{r}}{dt}$$

An object's state vector can be used to compute its classical or Keplerian orbital elements and vice versa. Each representation has its advantages. The elements are more descriptive of the size, shape and orientation of an orbit, and may be used to quickly and easily estimate the object's state at any arbitrary time provided its motion is accurately modeled by the two-body problem with only small perturbations.

On the other hand, the state vector is more directly useful in a numerical integration that accounts for significant, arbitrary, time-varying forces such as drag, thrust and gravitational perturbations from third bodies as well as the gravity of the primary body.

The state vectors ($\mathbf{r}$ and $\mathbf{v}$) can be easily used to compute the angular momentum vector as $\mathbf{h} = \mathbf{r} \times \mathbf{v}$.

Because even satellites in low Earth orbit experience significant perturbations (primarily from Earth's non-spherical shape), the Keplerian elements computed from the state vector at any moment are only valid at that time. Such element sets are known as osculating elements because they coincide with the actual orbit only at that moment.

4.3 See also

- ECEF
- Earth-centered inertial
- Orbital plane

Chapter 5

Semi-major axis

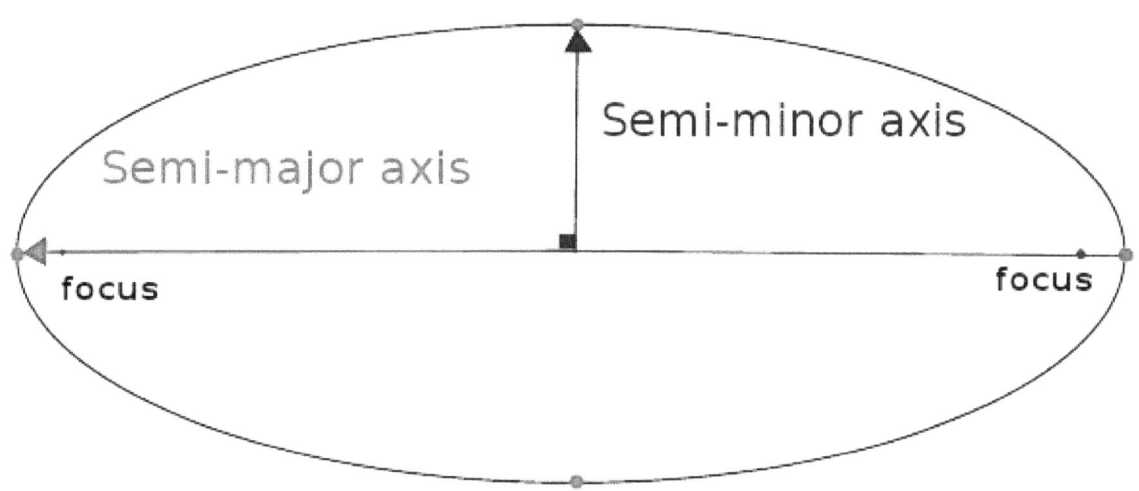

The semi-major and semi-minor axis of an ellipse

In geometry, the **major axis** of an ellipse is its longest diameter: a line segment that runs through the center and both foci, with ends at the widest points of the perimeter. The **semi-major axis** is one half of the major axis, and thus runs from the centre, through a focus, and to the perimeter. Essentially, it is the radius of an orbit at the orbit's two most distant points. For the special case of a circle, the semi-major axis is the radius. One can think of the semi-major axis as an ellipse's *long radius*.

The length of the semi-major axis a of an ellipse is related to the semi-minor axis's length b through the eccentricity e and the semi-latus rectum ℓ, as follows:

$$b = a\sqrt{1 - e^2},$$

$$\ell = a(1 - e^2),$$

$$a\ell = b^2.$$

The **semi-major axis of a hyperbola** is, depending on the convention, plus or minus one half of the distance between the two branches. Thus it is the distance from the center to either vertex (turning point) of the hyperbola.

A parabola can be obtained as the limit of a sequence of ellipses where one focus is kept fixed as the other is allowed to move arbitrarily far away in one direction, keeping ℓ fixed. Thus a and b tend to infinity, a faster than b.

5.1 Ellipse

The semi-major axis is the mean value of the smallest and largest distances from one focus to the ellipse. Now consider the equation in polar coordinates, with one focus at the origin and the other on the negative *x*-axis,

$$r(1 + e \cos \theta) = \ell.$$

The mean value of $r = \frac{\ell}{1-e}$ and $r = \frac{\ell}{1+e}$, (for $\theta = \pi$ and $\theta = 0$) is

$$a = \frac{\ell}{1 - e^2}.$$

In an ellipse, the semimajor axis is the geometric mean of the distance from the center to either focus and the distance from the center to either directrix.

5.2 Hyperbola

The **semi-major axis of a hyperbola** is, depending on the convention, plus or minus one half of the distance between the two branches; if this is *a* in the *x*-direction the equation is:

$$\frac{(x-h)^2}{a^2} - \frac{(y-k)^2}{b^2} = 1.$$

In terms of the semi-latus rectum and the eccentricity we have

$$a = \frac{\ell}{1 - e^2}.$$

The **transverse axis** of a hyperbola coincides with the semi-major axis.[1]

5.3 Astronomy

5.3.1 Orbital period

In astrodynamics the orbital period *T* of a small body orbiting a central body in a circular or elliptical orbit is:

$$T = 2\pi \sqrt{\frac{a^3}{\mu}}$$

where:

> *a* is the length of the orbit's semi-major axis

> μ is the standard gravitational parameter of the central body

Note that for all ellipses with a given semi-major axis, the orbital period is the same, regardless of eccentricity.

The specific angular momentum H of a small body orbiting a central body in a circular or elliptical orbit is:

$$H = \sqrt{a \cdot \mu \cdot (1 - e^2)}$$

where:

a and μ are as defined above

e is the eccentricity of the orbit

In astronomy, the semi-major axis is one of the most important orbital elements of an orbit, along with its orbital period. For Solar System objects, the semi-major axis is related to the period of the orbit by Kepler's third law (originally empirically derived),

$$T^2 \propto a^3$$

where T is the period and a is the semimajor axis. This form turns out to be a simplification of the general form for the two-body problem, as determined by Newton:

$$T^2 = \frac{4\pi^2}{G(M+m)} a^3$$

where G is the gravitational constant, M is the mass of the central body, and m is the mass of the orbiting body. Typically, the central body's mass is so much greater than the orbiting body's, that m may be ignored. Making that assumption and using typical astronomy units results in the simpler form Kepler discovered.

The orbiting body's path around the barycentre and its path relative to its primary are both ellipses. The semi-major axis is sometimes used in astronomy as the primary-to-secondary distance when the mass ratio of the primary to the secondary is significantly large (M»m); thus, the orbital parameters of the planets are given in heliocentric terms. The difference between the primocentric and "absolute" orbits may best be illustrated by looking at the Earth–Moon system. The mass ratio in this case is 81.30059. The Earth–Moon characteristic distance, the semi-major axis of the *geocentric* lunar orbit, is 384,400 km. The *barycentric* lunar orbit, on the other hand, has a semi-major axis of 379,700 km, the Earth's counter-orbit taking up the difference, 4,700 km. The Moon's average barycentric orbital speed is 1.010 km/s, whilst the Earth's is 0.012 km/s. The total of these speeds gives a geocentric lunar average orbital speed of 1.022 km/s; the same value may be obtained by considering just the geocentric semi-major axis value.

5.3.2 Average distance

It is often said that the semi-major axis is the "average" distance between the primary focus of the ellipse and the orbiting body. This is not quite accurate, because it depends on what the average is taken over.

- averaging the distance over the eccentric anomaly (q.v.) indeed results in the semi-major axis.
- averaging over the true anomaly (the true orbital angle, measured at the focus) results, oddly enough, in the semi-minor axis $b = a\sqrt{1 - e^2}$.
- averaging over the mean anomaly (the fraction of the orbital period that has elapsed since pericentre, expressed as an angle), finally, gives the time-average

$$a\left(1 + \frac{e^2}{2}\right).$$

The time-averaged value of the reciprocal of the radius, r^{-1}, is a^{-1}.

5.3.3 Energy; calculation of semi-major axis from state vectors

In astrodynamics, the **semi-major axis** a can be calculated from orbital state vectors:

$$a = -\frac{\mu}{2\varepsilon}$$

for an elliptical orbit and, depending on the convention, the same or

$$a = \frac{\mu}{2\varepsilon}$$

for a hyperbolic trajectory

and

$$\varepsilon = \frac{v^2}{2} - \frac{\mu}{|\mathbf{r}|}$$

(specific orbital energy)

and

$$\mu = G(M + m)$$

(standard gravitational parameter), where:

- v is orbital velocity from velocity vector of an orbiting object,
- $\mathbf{r}$ is a cartesian position vector of an orbiting object in coordinates of a reference frame with respect to which the elements of the orbit are to be calculated (e.g. geocentric equatorial for an orbit around Earth, or heliocentric ecliptic for an orbit around the Sun),
- G is the gravitational constant,
- M and m are the masses of the bodies, and
- ε, is the energy of the orbiting body.

Note that for a given amount of total mass, the specific energy and the semi-major axis are always the same, regardless of eccentricity or the ratio of the masses. Conversely, for a given total mass and semi-major axis, the total specific orbital energy is always the same. This statement will always be true under any given conditions.

5.4 References

[1] 7.1 Alternative Characterization

5.5 External links

- Semi-major and semi-minor axes of an ellipse With interactive animation

Chapter 6

Orbital eccentricity

This article is about eccentricity in astrodynamics. For other uses, see Eccentricity (disambiguation).

The **orbital eccentricity** of an astronomical object is a parameter that determines the amount by which its orbit around another body deviates from a perfect circle. A value of 0 is a circular orbit, values between 0 and 1 form an elliptical orbit, 1 is a parabolic escape orbit, and greater than 1 is a hyperbola. The term derives its name from the parameters of conic sections, as every Kepler orbit is a conic section. It is normally used for the isolated two-body problem, but extensions exist for objects following a rosette orbit through the galaxy.

6.1 Definition

In a two-body problem with inverse-square-law force, every orbit is a Kepler orbit. The eccentricity of this Kepler orbit is a non-negative number that defines its shape.

The eccentricity may take the following values:

- circular orbit: $e = 0$

- elliptic orbit: $0 < e < 1$ (see Ellipse)

- parabolic trajectory: $e = 1$ (see Parabola)

- hyperbolic trajectory: $e > 1$ (see Hyperbola)

The eccentricity e is given by

$$e = \sqrt{1 + \frac{2EL^2}{m_{\text{red}}\alpha^2}}$$

where E is the total orbital energy, L is the angular momentum, m_{red} is the reduced mass, and α the coefficient of the inverse-square law central force such as gravity or electrostatics in classical physics:

$$F = \frac{\alpha}{r^2}$$

(α is negative for an attractive force, positive for a repulsive one) (see also Kepler problem).

or in the case of a gravitational force:

6.1. DEFINITION

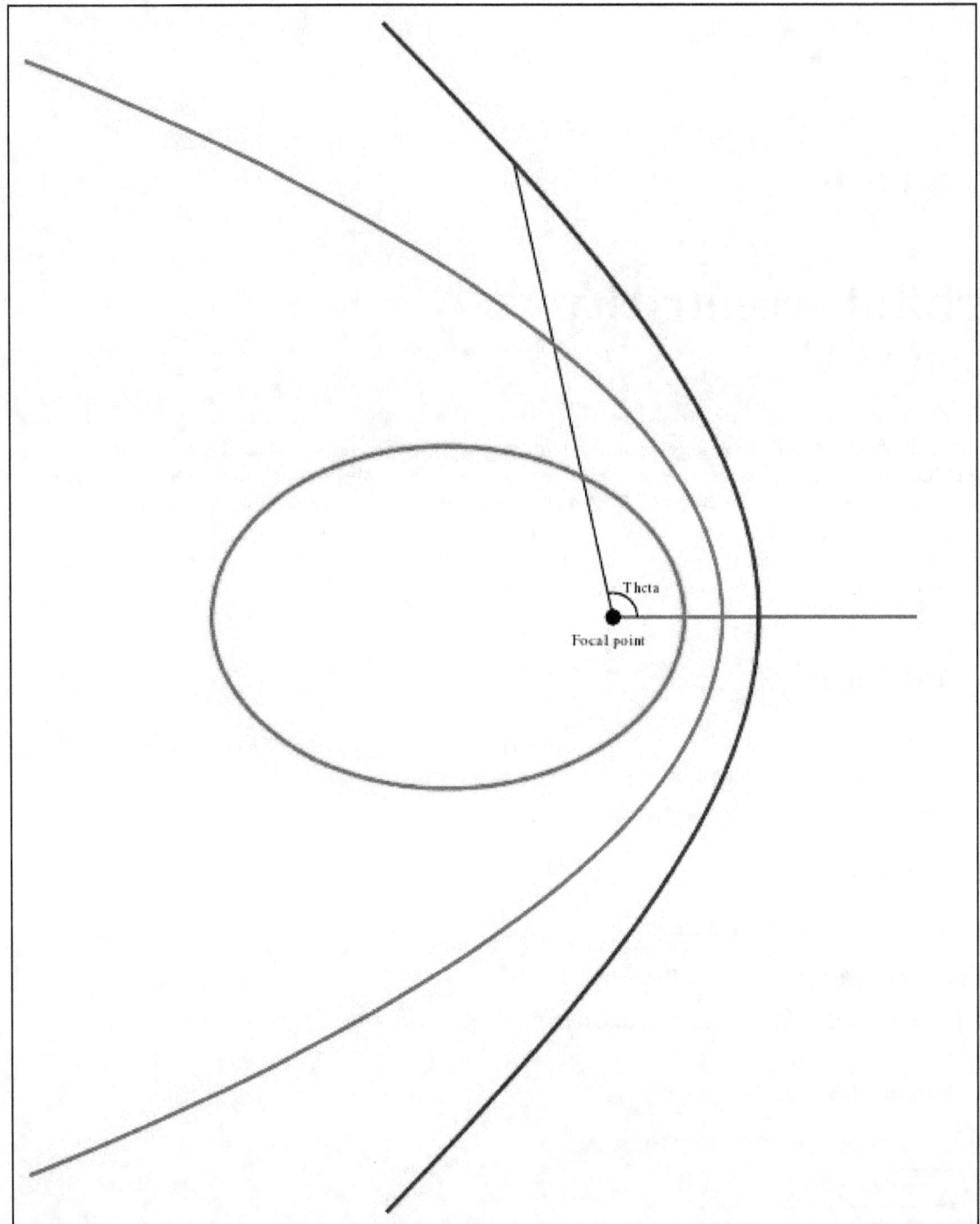

An elliptic, parabolic and hyperbolic Kepler orbit:
elliptic (eccentricity = 0.7)
parabolic (eccentricity = 1)
hyperbolic orbit (eccentricity = 1.3)

$$e = \sqrt{1 + \frac{2\epsilon h^2}{\mu^2}}$$

where ϵ is the specific orbital energy (total energy divided by the reduced mass), μ the standard gravitational parameter based on the total mass, and h the specific relative angular momentum (angular momentum divided by the reduced mass).

For values of e from 0 to 1 the orbit's shape is an increasingly elongated (or flatter) ellipse; for values of e from 1 to infinity the orbit is a hyperbola branch making a total turn of $2\,\mathrm{arccsc}\,e$, decreasing from 180 to 0 degrees. The limit case between an ellipse and a hyperbola, when e equals 1, is parabola.

Radial trajectories are classified as elliptic, parabolic, or hyperbolic based on the energy of the orbit, not the eccentricity. Radial orbits have zero angular momentum and hence eccentricity equal to one. Keeping the energy constant and reducing the angular momentum, elliptic, parabolic, and hyperbolic orbits each tend to the corresponding type of radial trajectory while e tends to 1 (or in the parabolic case, remains 1).

For a repulsive force only the hyperbolic trajectory, including the radial version, is applicable.

For elliptical orbits, a simple proof shows that $\arcsin(\epsilon)$ yields the projection angle of a perfect circle to an ellipse of eccentricity ϵ. For example, to view the eccentricity of the planet Mercury ($\epsilon = 0.2056$), one must simply calculate the inverse sine to find the projection angle of 11.86 degrees. Next, tilt any circular object (such as a coffee mug viewed from the top) by that angle and the apparent ellipse projected to your eye will be of that same eccentricity.

6.2 Etymology

From Medieval Latin *eccentricus*, derived from Greek *ekkentros* "out of the center", from *ek-*, *ex-* "out of" + *kentron* "center". Eccentric first appeared in English in 1551, with the definition "a circle in which the earth, sun, etc. deviates from its center." Five years later, in 1556, an adjective form of the word was added.

6.3 Calculation

The **eccentricity** of an orbit can be calculated from the orbital state vectors as the magnitude of the eccentricity vector:

$$e = |\mathbf{e}|$$

where:

- $\mathbf{e}$ is the eccentricity vector.

For elliptical orbits it can also be calculated from the periapsis and apoapsis since $r_p = a(1-e)$ and $r_a = a(1+e)$, where a is the semimajor axis.

$$e = \frac{r_a - r_p}{r_a + r_p}$$
$$= 1 - \frac{2}{(r_a/r_p) + 1}$$

where:

- r_a is the radius at apoapsis (i.e., the farthest distance of the orbit to the center of mass of the system, which is a focus of the ellipse).
- r_p is the radius at periapsis (the closest distance).

The eccentricity of an elliptical orbit can also be used to obtain the ratio of the periapsis to the apoapsis:

$$\frac{r_p}{r_a} = \frac{1-e}{1+e}$$

6.4 Examples

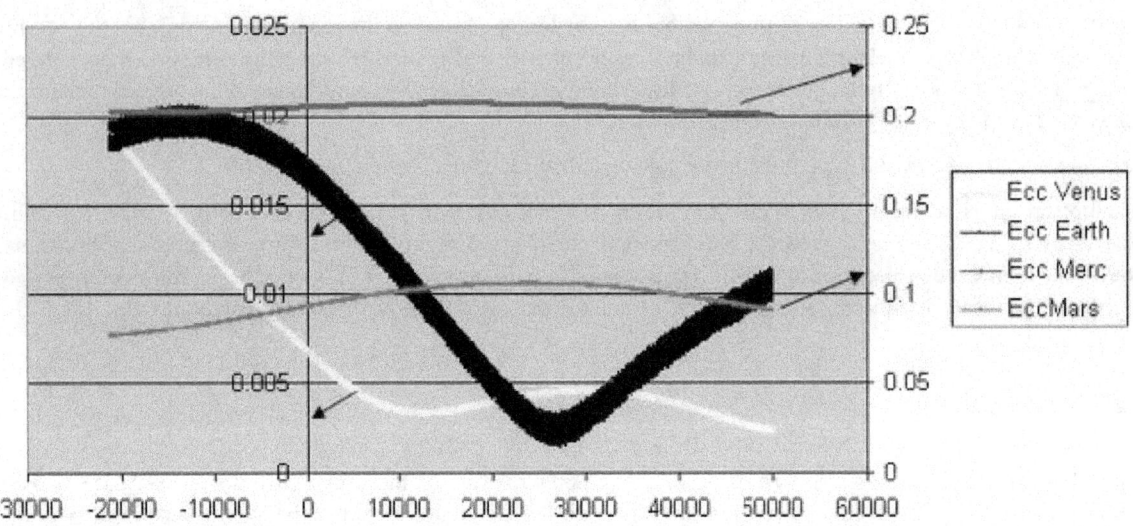

Gravity Simulator plot of the changing orbital eccentricity of Mercury, Venus, Earth, and Mars over the next 50,000 years. The arrows indicate the different scales used. The 0 point on this plot is the year 2007.

The eccentricity of the Earth's orbit is currently about 0.0167; the Earth's orbit is nearly circular. Over hundreds of thousands of years, the eccentricity of the Earth's orbit varies from nearly 0.0034 to almost 0.058 as a result of gravitational attractions among the planets (see graph).[1]

Mercury has the greatest orbital eccentricity of any planet in the Solar System (e=0.2056). Before 2006, Pluto was considered to be the planet with the most eccentric orbit (e=0.248). The Moon's value is 0.0549. For the values for all planets and other celestial bodies in one table, see *List of gravitationally rounded objects of the Solar System*. Sedna the most distant known Trans-Neptunian object in the Solar System has an extremely high eccentricity of 0.85491 due to which it has an aphelion estimated at 937 AU and a perihelion at about 76 AU.

Most of the Solar System's asteroids have orbital eccentricities between 0 and 0.35 with an average value of 0.17.[2] Their comparatively high eccentricities are probably due to the influence of Jupiter and to past collisions.

Comets have very different values of eccentricity. Periodic comets have eccentricities mostly between 0.2 and 0.7,[3] but some of them have highly eccentric elliptical orbits with eccentricities just below 1, for example, Halley's Comet has a value of 0.967. Non-periodic comets follow near-parabolic orbits and thus have eccentricities even closer to 1. Examples include Comet Hale–Bopp with a value of 0.995[4] and comet C/2006 P1 (McNaught) with a value of 1.000019.[5] As Hale–Bopp's value is less than 1, its orbit is elliptical and it will in fact return.[4] Comet McNaught has a hyperbolic orbit while within the influence of the planets, but is still bound to the Sun with an orbital period of about 10^5 years.[6] As of a 2010 Epoch, Comet C/1980 E1 has the largest eccentricity of any known hyperbolic comet with an eccentricity of 1.057,[7] and will leave the Solar System indefinitely.

Neptune's largest moon Triton has an eccentricity of 1.6×10^{-5},[8] the smallest eccentricity of any known body in the Solar System; its orbit is as close to a perfect circle as can be currently measured.

6.5 Mean eccentricity

The mean eccentricity of an object is the average eccentricity as a result of perturbations over a given time period. Neptune currently has an instant (current Epoch) eccentricity of 0.0113,[9] but from 1800 A.D. to 2050 A.D. has a *mean* eccentricity of 0.00859.[10]

6.6 Climatic effect

Orbital mechanics require that the duration of the seasons be proportional to the area of the Earth's orbit swept between the solstices and equinoxes, so when the orbital eccentricity is extreme, the seasons that occur on the far side of the orbit (aphelion) can be substantially longer in duration. Today, northern hemisphere fall and winter occur at closest approach (perihelion), when the earth is moving at its maximum velocity—while the opposite occurs in the southern hemisphere. As a result, in the northern hemisphere, fall and winter are slightly shorter than spring and summer—but in global terms this is balanced with them being longer below the equator. In 2006, the northern hemisphere summer was 4.66 days longer than winter and spring was 2.9 days longer than fall.[11] Apsidal precession slowly changes the place in the Earth's orbit where the solstices and equinoxes occur (this is not the precession of the axis). Over the next 10,000 years, northern hemisphere winters will become gradually longer and summers will become shorter. Any cooling effect in one hemisphere is balanced by warming in the other—and any overall change will, however, be counteracted by the fact that the eccentricity of Earth's orbit will be almost halved, reducing the mean orbital radius and raising temperatures in both hemispheres closer to the mid-interglacial peak.

6.7 See also

- Eccentricity (mathematics)
- Eccentricity vector
- Equation of time
- Milankovitch cycles
- Orbits

6.8 References

[1] A. Berger & M.F. Loutre (1991). "Graph of the eccentricity of the Earth's orbit". Illinois State Museum (Insolation values for the climate of the last 10 million years). Retrieved 2009-12-17.

[2] Asteroids

[3] Lewis, John (2 December 2012). *Physics and Chemistry of the Solar System*. Academic Press. Retrieved 2015-03-29.

[4] "JPL Small-Body Database Browser: C/1995 O1 (Hale-Bopp)" (2007-10-22 last obs). Retrieved 2008-12-05.

[5] "JPL Small-Body Database Browser: C/2006 P1 (McNaught)" (2007-07-11 last obs). Retrieved 2009-12-17.

[6] "Comet C/2006 P1 (McNaught) - facts and figures". Perth Observatory in Australia. 2007-01-22. Retrieved 2011-02-01.

[7] "JPL Small-Body Database Browser: C/1980 E1 (Bowell)" (1986-12-02 last obs). Retrieved 2010-03-22.

[8] David R. Williams (22 January 2008). "Neptunian Satellite Fact Sheet". NASA. Retrieved 2009-12-17.

[9] Williams, David R. (2007-11-29). "Neptune Fact Sheet". NASA. Retrieved 2009-12-17.

[10] "Keplerian elements for 1800 A.D. to 2050 A.D.". JPL Solar System Dynamics. Retrieved 2009-12-17.

[11] This information is concerning the summer of the year 2006 not the current year we are in now.

Prussing, John E., and Bruce A. Conway. Orbital Mechanics. New York: Oxford University Press, 1993.

6.9 External links

- World of Physics: Eccentricity
- The NOAA page on Climate Forcing Data includes (calculated) data from Berger (1978), Berger and Loutre (1991). Laskar et al. (2004) on Earth orbital variations, Includes eccentricity over the last 50 million years and for the coming 20 million years.
- The orbital simulations by Varadi, Ghil and Runnegar (2003) provides series for Earth orbital eccentricity and orbital inclination.
- Kepler's Second law's simulation

Chapter 7

Orbital inclination

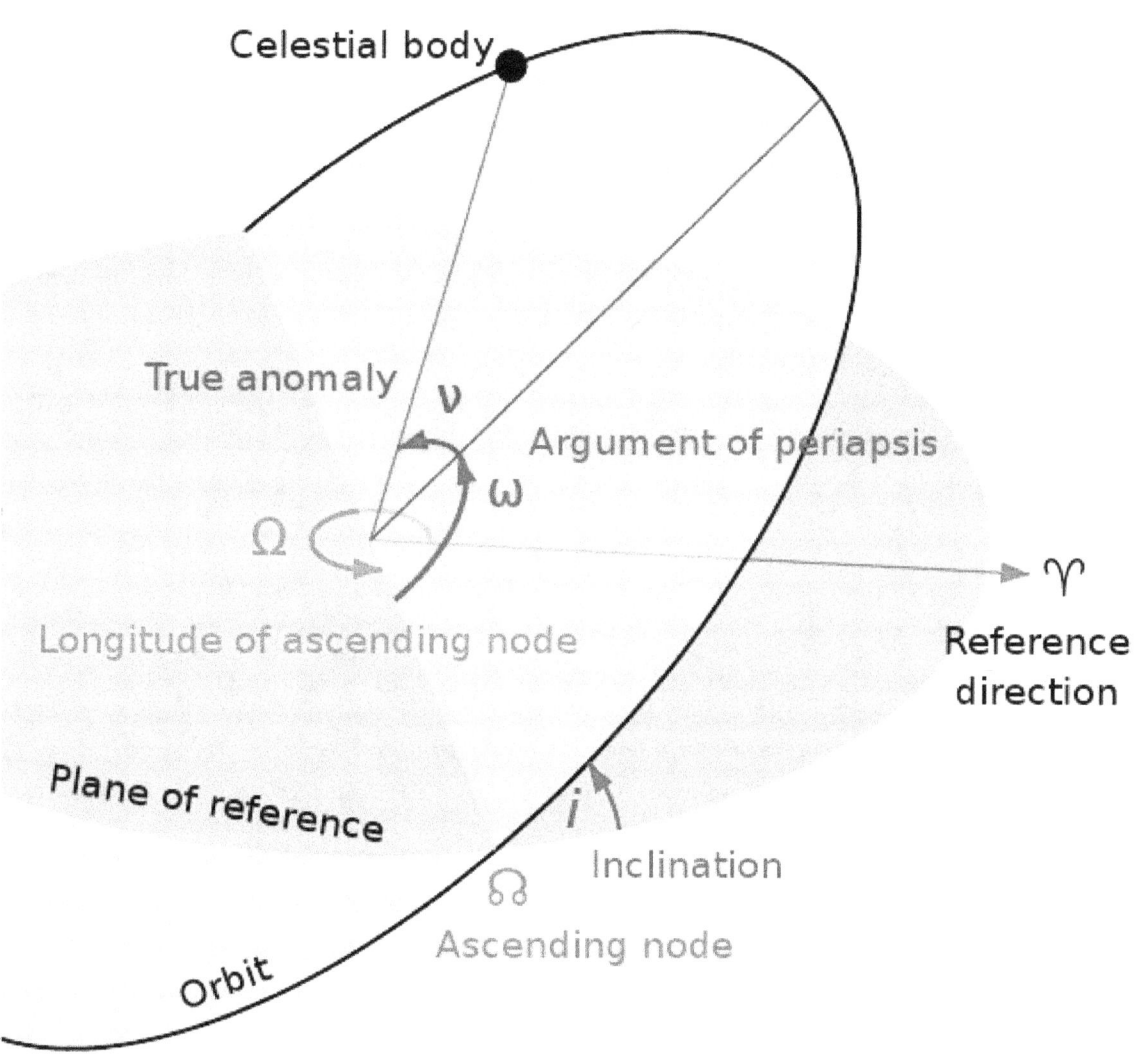

Fig. 1: One view of inclination i (green) and other orbital parameters

"Inclination" redirects here. For other uses, see Inclination (disambiguation).

Orbital inclination is the angle between a reference plane and the orbital plane or axis of direction of an object in orbit around another object.

7.1 Orbits

The inclination is one of the six orbital parameters describing the shape and orientation of a celestial orbit. It is the angular distance of the orbital plane from the plane of reference (usually the primary's equator or the ecliptic), normally stated in degrees.[1]

In the Solar System, the inclination of the orbit of a planet is defined as the angle between the plane of the orbit of the planet and the ecliptic.[2] Therefore Earth's inclination is, by definition, zero. Inclination could instead be measured with respect to another plane, such as the Sun's equator or even Jupiter's orbital plane, but the ecliptic is more practical for Earth-bound observers. Most planetary orbits in the Solar System have relatively small inclinations, both in relation to each other and to the Sun's equator. On the other hand, the dwarf planets Pluto and Eris have inclinations to the ecliptic of 17 degrees and 44 degrees respectively, and the large asteroid Pallas is inclined at 34 degrees.

7.1.1 Natural and artificial satellites

The inclination of orbits of natural or artificial satellites is measured relative to the equatorial plane of the body they orbit if they do so close enough. The equatorial plane is the plane perpendicular to the axis of rotation of the central body.

- an inclination of 0° means the orbiting body orbits the planet in its equatorial plane, in the same direction as the planet rotates;

- an inclination greater than 0° and less than 90° is a prograde orbit.

- an inclination greater than 90° and less than 180° is a retrograde orbit.

- an inclination of exactly 90° is a polar orbit, in which the spacecraft passes over the north and south poles of the planet; and

- an inclination of exactly 180° is a retrograde equatorial orbit.

For impact-generated moons of terrestrial planets not too far from their star, with a large planet–moon distance, it is expected that the orbital planes of moons will tend to be aligned with the planet's orbit around the star due to tides from the star, but if the planet–moon distance is small it may be inclined. For gas giants, the orbits of moons will tend to be aligned with the giant planet's equator because these formed in circumplanetary disks.[4]

The term "critical inclination" is often used when describing artificial satellites in orbit around Earth. This term refers to a satellite orbiting with an inclination of 63.4°. This inclination is described as critical as there is zero apogee drift for satellites in elliptical orbits at this inclination.[5]

7.1.2 Exoplanets and multiple star systems

The inclination of exoplanets or members of multiple stars is the angle of the plane of the orbit relative to the plane perpendicular to the line-of-sight from Earth to the object.

- An inclination of 0° is a face-on orbit, meaning the plane of its orbit is parallel to the sky.

- An inclination of 90° is an edge-on orbit, meaning the plane of its orbit is perpendicular to the sky.

Since the word 'inclination' is used in exoplanet studies for this line-of-sight inclination then the angle between the planet's orbit and the star's rotation must use a different word and is termed the spin-orbit angle or spin-orbit alignment. In most cases the orientation of the star's rotational axis is unknown.

Because the radial-velocity method more easily finds planets with orbits closer to edge-on, most exoplanets found by this method have inclinations between 45° and 135°, although in most cases the inclination is not known. Consequently, most exoplanets found by radial velocity have true masses no more than 70% greater than their minimum masses. If the orbit is almost face-on, especially for superjovians detected by radial velocity, then those objects may actually be brown dwarfs or even red dwarfs. One particular example is HD 33636 B, which has true mass 142 MJ, corresponding to an M6V star, while its minimum mass was 9.28 MJ.

If the orbit is almost edge-on, then the planet can be seen transiting its star.

7.2 Other meanings

- For planets and other rotating celestial bodies, the angle of the axis of rotation with respect to the normal to plane of the orbit is sometimes also called inclination or axial inclination, but to avoid ambiguity can be called axial tilt or **obliquity**.

- In geology, the magnetic inclination is the angle made by a compass needle with respect to the horizontal surface of Earth at a given latitude.

7.3 Calculation

In astrodynamics, the inclination i can be computed from the orbital momentum vector **h** (or any vector perpendicular to the orbital plane) as $i = \arccos \frac{h_z}{|\mathbf{h}|}$, where h_z is the z-component of **h**.

Mutual inclination of two orbits may be calculated from their inclinations to another plane using cosine rule for angles.

7.4 See also

- Altitude (astronomy)
- Axial tilt
- Azimuth
- Beta Angle
- Kepler orbits
- Kozai effect
- Orbital inclination change

7.5 References

[1] Chobotov, Vladimir A. (2002). *Orbital Mechanics* (3rd ed.). AIAA. pp. 28–30;. ISBN 1-56347-537-5.

[2] McBride, Neil; Bland, Philip A.; Gilmour, Iain (2004). *An Introduction to the Solar System*. Cambridge University Press. p. 248. ISBN 0-521-54620-6.

[3] "The MeanPlane (Invariable plane) of the Solar System passing through the barycenter". 2009-04-03. Retrieved 2009-04-10. (produced with Solex 10 written by Aldo Vitagliano)

7.5. REFERENCES

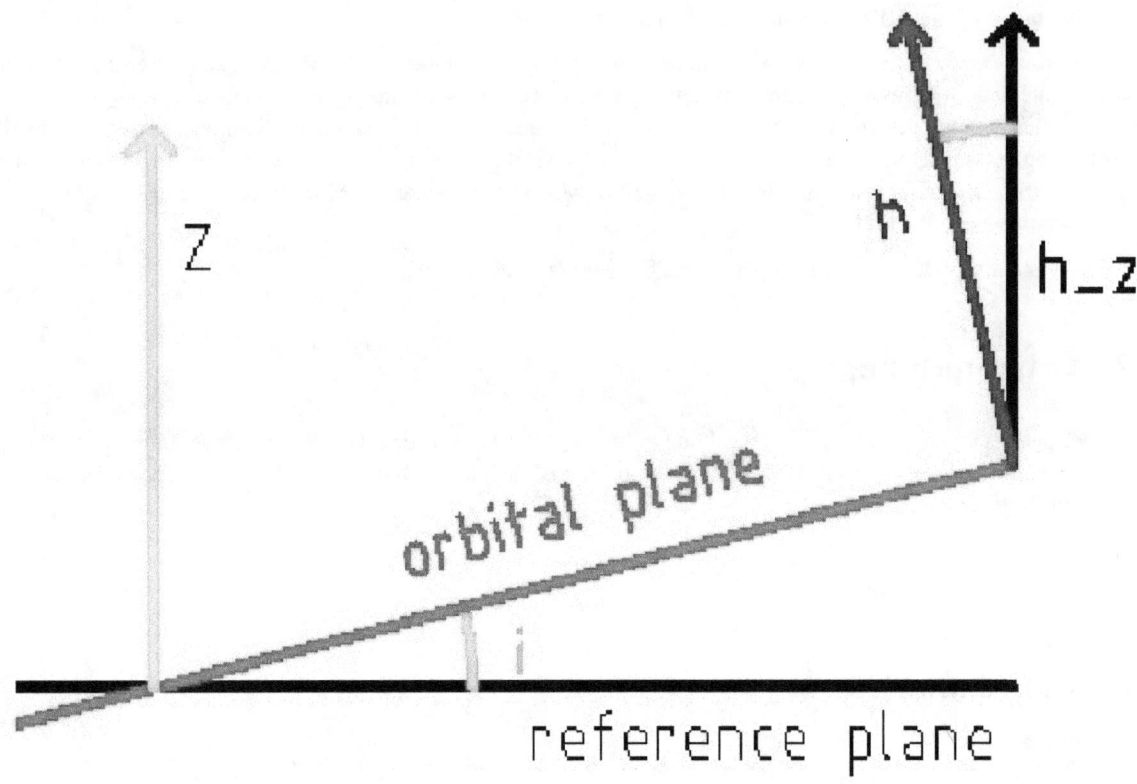

components of the calculation of the orbital inclination from the momentum vector

[4] Moon formation and orbital evolution in extrasolar planetary systems-A literature review, K Lewis – EPJ Web of Conferences, 2011 – epj-conferences.org

[5] Arctic Communications System Utilizing Satellites in Highly Elliptical Orbits, Lars Løge – Section 3.1, Page 17

Chapter 8

Argument of periapsis

The **argument of periapsis** (also called **argument of perifocus** or **argument of pericenter**), symbolized as ω, is one of the orbital elements of an orbiting body. Specifically, ω is the angle from the body's ascending node to its periapsis, measured in the direction of motion. For specific types of orbits, words such as *perihelion* (for Sun-centered orbits), *perigee* (for Earth-centered orbits), *periastron* (for orbits around stars) and so on may replace the word *periapsis*. See apsis for more information.

An argument of periapsis of 0° means that the orbiting body will be at its closest approach to the central body at the same moment that it crosses the plane of reference from South to North. An argument of periapsis of 90° means that the orbiting body will reach periapsis at its northmost distance from the plane of reference.

Adding the argument of periapsis to the longitude of the ascending node gives the longitude of the periapsis. However, especially in discussions of binary stars and exoplanets, the terms "longitude of periapsis" or "longitude of periastron" are often used synonymously with "argument of periapsis".

8.1 Calculation

In astrodynamics the **argument of periapsis** ω can be calculated as follows:

$$\omega = \arccos \frac{\mathbf{n} \cdot \mathbf{e}}{|\mathbf{n}||\mathbf{e}|}$$

(if $e_z < 0$ then $\omega = 2\pi - \omega$)

where:

- $\mathbf{n}$ is a vector pointing towards the ascending node (i.e. the z-component of $\mathbf{n}$ is zero).
- $\mathbf{e}$ is the eccentricity vector (a vector pointing towards the periapsis).

In the case of equatorial orbits (which have no ascending node), the argument is strictly undefined. However, if the convention of setting the longitude of the ascending node Ω to 0 is followed, then the value of ω follows from the two-dimensional case:

$$\omega = \arctan 2(e_y, e_x)$$

(if the orbit is clockwise (i.e. $(\mathbf{r} \times \mathbf{v})_z < 0$) then $\omega = 2\pi - \omega$)

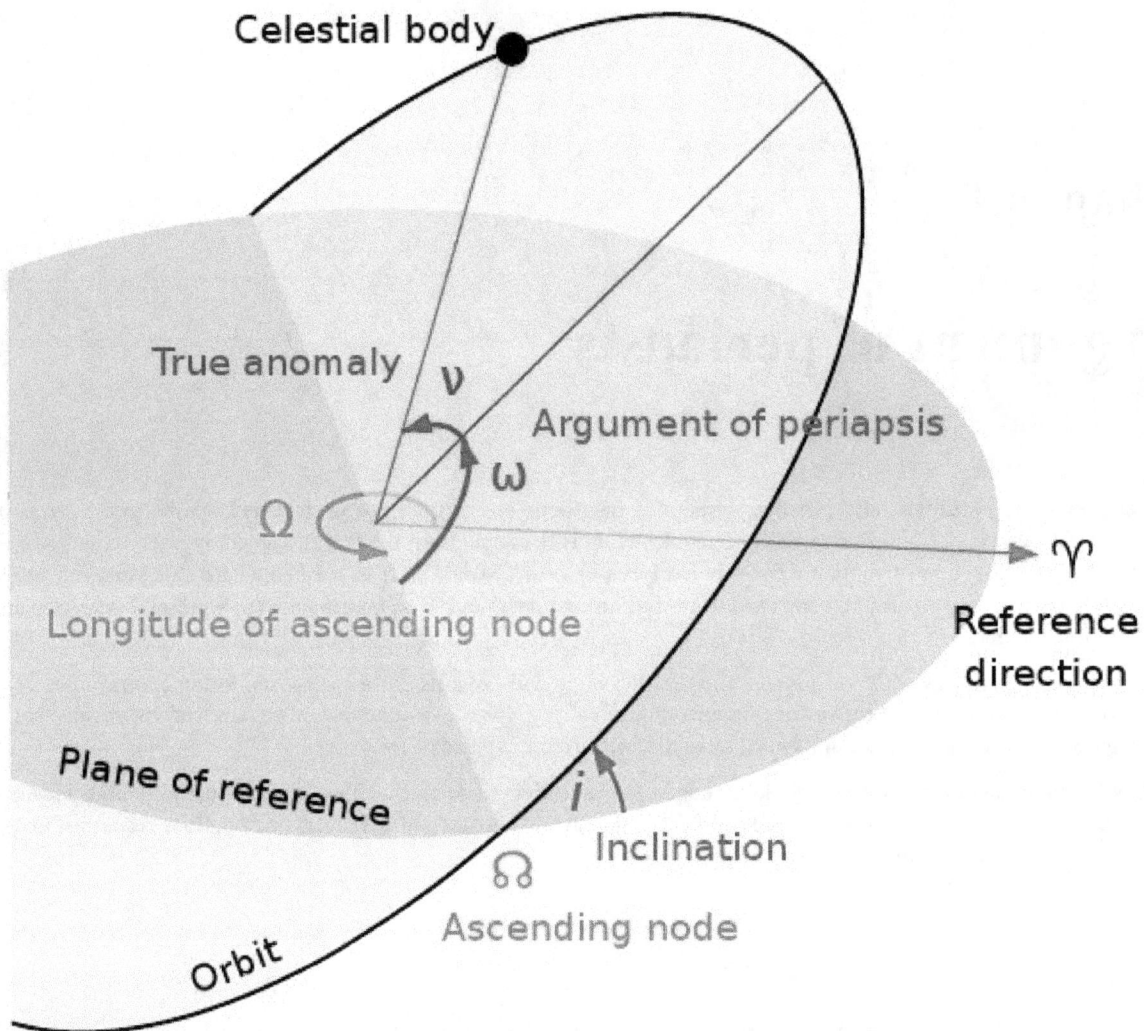

Fig. 1: *Diagram of orbital elements, including the argument of periapsis (ω).*

where:

- e_x and e_y are the *x* and *y* components of the eccentricity vector **e**.

In the case of circular orbits it is often assumed that the periapsis is placed at the ascending node and therefore $\varpi = 0$.

8.2 See also

- Kepler orbit
- Orbital mechanics
- Orbital node

8.3 References

8.4 External links

Chapter 9

Longitude of the ascending node

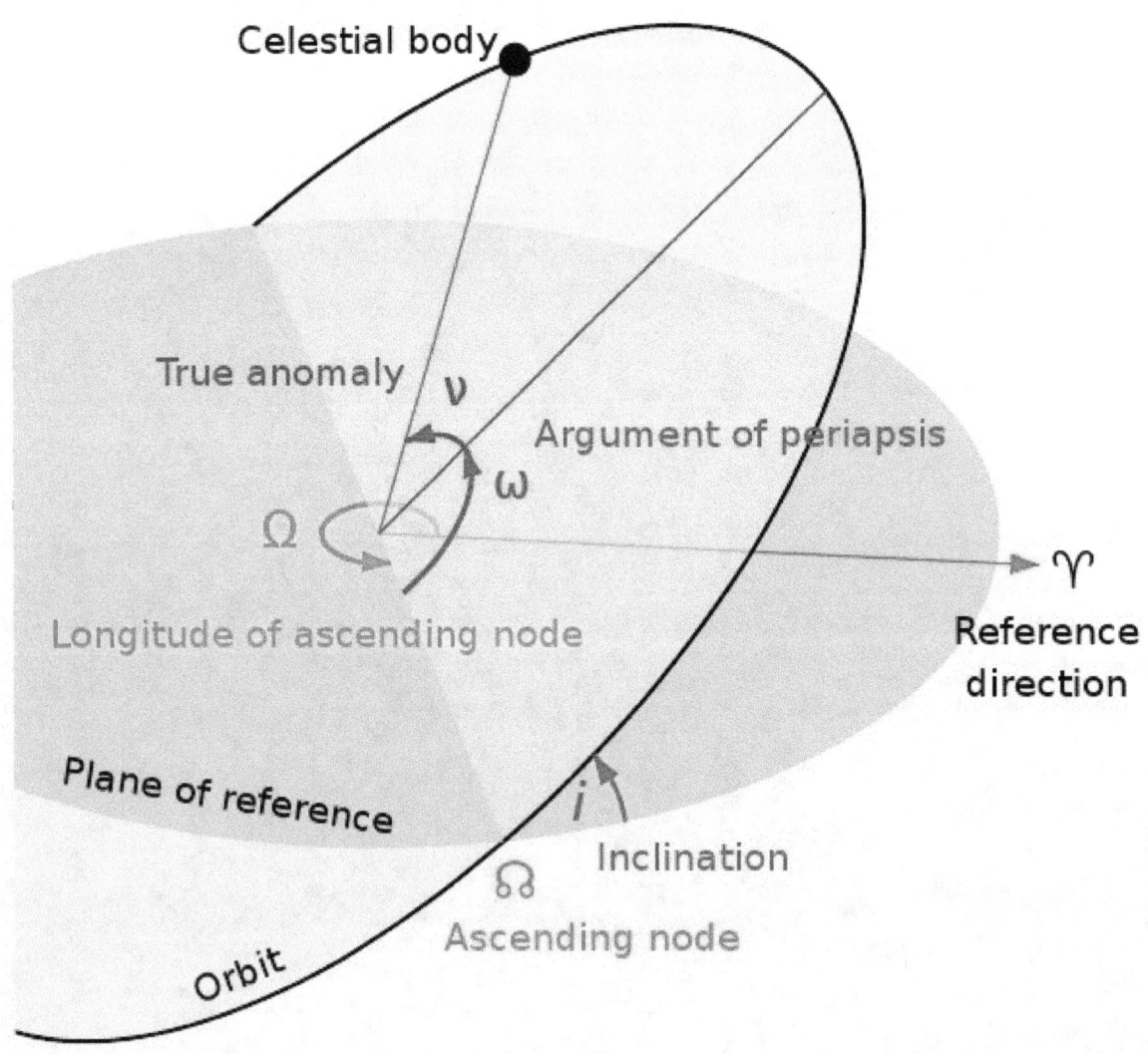

The longitude of the ascending node.

The **longitude of the ascending node** (☊ or Ω) is one of the orbital elements used to specify the orbit of an object in space. It is the angle from a reference direction, called the *origin of longitude*, to the direction of the ascending node, measured in a reference plane.[1] Commonly used reference planes and origins of longitude include:

- For a geocentric orbit, Earth's equatorial plane as the reference plane, and the First Point of Aries as the origin of longitude. In this case, the longitude is also called the **right ascension of the ascending node**, or **RAAN**. The angle is measured eastwards (or, as seen from the north, counterclockwise) from the First Point of Aries to the node.[2][3]

- For a heliocentric orbit, the ecliptic as the reference plane, and the First Point of Aries as the origin of longitude. The angle is measured counterclockwise (as seen from north of the ecliptic) from the First Point of Aries to the node.[2]

- For an orbit outside the Solar System, the plane through the primary perpendicular to a line through the observer and the primary (called the *plane of the sky*) as the reference plane, and north, i.e., the perpendicular projection of the direction from the observer to the North Celestial Pole onto the plane of the sky, as the origin of longitude. The angle is measured eastwards (or, as seen by the observer, counterclockwise) from north to the node.[4], pp. 40, 72, 137; [5], chap. 17.

In the case of a binary star known only from visual observations, it is not possible to tell which node is ascending and which is descending. In this case the orbital parameter which is recorded is the **longitude of the node**, Ω, which is the longitude of whichever node has a longitude between 0 and 180 degrees.[5], chap. 17;[4], p. 72.

9.1 Calculation from state vectors

In astrodynamics, the longitude of the ascending node can be calculated from the specific relative angular momentum vector h as follows:

$$\mathbf{n} = \mathbf{k} \times \mathbf{h} = (-h_y, h_x, 0)$$

$$\Omega = \arccos \frac{n_x}{|\mathbf{n}|} \quad (n_y \geq 0);$$

$$\Omega = 2\pi - \arccos \frac{n_x}{|\mathbf{n}|} \quad (n_y < 0).$$

Here, $n=<n_x, n_y, n_z>$ is a vector pointing towards the ascending node. The reference plane is assumed to be the *xy*-plane, and the origin of longitude is taken to be the positive *x*-axis. k is the unit vector (0, 0, 1), which is the normal vector to the *xy* reference plane.

For non-inclined orbits (with inclination equal to zero), Ω is undefined. For computation it is then, by convention, set equal to zero; that is, the ascending node is placed in the reference direction, which is equivalent to letting n point towards the positive *x*-axis.

9.2 See also

- Kepler orbits
- Equinox
- Orbital node
- perturbation of the orbital plane can cause revolution of the ascending node

9.3 References

[1] Parameters Describing Elliptical Orbits, web page, accessed May 17, 2007.

[2] Orbital Elements and Astronomical Terms, Robert A. Egler, Dept. of Physics, North Carolina State University. Web page, accessed May 17, 2007.

[3] Keplerian Elements Tutorial, amsat.org, accessed May 17, 2007.

[4] *The Binary Stars*, R. G. Aitken, New York: Semi-Centennial Publications of the University of California, 1918.

[5] *Celestial Mechanics*, Jeremy B. Tatum, on line, accessed May 17, 2007.

Chapter 10

Longitude of the periapsis

In celestial mechanics, the **longitude of the periapsis** (symbolized ϖ) of an orbiting body is the longitude (measured from the point of the vernal equinox) at which the periapsis (closest approach to the central body) would occur if the body's inclination were zero. For motion of a planet around the Sun, this position could be called **longitude of perihelion**. The longitude of periapsis is a compound angle, with part of it being measured in the plane of reference and the rest being measured in the plane of the orbit. Likewise, any angle derived from the longitude of periapsis (e.g. mean longitude and true longitude) will also be compound.

Sometimes, the term *longitude of periapsis* is used to refer to ω, the angle between the ascending node and the periapsis. That usage of the term is especially common in discussions of binary stars and exoplanets.[1] However, the angle ω is less ambiguously known as the argument of periapsis.

10.1 Calculation from state vectors

ϖ is the sum of the longitude of ascending node Ω and the argument of periapsis ω :

$$\varpi = \Omega + \omega$$

which are derived from the orbital state vectors.

10.2 References

[1] See e.g. p. 201, *The Binary Stars*, Robert Grant Aitken, Semicentennial Publications of the University of California, 1918, or Format, *Sixth Catalog of Orbits of Visual Binary Stars*, William I. Hartkopf & Brian D. Mason, U. S. Naval Observatory, Washington, D.C. Accessed on line October 25, 2008

10.3 External links

- Determination of the Earth's Orbital Parameters Past and future longitude of perihelion for Earth.

Chapter 11

True anomaly

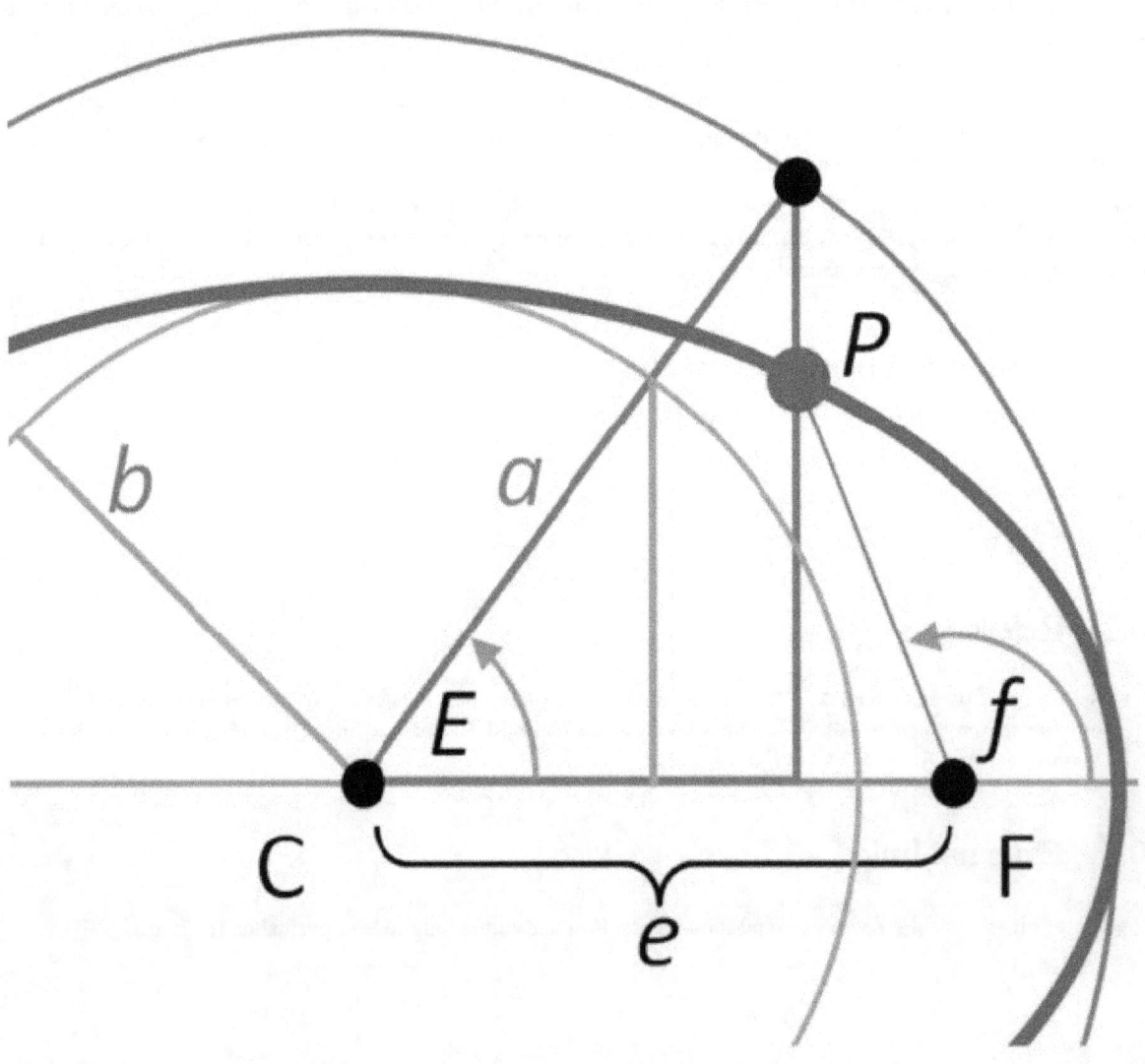

The true anomaly of point P is the angle f. The center of the ellipse is point C, and the focus is point F.

In celestial mechanics, **true anomaly** is an angular parameter that defines the position of a body moving along a Keplerian orbit. It is the angle between the direction of periapsis and the current position of the body, as seen from the main focus of the ellipse (the point around which the object orbits).

The true anomaly is usually denoted by the Greek letters ν or θ, or the Latin letter f.

The true anomaly is one of three angular parameters ("anomalies") that define a position along an orbit, the other two being the eccentric anomaly and the mean anomaly.

11.1 Formulas

11.1.1 From state vectors

For elliptic orbits **true anomaly** ν can be calculated from orbital state vectors as:

$$\nu = \arccos \frac{\mathbf{e} \cdot \mathbf{r}}{|\mathbf{e}||\mathbf{r}|} \text{ (if } \mathbf{r} \cdot \mathbf{v} < 0 \text{ then replace } \nu \text{ by } 2\pi - \nu \text{)}$$

where:

- $\mathbf{v}$ is orbital velocity vector of the orbiting body,
- $\mathbf{e}$ is eccentricity vector,
- $\mathbf{r}$ is orbital position vector (segment fp) of the orbiting body.

Circular orbit

For circular orbits the true anomaly is undefined because circular orbits do not have a uniquely determined periapsis. Instead one uses the argument of latitude u :

$$u = \arccos \frac{\mathbf{n} \cdot \mathbf{r}}{|\mathbf{n}||\mathbf{r}|} \text{ (if } \mathbf{n} \cdot \mathbf{v} > 0 \text{ then replace } u \text{ by } 2\pi - u \text{)}$$

where:

- $\mathbf{n}$ is vector pointing towards the ascending node (i.e. the z-component of $\mathbf{n}$ is zero).

Circular orbit with zero inclination

For circular orbits with zero inclination the argument of latitude is also undefined, because there is no uniquely determined line of nodes. One uses the true longitude instead:

$$l = \arccos \frac{r_x}{|\mathbf{r}|} \text{ (if } v_x > 0 \text{ then replace } l \text{ by } 2\pi - l \text{)}$$

where:

- r_x is x-component of orbital position vector $\mathbf{r}$,
- v_x is x-component of orbital velocity vector $\mathbf{v}$.

11.1.2 From the eccentric anomaly

The relation between the true anomaly ν and the eccentric anomaly E is:

$$\cos \nu = \frac{\cos E - e}{1 - e \cdot \cos E}$$

or using sin[1] and tan

$$\sin \nu = \frac{\sqrt{1 - e^2} \sin E}{1 - e \cos E}$$

$$\tan \nu = \frac{\sin \nu}{\cos \nu} = \frac{\sqrt{1 - e^2} \sin E}{\cos E - e}$$

or equivalently

$$\tan \frac{\nu}{2} = \sqrt{\frac{1 + e}{1 - e}} \tan \frac{E}{2}.$$

Therefore

$$\nu = 2 \arg\left(\sqrt{1 - e} \cos \frac{E}{2}, \sqrt{1 + e} \sin \frac{E}{2} \right)$$

where $\arg(x, y)$ is the polar argument of the vector (x, y) (available in many programming languages as the library function atan2(y, x) in Fortran and MATLAB, or as ArcTan(x, y) in Wolfram Mathematica).

11.1.3 Radius from true anomaly

The radius (distance from the focus of attraction and the orbiting body) is related to the true anomaly by the formula

$$r = a \cdot \frac{1 - e^2}{1 + e \cdot \cos \nu}$$

where a is the orbit's semi-major axis (segment cz).

11.2 Notes

[1] Fundamentals of Astrodynamics and Applications by David A. Vallado

11.3 See also

- Kepler's laws of planetary motion
- Eccentric anomaly
- Mean anomaly
- Ellipse
- Hyperbola

11.4 References

- Murray, C. D. & Dermott, S. F. 1999, *Solar System Dynamics*, Cambridge University Press, Cambridge. ISBN 0-521-57597-4

- Plummer, H.C., 1960, *An Introductory treatise on Dynamical Astronomy*, Dover Publications, New York. OCLC 1311887 (Reprint of the 1918 Cambridge University Press edition.)

Chapter 12

Mean anomaly

In celestial mechanics, **mean anomaly** is a parameter relating position and time for a body moving in a Kepler orbit. It is based on equal areas being swept in equal intervals of time by a line joining the focus and the orbiting body (Kepler's second law).

The mean anomaly increases uniformly from 0 to 2π radians during each orbit. However, it is not an angle. Due to Kepler's second law, the mean anomaly is proportional to the area swept by the focus-to-body line since the last periapsis.

The mean anomaly is usually denoted by the letter M, and is given by the formula:

$$M = nt = \sqrt{\frac{G(M_\star + m)}{a^3}}\, t$$

where n is the mean motion, a is the length of the orbit's semi-major axis, $M_\star$ and m are the orbiting masses, and G is the gravitational constant.

The mean anomaly is the time since the last periapsis multiplied by the mean motion, and the mean motion is 2π radians divided by the duration of a full orbit.

The mean anomaly is one of three angular parameters ("anomalies") that define a position along an orbit, the other two being the eccentric anomaly and the true anomaly. If the mean anomaly is known at any given instant, it can be calculated at any later (or prior) instant by simply adding (or subtracting) $\sqrt{\frac{G(M_\star + m)}{a^3}}\, \delta t$ where δt represents the time difference. The other anomalies can hence be calculated.

12.1 Formulas

The mean anomaly M can be computed from the eccentric anomaly E and the eccentricity e with Kepler's Equation:

$$M = E - e \cdot \sin E$$

To find the position of the object in an elliptic Kepler orbit at a given time t, the mean anomaly is found by multiplying the time and the mean motion, then it is used to find the eccentric anomaly by solving Kepler's equation.

It is also frequently seen:

$$M = M_0 + nt$$

Again n is the mean motion. However, t, in this instance, is the *time since epoch*, which is how much time has passed since the measurement of M_0 was taken. The value M_0 denotes the *mean anomaly at epoch*, which is the mean anomaly at the time the measurement was taken.

12.2 See also

- Kepler orbit
- Ellipse
- Eccentric anomaly
- True anomaly

12.3 References

- Murray, C. D. & Dermott, S. F. 1999, *Solar System Dynamics*, Cambridge University Press, Cambridge.
- Plummer, H.C., 1960, *An Introductory treatise on Dynamical Astronomy*, Dover Publications, New York. (Reprint of the 1918 Cambridge University Press edition.)

Chapter 13

Eccentric anomaly

In orbital mechanics, **eccentric anomaly** is an angular parameter that defines the position of a body that is moving along an elliptic Kepler orbit. The eccentric anomaly is one of three angular parameters ("anomalies") that define a position along an orbit, the other two being the true anomaly and the mean anomaly.

13.1 Graphical Representation

Consider the ellipse with equation given by:

$$\frac{x^2}{a^2} + \frac{y^2}{b^2} = 1,$$

where a is the *semi-major* axis and b is the *semi-minor* axis.

For a point on the ellipse, $P=P(x, y)$, representing the position of an orbiting body in an elliptical orbit, the eccentric anomaly is the angle E in the figure to the right. The eccentric anomaly, E, is observed by drawing a right triangle with one vertex at the center of the ellipse, having hypotenuse a (equal to the *semi-major* axis of the ellipse), and opposite side (perpendicular to the *major-axis* and touching the point P' on the auxiliary circle of radius a) that crosses through the point P. The eccentric anomaly is measured in the same direction as the true anomaly, shown in the figure as f. The eccentric anomaly E in terms of these coordinates is given by:[1]

$$\cos E = \frac{x}{a},$$

$$\sin E = \frac{y}{b}$$

The second equation is established using the relationship $\left(\frac{y}{b}\right)^2 = 1 - \cos^2 E = \sin^2 E$, which implies that $\sin E = \pm\frac{y}{b}$. The equation $\sin E = -\frac{y}{b}$ is immediately able to be ruled out since it traverses the ellipse in the wrong direction. It can also be noted that the second equation can be viewed as being the similar triangle with adjacent side through P and the minor auxiliary circle, hypotenuse b, and whose opposite side is y.

13.2 Formulas

13.2.1 Radius and eccentric anomaly

The eccentricity e is defined as:

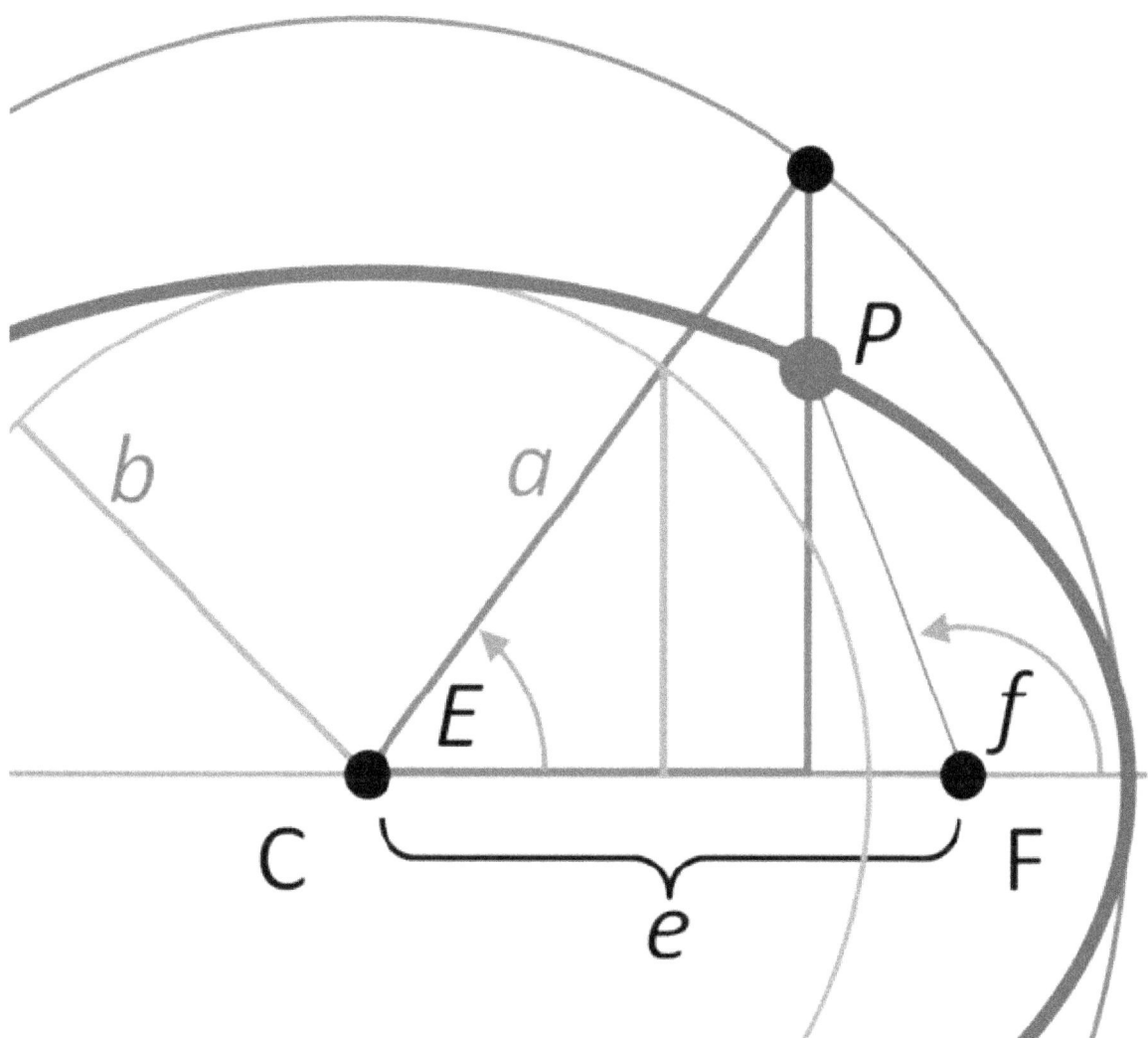

The eccentric anomaly of point P is the angle E. The center of the ellipse is point C, and the focus is point F.

$$e = \sqrt{1 - \left(\frac{b}{a}\right)^2}.$$

From Pythagoras' theorem applied to the triangle with r as hypotenuse:

$$\begin{aligned} r^2 &= b^2 \sin^2 E + (ae - a\cos E)^2 \\ &= a^2(1-e^2)(1-\cos^2 E) + a^2(e^2 - 2e\cos E + \cos^2 E) \\ &= a^2 - 2a^2 e \cos E + a^2 e^2 \cos^2 E \\ &= a^2(1 - e\cos E)^2 \end{aligned}$$

Thus, the radius (distance from the focus to point P) is related to the eccentric anomaly by the formula

$$r = a(1 - e \cdot \cos E).$$

With this result the eccentric anomaly can be determined from the true anomaly as shown next.

13.2.2 From the true anomaly

The *true anomaly* is the angle labeled f in the figure, located at the focus of the ellipse; it is often referred to as θ as in the calculations below. The true anomaly and the eccentric anomaly are related as follows:[2]

Using the formula for r above, the sine and cosine of E are found in terms of θ:

$$\cos E = \frac{x}{a} = \frac{ae + r\cos\theta}{a} = e + (1 - e\cos E)\cos\theta \rightarrow \cos E = \frac{e + \cos\theta}{1 + e\cos\theta}$$

$$\sin E = \sqrt{1 - \cos^2 E} = \frac{\sqrt{1 - e^2}\sin\theta}{1 + e\cos\theta}.$$

Hence,

$$\tan E = \frac{\sin E}{\cos E} = \frac{\sqrt{1 - e^2}\sin\theta}{e + \cos\theta}.$$

Angle E is therefore the adjacent angle of a right triangle with hypotenuse $1 + e\cos\theta$, adjacent side $e + \cos\theta$, and opposite side $\sqrt{(1-e^2)}\sin\theta$.

Also,

$$\tan\frac{\theta}{2} = \sqrt{\frac{1+e}{1-e}} \cdot \tan\frac{E}{2}$$

Substituting $\cos E$ as found above into the expression for r, the radial distance from the focal point to the point P, can be found in terms of the true anomaly as well:[2]

$$r = \frac{a\left(1 - e^2\right)}{1 + e\cos\theta}.$$

13.2.3 From the mean anomaly

The eccentric anomaly E is related to the mean anomaly M by Kepler's equation:[3]

$$M = E - e \cdot \sin E$$

This equation does not have a closed-form solution for E given M. It is usually solved by numerical methods, e.g. Newton-Raphson method.

13.3 In-line references and notes

[1] George Albert Wentworth (1914). "The ellipse §126". *Elements of analytic geometry* (2nd ed.). Ginn & Co. p. 141.

[2] James Bao-yen Tsui (2000). *Fundamentals of global positioning system receivers: a software approach* (3rd ed.). John Wiley & Sons. p. 48. ISBN 0-471-38154-3.

[3] Michel Capderou (2005). "Definition of the mean anomaly, Eq. 1.68". *Satellites: orbits and missions*. Springer. p. 21. ISBN 2-287-21317-1.

13.4 Background references

- Murray, Carl D.; & Dermott, Stanley F. (1999); *Solar System Dynamics*, Cambridge University Press, Cambridge, GB

- Plummer, Henry C. K. (1960); *An Introductory Treatise on Dynamical Astronomy*, Dover Publications, New York, NY (Reprint of the 1918 Cambridge University Press edition)

13.5 See also

- Eccentricity vector
- Orbital eccentricity

Chapter 14

Epoch (astronomy)

In astronomy, an **epoch** is a moment in time used as a reference point for some time-varying astronomical quantity, such as the celestial coordinates or elliptical orbital elements of a celestial body, because these are subject to perturbations and vary with time.[1] These time-varying astronomical quantities might include, for example, the mean longitude or mean anomaly of a body, the node of its orbit relative to a reference plane, the direction of the apogee or aphelion of its orbit, or the size of the major axis of its orbit.

The main use of astronomical quantities specified in this way is to calculate other relevant parameters of motion, in order to predict future positions and velocities. The applied tools of the disciplines of celestial mechanics or its subfield orbital mechanics (for predicting orbital paths and positions for bodies in motion under the gravitational effects of other bodies) can be used to generate an ephemeris, a table of values giving the positions and velocities of astronomical objects in the sky at a given time or times.

Astronomical quantities can be specified in any of several ways, for example, as a polynomial function of the time-interval, with an epoch as a temporal point of origin (this is a common current way of using an epoch). Alternatively, the time-varying astronomical quantity can be expressed as a constant, equal to the measure that it had at the epoch, leaving its variation over time to be specified in some other way—for example, by a table, as was common during the 17th and 18th centuries.

The word **epoch** was often used in a different way in older astronomical literature, e.g. during the 18th century, in connection with astronomical tables. At that time, it was customary to denote as "epochs", not the standard date and time of origin for time-varying astronomical quantities, but rather the values at that date and time *of those time-varying quantities themselves*.[2] In accordance with that alternative historical usage, an expression such as 'correcting the epochs' would refer to the adjustment, usually by a small amount, of the values of the tabulated astronomical quantities applicable to a fixed standard date and time of reference (and not, as might be expected from current usage, to a change from one date and time of reference to a different date and time).

14.1 Epoch versus equinox

Astronomical data are often specified not only in their relation to an epoch or date of reference, but also in their relations to other conditions of reference, such as coordinate systems specified by "equinox", or "equinox and equator", or "equinox and ecliptic" – when these are needed for fully specifying astronomical data of the considered type.

14.1.1 Date-references for coordinate systems

When the data are dependent for their values on a particular coordinate system, the date of that coordinate system needs to be specified directly or indirectly.

Celestial coordinate systems most commonly used in astronomy are equatorial coordinates and ecliptic coordinates. These are defined relative to the (moving) vernal equinox position, which itself is determined by the orientations of the Earth's rotation axis and orbit around the Sun. Their orientations vary (though slowly, e.g. due to precession), and there is an infinity of such coordinate systems possible. Thus the coordinate systems most used in astronomy need their own date-reference because the coordinate systems of that type are themselves in motion, e.g. by the precession of the equinoxes, nowadays often resolved into precessional components, separate precessions of the equator and of the ecliptic.

The epoch of the coordinate system need not be the same, and often in practice is not the same, as the epoch for the data themselves.

The difference between reference to an epoch alone, and a reference to a certain equinox with equator or ecliptic, is therefore that the reference to the epoch contributes to specifying the date of the values of astronomical variables themselves; while the reference to an equinox along with equator/ecliptic, of a certain date, addresses the identification of, or changes in, the coordinate system in terms of which those astronomical variables are expressed. (Sometimes the word 'equinox' may be used alone, e.g. where it is obvious from the context to users of the data in which form the considered astronomical variables are expressed, in equatorial form or ecliptic form.)

The equinox with equator/ecliptic of a given date defines which coordinate system is used. Most standard coordinates in use today refer to 2000 Jan 1.5 **TT** (i.e. to 12h on the Terrestrial Time scale on 2000 Jan 1), which occurred about 64 seconds sooner than noon UT1 on the same date (see ΔT). Before about 1984, coordinate systems dated to 1950 or 1900 were commonly used.

There is a special meaning of the expression "equinox (and ecliptic/equator) **of date**". When coordinates are expressed as polynomials in time relative to a reference frame defined in this way, that means the values obtained for the coordinates in respect of any interval t after the stated epoch, are in terms of the coordinate system of the same date as the obtained values themselves, i.e. the date of the coordinate system is equal to (epoch + t).[3]

It can be seen that the date of the coordinate system need not be the same as the epoch of the astronomical quantities themselves. But in that case (apart from the "equinox of date" case described above), two dates will be associated with the data: one date is the epoch for the time-dependent expressions giving the values, and the other date is that of the coordinate system in which the values are expressed.

For example, orbital elements, especially osculating elements for minor planets, are routinely given with reference to two dates: first, relative to a recent epoch for all of the elements; but some of the data are dependent on a chosen coordinate system, and then it is usual to specify the coordinate system of a standard epoch which often is not the same as the epoch of the data. An example is as follows: For minor planet (5145) Pholus, orbital elements have been given including the following data:[4]

Epoch 2010 Jan. 4.0 TT . . . = JDT 2455200.5
M 72.00071 . . . .(2000.0)
n. 0.01076162 .. . . . Peri . 354.75938
a 20.3181594 . . . . . Node . 119.42656
e. 0.5715321 . . . . . Incl .. 24.66109

where the epoch is expressed in terms of Terrestrial Time, with an equivalent Julian date. Four of the elements are independent of any particular coordinate system: M is mean anomaly (deg), n: mean daily motion (deg/d), a: size of semi-major axis (AU), e: eccentricity (dimensionless). But the argument of perihelion, longitude of the ascending node and the inclination are all coordinate-dependent, and are specified relative to the reference frame of the equinox and ecliptic of another date "2000.0", otherwise known as J2000, i.e. 2000 Jan 1.5 (12h on January 1) or JD 2451545.0.[5]

14.1.2 Epochs and periods of validity

In the particular set of coordinates exampled above, much of the time-dependence of the elements has been omitted as unknown or undetermined, for example, the element n allows an approximate time-dependence of the element M to be calculated, but the other elements and n itself are treated as constant, which represents a temporary approximation, see Osculating elements.

Thus a particular coordinate system (equinox and equator/ecliptic of a particular date, such as J2000.0) could be used forever, but a set of osculating elements for a particular epoch may only be (approximately) valid for a rather limited time, because osculating elements such as those exampled above do not show the effect of future perturbations which will change the values of the elements.

Nevertheless, the period of validity is a different matter in principle, and not the result of the use of an epoch to express the data. In other cases, e.g. the case of a complete analytical theory of the motion of some astronomical body, all of the elements will usually be given in the form of polynomials in interval of time from the epoch, and they will also be accompanied by trigonometrical terms of periodical perturbations specified appropriately. In that case, their period of validity may stretch over several centuries or even millennia on either side of the stated epoch.

Some data and some epochs have a long period of use for other reasons. For example, the boundaries of the IAU constellations are specified relative to an equinox from near the beginning of the year 1875. This is a matter of convention, but the convention is defined in terms of the equator and ecliptic as they were in 1875. To find out in which constellation a particular comet stands today, the current position of that comet must be expressed in the coordinate system of 1875 (equinox/equator of 1875). Thus that coordinate system can still be used today, even though most comet predictions made originally for 1875 (epoch = 1875) would no longer, because of the lack of information about their time-dependence and perturbations, be useful today.

14.2 Changing the standard equinox and epoch

To calculate the visibility of a celestial object for an observer at a specific time and place on the Earth, the coordinates of the object are needed relative to a coordinate system of current date. If coordinates relative to some other date are used, then that will cause errors in the results. The magnitude of those errors increases with the time difference between the date and time of observation and the date of the coordinate system used, because of precession of the equinoxes. If the time difference is small, then fairly easy and small corrections for the precession may well suffice. If the time difference gets large, then fuller and more accurate corrections must be applied. For this reason, a star position read from a star atlas or catalog based on a sufficiently old equinox and equator cannot be used without corrections, if reasonable accuracy is required.

Additionally, stars move relative to each other through space. Apparent motion across the sky relative to other stars is called proper motion. Most stars have very small proper motions, but a few have proper motions that accumulate to noticeable distances after a few tens of years. So, some stellar positions read from a star atlas or catalog for a sufficiently old epoch require proper motion corrections as well, for reasonable accuracy.

Due to precession and proper motion, star data become less useful as the age of the observations and their epoch, and the equinox and equator to which they are referred, get older. After a while, it is easier or better to switch to newer data, generally referred to a newer epoch and equinox/equator, than to keep applying corrections to the older data.

14.3 Specifying an epoch or equinox

Epochs and equinoxes are moments in time, so they can be specified in the same way as moments that indicate things other than epochs and equinoxes. The following standard ways of specifying epochs and equinoxes seem most popular:

- Julian days, e.g., JD 2433282.4235 for 1950 January 0.9235 TT
- Besselian years (see below), e.g., 1950.0 or B1950.0 for 1950 January 0.9235 TT
- Julian years, e.g., J2000.0 for 2000 January 1.5000 TT

All three of these are expressed in TT = Terrestrial Time.

Besselian years, used mostly for star positions, can be encountered in older catalogs but are now becoming obsolete. The Hipparcos catalog summary,[6] for example, defines the "catalog epoch" as J1991.25 (one quarter-year after the start of calendar year 1991).

14.4 Besselian years

A Besselian year is named after the German mathematician and astronomer Friedrich Bessel (1784–1846). Meeus[7] defines the beginning of a Besselian year to be the moment at which the mean longitude of the Sun, including the effect of aberration and measured from the mean equinox of the date, is exactly 280 degrees. This moment falls near the beginning of the corresponding Gregorian year. The definition depended on a particular theory of the orbit of the Earth around the Sun, that of Newcomb (1895), which is now obsolete; for that reason among others, the use of Besselian years has also become or is becoming obsolete.

Lieske[8] says that a "Besselian epoch" can be calculated from the Julian date according to

$$B = 1900.0 + (\text{Julian date} - 2415020.31352) / 365.242198781$$

This relationship is included in the SOFA software library.[9]

Lieske's definition is not exactly consistent with the earlier definition in terms of the mean longitude of the Sun. When using Besselian years, specify which definition is being used.

To distinguish between calendar years and Besselian years, it became customary to add ".0" to the Besselian years. Since the switch to Julian years in the mid-1980s, it has become customary to prefix "B" to Besselian years. So, "1950" is the calendar year 1950, and "1950.0" = "B1950.0" is the beginning of Besselian year 1950.

- The IAU constellation boundaries are defined in the equatorial coordinate system relative to the equinox of B1875.0.
- The Henry Draper Catalog uses the equinox B1900.0.
- The classical star atlas Tabulae Caelestes used B1925.0 as its equinox.

According to Meeus, and also according to the formula given above,

- B1900.0 = JDE 2415020.3135 = 1900 January 0.8135 TT
- B1950.0 = JDE 2433282.4235 = 1950 January 0.9235 TT

14.5 Julian years and J2000

A Julian year is an interval with the length of a mean year in the Julian calendar, i.e. 365.25 days. This interval measure does not itself define any epoch: the Gregorian calendar is in general use for dating. But, standard conventional epochs which are not Besselian epochs have been often designated nowadays with a prefix "J", and the calendar date to which they refer is widely known, although not always the same date in the year: thus "J2000" refers to the instant of 12h on 1 January 2000, and J1900 refers to the instant of 12h (midday) on 0 January 1900, equal to 31 Dec 1899.[10] It is also usual now to specify on what time scale the time of day is expressed in that epoch-designation, e.g. often Terrestrial Time.

In addition, an epoch optionally prefixed by "J" and designated as a year with decimals (2000 +x), where x is positive or negative and quoted to 1 or 2 decimal places, has come to mean a date that is an interval of x Julian years of 365.25 days away from the epoch J2000 = JD 2451545.0 (TT), still corresponding (in spite of the use of the prefix "J" or word "Julian") to the Gregorian calendar date of 2000 Jan 1 at 12h TT (about 64 seconds before noon UTC on the same calendar day).[9] (See also Julian year (astronomy).) Like the Besselian epoch, an arbitrary Julian epoch is therefore related to the Julian date by

$$J = 2000.0 + (\text{Julian date} - 2451545.0)/365.25 \ .$$

The IAU decided at their General Assembly of 1976[11] that the new standard equinox of J2000.0 should be used starting in 1984. Before that, the equinox of B1950.0 seems to have been the standard.

Different astronomers or groups of astronomers used to define epochs to suit themselves, but nowadays standard epochs are generally defined by international agreement through the IAU, so astronomers worldwide can collaborate more effectively. It is inefficient and error-prone if data or observations of one group have to be translated in non-standard ways so that other groups could compare the data with information from other sources. An example of how this works: if a star's position is measured by someone today, he/she then uses a standard transformation to obtain the position expressed in terms of the standard reference frame of J2000, and it is often then this J2000 position which is shared with others.

On the other hand, there has also been an astronomical tradition of retaining observations in just the form in which they were made, so that others can later correct the reductions to standard if that proves desirable, as has sometimes occurred.

The currently-used standard epoch "J2000" is defined by international agreement to be equivalent to:

1. The Gregorian date January 1, 2000 at approximately 12:00 GMT (Greenwich Mean Time).
2. The Julian date 2451545.0 TT (Terrestrial Time).[12]
3. January 1, 2000, 11:59:27.816 TAI (International Atomic Time).[13]
4. January 1, 2000, 11:58:55.816 UTC (Coordinated Universal Time).[14]

14.6 Epoch of the day

In addition to its usual application concerning a reference point for long term astronomical calculations, the term Epoch has also been used to refer to the time of the beginning of the day.

In ordinary usage, the civil day is reckoned by the midnight epoch, that is, the civil day begins at midnight. In older astronomical usage, it was usual, until 1 January 1925, to reckon by a noon epoch, 12 hours after the start of the civil day of the same denomination, so that the day began when the mean sun crossed the meridian at noon.[15]

In traditional cultures and in antiquity other epochs were used. In ancient Egypt days were reckoned from sunrise to sunrise, following a morning epoch. It has been suggested that this may be related to the fact that the Egyptians regulated their year by the heliacal rising of the star Sirius, a phenomenon which occurs in the morning just before dawn.[16]

In cultures following a lunar or lunisolar calendar, in which the beginning of the month is determined by the appearance of the New Moon in the evening, the beginning of the day was reckoned from sunset to sunset, following an evening epoch. This practice was followed in the Jewish and Islamic calendars[17] and in Medieval Western Europe in reckoning the dates of religious festivals.[18]

14.7 See also

- Astrometry
- Epoch (reference date)
- International Celestial Reference System
- International Celestial Reference Frame

14.8 References

[1] Soop, E. M. (1994). *Handbook of Geostationary Orbits*. Springer. ISBN 978-0-7923-3054-7.

[2] M Chapront-Touzé (ed.), *Jean le Rond d'Alembert, Oeuvres Complètes: Ser.1, Vol.6*, Paris (CNRS) (2002), p.xxx, n.50.

[3] Examples of this usage are seen in: J L Simon et al., "Numerical expressions for precession formulae and mean elements for the Moon and the planets", *Astronomy and Astrophysics* 282 (1994), pp. 663-683.

[4] Harvard Minor Planet Center, data for *Pholus*

[5] See Explanation of Orbital Elements.

[6] "The Hipparcos and Tycho Catalogues", ESA SP-1200, Vol. 1, page XV. ESA, 1997

[7] Meeus, J.: "Astronomical Algorithms", page 125. Willmann-Bell, 1991

[8] Lieske, J.H.: "Precession Matrix Based on IAU (1976) System of Astronomical Constants", page 282. Astronomy & Astrophysics, 73, 282-284 (1979)

[9] "SOFA Libraries Issue 2007-08-10". 2007-08-18. Retrieved 2008-10-01.

[10] See NASA Jet Propulsion Laboratory 'spice' toolkit documentation, function J1900.

[11] Aoki, S.; Soma, M.; Kinoshita, H.; Inoue, K. (December 1983). "Conversion matrix of epoch B 1950.0 FK 4-based positions of stars to epoch J 2000.0 positions in accordance with the new IAU resolutions". *Astronomy and Astrophysics* **128** (3): 263–267. Bibcode:1983A&A...128..263A. ISSN 0004-6361.

[12] Seidelmann, P. K., Ed. (1992). *Explanatory Supplement to the Astronomical Almanac*. Sausalito, CA: University Science Books. p. 8.

[13] Seidelmann, P. K., Ed. (1992). *Explanatory Supplement to the Astronomical Almanac*. Sausalito, CA: University Science Books. Glossary, s.v. Terrestrial Dynamical Time.

[14] This article uses a 24-hour clock, so 11:59:27.816 is equivalent to 11:59:27.816 AM.

[15] H. C. Wilson, "Change of astronomical time", *Popular Astronomy*, 33 (1925): 1-2.

[16] Otto Neugebauer, *A History of Ancient Mathematical Astronomy*, (New York: Springer, 1975), p. 1067. ISBN 0-387-06995-X

[17] Otto Neugebauer, *A History of Ancient Mathematical Astronomy*, (New York: Springer, 1975), pp. 1067-1069. ISBN 0-387-06995-X

[18] Bede, *The Reckoning of Time*, 5, trans. Faith Wallis, (Liverpool: Liverpool University Press, 2004), pp. 22-24. ISBN 0-85323-693-3

14.9 External links

- Standish, E. M., Jr. (November 1982). "Conversion of positions and proper motions from B1950.0 to the IAU system at J2000.0". *Astronomy and Astrophysics* **115** (1): 20–22. Bibcode:1982A&A...115...20S.

- What is Terrestrial Time? - U.S. Naval Observatory

- International Celestial Reference System, or ICRS - U.S. Naval Observatory

- IERS Conventions 2003 (defines ICRS and other related standards)

Chapter 15

Orbital node

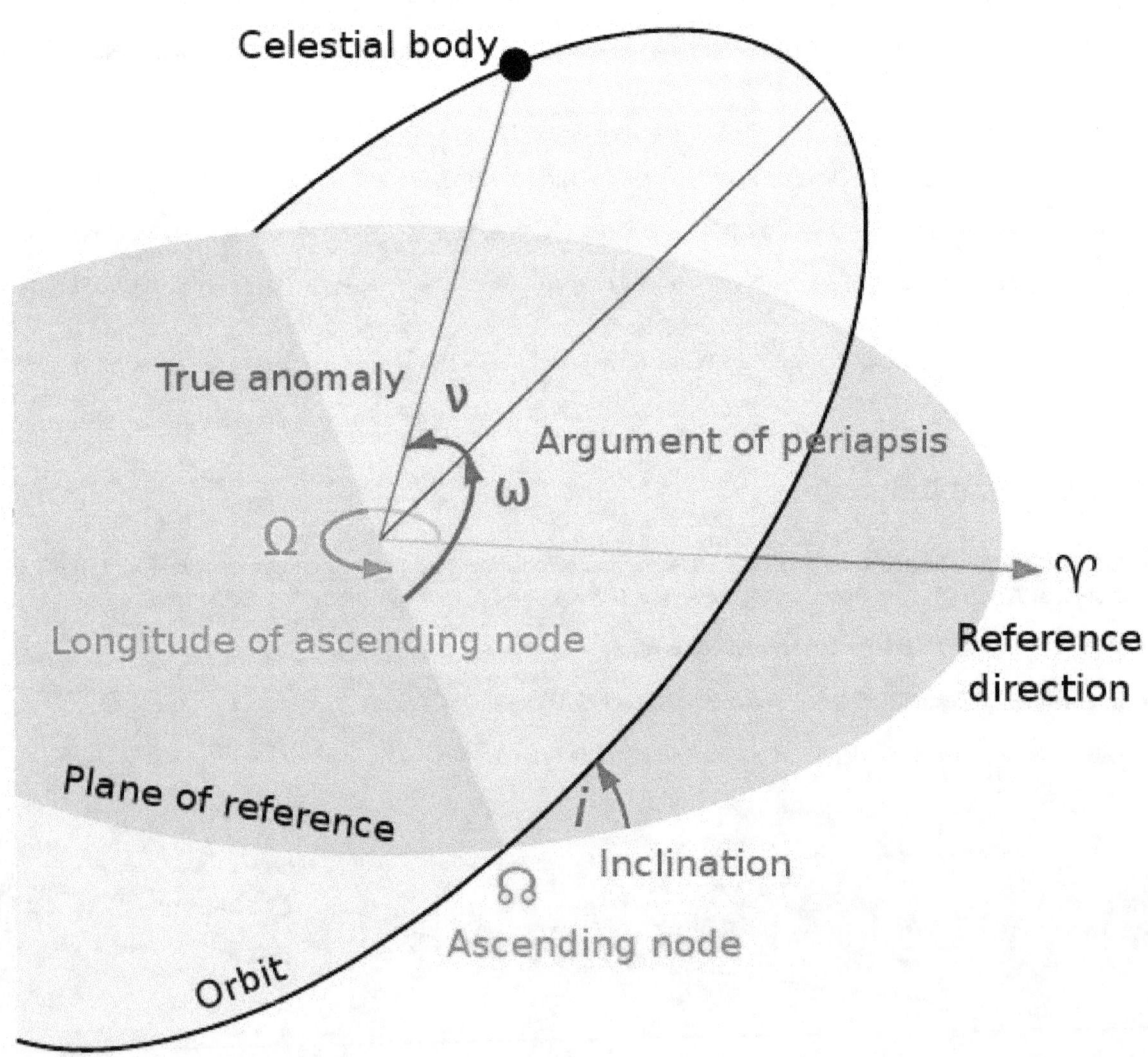

The ascending node

An **orbital node** is one of the two points where an orbit crosses a plane of reference to which it is inclined.[1] An orbit that is contained in the plane of reference (called *non-inclined*) has no nodes.

15.1 Planes of reference

Common planes of reference include:

- For a geocentric orbit, the Earth's equatorial plane. In this case, non-inclined orbits are called *equatorial*.[2]

- For a heliocentric orbit, the ecliptic. In this case, non-inclined orbits are called *ecliptic*.[2]

- For an orbit outside the Solar System, the plane through the primary perpendicular to a line through the observer and the primary (called the *plane of the sky*).[3], chap. 17.

15.2 Node distinction

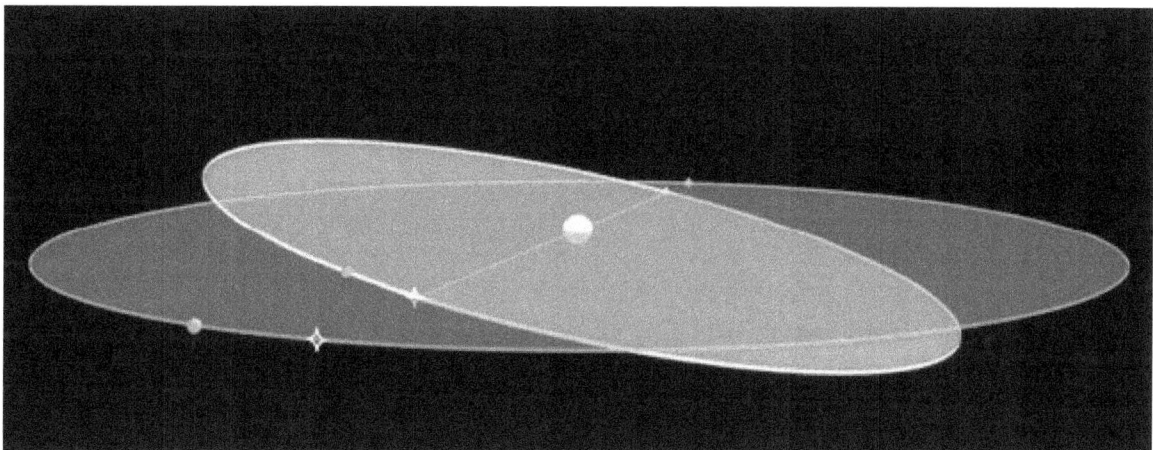

Animation about nodes of two elliptic trajectories. (Click onto image.)

If a reference direction from one side of the plane of reference to the other is defined, the two nodes can be distinguished. For geocentric and heliocentric orbits, the **ascending node** (or **north node**) is where the orbiting object moves north through the plane of reference, and the **descending node** (or **south node**) is where it moves south through the plane.[4] In the case of objects outside the Solar System, the ascending node is the node where the orbiting secondary passes away from the observer, and the descending node is the node where it moves towards the observer.[5], p. 137.

The position of the node may be used as one of a set of parameters, called *orbital elements*, which describe the orbit. This is done by specifying the longitude of the ascending node (or, sometimes, the longitude of the node.)

The **line of nodes** is the intersection of the object's orbital plane with the plane of reference. It passes through the two nodes.[2]

15.3 Symbols and nomenclature

The symbol of the ascending node is ☊ (Unicode: U+260A, ☊), and the symbol of the descending node is ☋ (Unicode: U+260B, ☋). In medieval and early modern times the ascending and descending nodes were called the *dragon's head* (Latin: *caput draconis*, Arabic: *ra's al-jauzahar*) and *dragon's tail* (Latin: *cauda draconis*), respectively.[6], p. 141; [7], p. 245. These terms originally referred to the times when the Moon crossed the apparent path of the sun in the sky. Also, corruptions of the Arabic term such as *ganzaar*, *genzahar*, *geuzaar* and *zeuzahar* were used in the medieval West to denote either of the nodes.[8], pp. 196–197; [9], p. 65; [10], pp. 95–96. The Greek terms αναβιβάζων and καταβιβάζων were also used for the ascending and descending nodes, giving rise to the English words *anabibazon* and *catabibazon*.[11]; [12], ¶27.

15.4 Earth orbit nodes

For the orbit of the Earth around the Sun is important the line of nodes formed by the ecliptic and the equator. There are two special days in which this line points to the sun. For these days, night and day take the same time. These two points in the sun orbit are named equinoxes.

15.5 Lunar nodes

Main article: Lunar node

For the orbit of the Moon around the Earth, the reference plane is taken to be the ecliptic, not the equatorial plane. The gravitational pull of the Sun upon the Moon causes its nodes, called the lunar nodes, to precess gradually westward, performing a complete circle in approximately 18.6 years.[1][13]

15.6 See also

- Eclipse
- Euler angles
- Longitude of the ascending node

15.7 References

[1] node, entry in *The Columbia Encyclopedia*, 6th ed., New York: Columbia University Press, 2001–04. Accessed on line May 17, 2007.

[2] line of nodes, entry in *The Encyclopedia of Astrobiology, Astronomy, and Spaceflight*, David Darling, on line, accessed May 17, 2007.

[3] *Celestial Mechanics*, Jeremy B. Tatum, on line, accessed May 17, 2007.

[4] ascending node, entry in *The Encyclopedia of Astrobiology, Astronomy, and Spaceflight*, David Darling, on line, accessed May 17, 2007.

[5] *The Binary Stars*, R. G. Aitken, New York: Semi-Centennial Publications of the University of California, 1918.

[6] Survey of Islamic Astronomical Tables, E. S. Kennedy, *Transactions of the American Philosophical Society*, new series, 46, #2 (1956), pp. 123–177.

[7] Cyclopædia, or, An universal dictionary of arts and sciences, Ephraim Chambers, London: Printed for J. and J. Knapton [and 18 others], 1728, vol. 1.

[8] Planetary Latitudes, the Theorica Gerardi, and Regiomontanus, Claudia Kren, *Isis*, 68, #2 (June 1977), pp. 194–205.

[9] Prophatius Judaeus and the Medieval Astronomical Tables, Richard I. Harper, *Isis* 62, #1 (Spring, 1971), pp. 61–68.

[10] Lexicographical Gleanings from the Philobiblon of Richard de Bury, Andrew F. West, *Transactions of the American Philological Association* (1869-1896), 22 (1891), pp. 93–104.

[11] anabibazon, entry in *Webster's third new international dictionary of the English language unabridged: with seven language dictionary*, Chicago: Encyclopaedia Britannica, 1986. ISBN 0-85229-503-0.

[12] New thoughts on the genesis of the mysteries of Mithras, Roger Beck, *Topoi* 11, #1 (2001), pp. 59–76.

[13] Introduction to Astronomy 250, Coordinates, Seasons, Eclipses, lecture notes, Astronomy 250, Marcia Rieke, University of Arizona, accessed on line May 17, 2007.

Chapter 16

Precession

For other uses, see Precession (disambiguation).

Precession is a change in the orientation of the rotational axis of a rotating body. In an appropriate reference frame it can be defined as a change in the first Euler angle, whereas the third Euler angle defines the rotation itself. In other words, the axis of rotation of a precessing body itself rotates around another axis. A motion in which the second Euler angle changes is called *nutation*. In physics, there are two types of precession: torque-free and torque-induced.

In astronomy, "precession" refers to any of several slow changes in an astronomical body's rotational or orbital parameters, and especially to Earth's precession of the equinoxes. *(See section Astronomy below.)*

16.1 Torque-free

In torque-free precession, the angular momentum remains fixed, but the angular velocity vector changes. What makes this possible is a time-varying moment of inertia, or more precisely, a time-varying inertia matrix. The inertia matrix is composed of moments of inertia calculated with respect to separate coordinate axes (e.g. x, y, z), or basis sets. If an object is asymmetric around its principal axis of rotation, the moment of inertia with respect to each basis will change with time, while preserving angular momentum. The result is that the component angular velocities around each axis will vary inversely to each axis' moment of inertia. Poinsot's ellipsoid is a geometrical analog of the functions that govern torque-free motion of a rotating rigid body.

The torque-free precession rate of an object with an axis of symmetry, such as a disk, spinning about an axis not aligned with that axis of symmetry can be calculated as follows:

$$\omega_p = \frac{I_s \omega_s}{I_p \cos(\alpha)} \quad [1]$$

where ω_p is the precession rate, ω_s is the spin rate about the axis of symmetry, I_s is the moment of inertia about the axis of symmetry, I_p is moment of inertia about either of the other two equal perpendicular principal axes, and α is the angle between the moment of inertia direction and the symmetry axis.[2]

When an object is not perfectly solid, internal vortices will tend to damp torque-free precession, and the rotation axis will align itself with one of the inertia axes of the body.

For a generic solid object without any axis of symmetry, the evolution of the object's orientation, represented (for example) by a rotation matrix R that transforms internal to external coordinates, may be numerically simulated. Given the object's fixed internal moment of inertia tensor I_0 and fixed external angular momentum L, the instantaneous angular velocity is $\omega(R) = R I_0^{-1} R^T L$. Precession occurs by repeatedly recalculating ω and applying a small rotation vector $\omega\, dt$ for the short time dt; e.g., $R_{new} = \exp([\omega(R_{old})]_\times \, dt) R_{old}$ for the skew-symmetric matrix $[\omega]_\times$. The errors induced by finite time steps tend to increase the rotational kinetic energy, $E(R) = \omega(R) \cdot L/2$; this unphysical tendency can be counter-acted by repeatedly applying a small rotation vector v perpendicular to both ω and L, noting that $E(\exp([v]_\times) R) \approx E(R) + (\omega(R) \times L) \cdot v$.

Precession of a gyroscope

Another type of torque-free precession can occur when there are multiple reference frames at work. For example, Earth is subject to local torque induced precession due to the gravity of the sun and moon acting on Earth's axis, but at the same time the solar system is moving around the galactic center. As a consequence, an accurate measurement of Earth's axial reorientation relative to objects outside the frame of the moving galaxy (such as distant quasars commonly used as precession measurement reference points) must account for a minor amount of non-local torque-free precession, due to the solar system's motion.

16.2 Torque-induced

Torque-induced precession (**gyroscopic precession**) is the phenomenon in which the axis of a spinning object (e.g.,a gyroscope) describes a cone in space when an external torque is applied to it. The phenomenon is commonly seen in a spinning toy top, but all rotating objects can undergo precession. If the speed of the rotation and the magnitude of the

16.2. TORQUE-INDUCED

external torque are constant, the spin axis will move at right angles to the direction that would intuitively result from the external torque. In the case of a toy top, its weight is acting downwards from its center of mass and the normal force (reaction) of the ground is pushing up on it at the point of contact with the support. These two opposite forces produce a torque which causes the top to precess.

The response of a rotating system to an applied torque. When the device swivels, and some roll is added, the wheel tends to pitch.

The device depicted on the right is gimbal mounted. From inside to outside there are three axes of rotation: the hub of the wheel, the gimbal axis, and the vertical pivot.

To distinguish between the two horizontal axes, rotation around the wheel hub will be called *spinning*, and rotation around the gimbal axis will be called *pitching*. Rotation around the vertical pivot axis is called *rotation*.

First, imagine that the entire device is rotating around the (vertical) pivot axis. Then, spinning of the wheel (around the wheelhub) is added. Imagine the gimbal axis to be locked, so that the wheel cannot pitch. The gimbal axis has sensors, that measure whether there is a torque around the gimbal axis.

In the picture, a section of the wheel has been named dm_1. At the depicted moment in time, section dm_1 is at the perimeter of the rotating motion around the (vertical) pivot axis. Section dm_1, therefore, has a lot of angular rotating velocity with respect to the rotation around the pivot axis, and as dm_1 is forced closer to the pivot axis of the rotation (by the wheel spinning further), because of the Coriolis effect, with respect to the vertical pivot axis, dm_1 tends to move in the direction of the top-left arrow in the diagram (shown at 45°) in the direction of rotation around the pivot axis.[3] Section dm_2 of the wheel is moving away from the pivot axis, and so a force (again, a Coriolis force) acts in the same direction as in the case of dm_1. Note that both arrows point in the same direction.

The same reasoning applies for the bottom half of the wheel, but there the arrows point in the opposite direction to that of the top arrows. Combined over the entire wheel, there is a torque around the gimbal axis when some spinning is added to rotation around a vertical axis.

It is important to note that the torque around the gimbal axis arises without any delay; the response is instantaneous.

In the discussion above, the setup was kept unchanging by preventing pitching around the gimbal axis. In the case of a spinning toy top, when the spinning top starts tilting, gravity exerts a torque. However, instead of rolling over, the spinning top just pitches a little. This pitching motion reorients the spinning top with respect to the torque that is being exerted. The result is that the torque exerted by gravity – via the pitching motion – elicits gyroscopic precession (which in turn yields a counter torque against the gravity torque) rather than causing the spinning top to fall to its side.

Precession or gyroscopic considerations have an effect on bicycle performance at high speed. Precession is also the mechanism behind gyrocompasses.

16.2.1 Classical (Newtonian)

Precession is the result of the angular velocity of rotation and the angular velocity produced by the torque. It is an angular velocity about a line that makes an angle with the permanent rotation axis, and this angle lies in a plane at right angles to the plane of the couple producing the torque. The permanent axis must turn towards this line, because the body cannot continue to rotate about any line that is not a principal axis of maximum moment of inertia; that is, the permanent axis turns in a direction at right angles to that in which the torque might be expected to turn it. If the rotating body is symmetrical and its motion unconstrained, and, if the torque on the spin axis is at right angles to that axis, the axis of precession will be perpendicular to both the spin axis and torque axis.

Under these circumstances the angular velocity of precession is given by:

$$\omega_p = \frac{mgr}{I_s \omega_s}$$

Where I_s is the moment of inertia, ω_s is the angular velocity of spin about the spin axis, m is the mass, g is the acceleration due to gravity and r is the perpendicular distance of the spin axis about the axis of precession. The torque vector originates at the center of mass. Using $\omega = \frac{2\pi}{T}$, we find that the period of precession is given by:

$$T_p = \frac{4\pi^2 I_s}{mgr T_s}$$

Where I_s is the moment of inertia, T_s is the period of spin about the spin axis, and τ is the torque. In general, the problem is more complicated than this, however.

There is a non-mathematical way of understanding the cause of gyroscopic precession. The behavior of spinning objects simply obeys the law of inertia by resisting any change in direction. If a force is applied to a spinning object to induce a change the direction of the spin axis, the object behaves as if that force was applied at a location exactly 90 degrees ahead, in the direction of rotation. This is why: A solid object can be thought of as an assembly of individual molecules. If the object is spinning, each molecule's direction of travel constantly changes as that molecule revolves around the object's spin axis. When a force is applied that is parallel to the axis, molecules are being forced to move in new directions at certain places during their path around the axis. These new changes in direction are resisted by inertia.

Imagine the object to be a spinning bicycle wheel, held at both ends of its axle in the hands of a subject. The wheel is spinning clock-wise as seen from a viewer to the subject's right. Clock positions on the wheel are given relative to

16.2. TORQUE-INDUCED

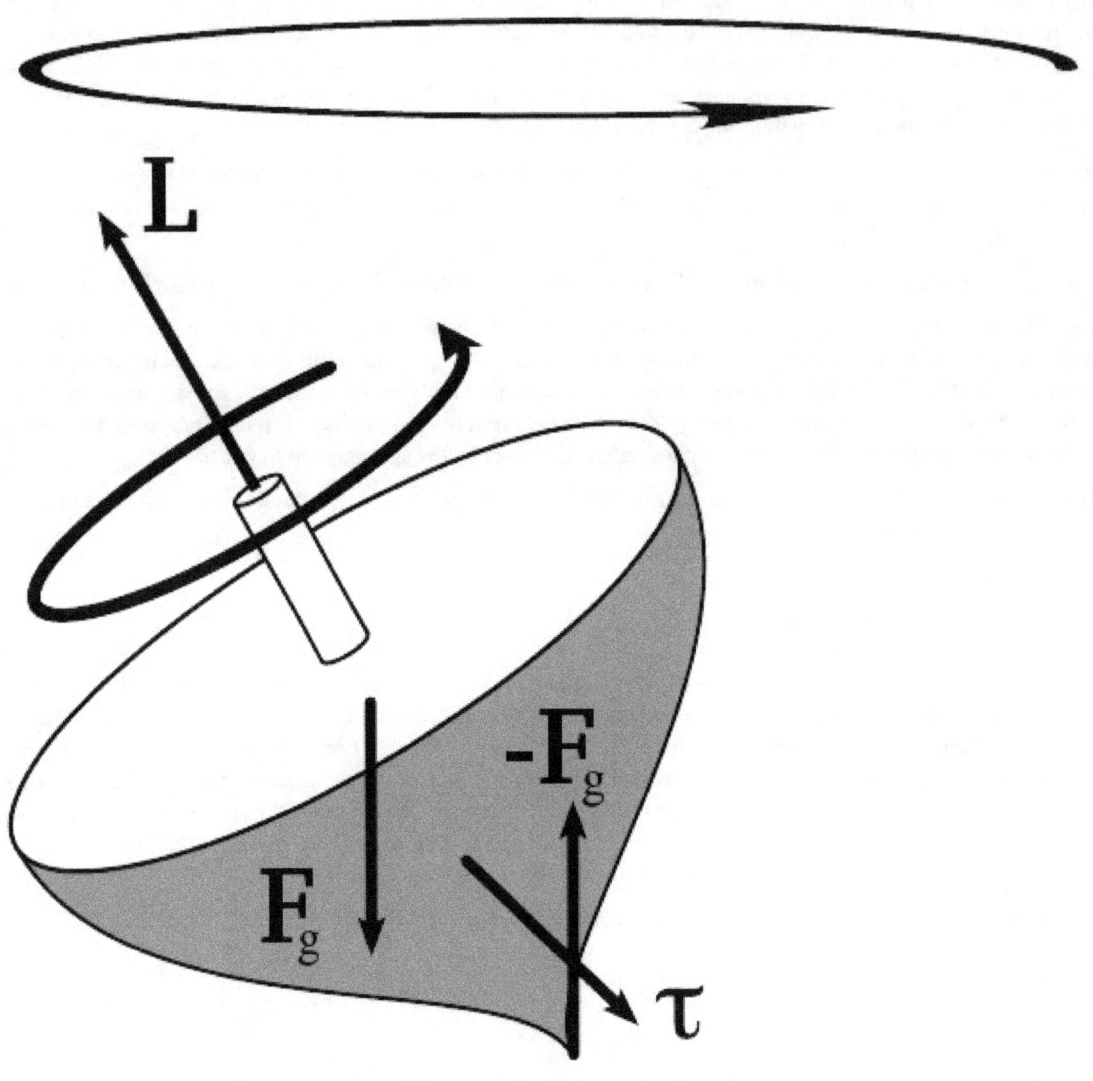

The torque caused by the normal force $-F_g$ and the weight of the top causes a change in the angular momentum L in the direction of that torque. This causes the top to precess.

this viewer. As the wheel spins, the molecules comprising it are travelling vertically downward the instant they pass the 3-o'clock position, horizontally to the left the instant they pass 6 o'clock, vertically upward at 9 o'clock, and horizontally to the right at 12 o'clock. Between these positions, each molecule travels components of these directions, which should be kept in mind as you read ahead. The viewer then applies a force to the wheel at the 3-o'clock position in a direction away from himself. The molecules at the 3-o'clock position are not being forced to change their direction when this happens; they still travel vertically downward. Actually, the force attempts to displace them some amount horizontally at that moment, but the ostensible component of that motion, attributed to the horizontal force, never occurs, as it would if the wheel was not spinning. Therefore, neither the horizontal nor downward components of travel are affected by the horizontally-applied force. The horizontal component started at zero and remains at zero, and the downward component is at its maximum and remains at maximum. The same holds true for the molecules located at 9 o'clock; they still travel vertically upward and not at all horizontally, thus are unaffected by the force that was applied. However, molecules at 6 and 12 o'clock are being forced to change direction. At 6 o'clock, molecules are forced to veer toward the viewer. At the same time, molecules that are passing 12 o'clock are being forced to veer away from the viewer. The inertia of those molecules resists this change in direction. The result is that they apply an equal and opposite reactive force in response. At 6 o'clock, molecules exert a push directly away from the viewer, while molecules at 12 o'clock push directly toward the viewer. This all happens instantaneously as the force is applied at 3 o'clock. Since no physical force was actually applied at 6 or 12 o'clock, there is nothing to oppose these reactive forces; therefore, the reaction is free to take place. This makes the wheel as a whole tilt toward the viewer. Thus, when the force was applied at 3 o'clock, the wheel behaved as if that force was applied at 6 o'clock, which is 90 degrees ahead in the direction of rotation. This principle is demonstrated in helicopters. Helicopter controls are rigged so that inputs to them are transmitted to the rotor blades at points 90 degrees prior to the point where the change in aircraft attitude is desired.

Precession causes another phenomenon for spinning objects such as the bicycle wheel in this scenario. If the subject holding the wheel removes one hand from the end of the axle, the wheel will not topple over, but will remain upright, supported at just the other end of its axle. However, it will immediately take on an additional motion; it will begin to rotate about a vertical axis, pivoting at the point of support as it continues spinning. If the wheel was not spinning, it would topple over and fall when one hand is removed. The ostensible action of the wheel beginning to topple over is equivalent to applying a force to it at 12 o'clock in the direction of the unsupported side (or a force at 6 o'clock toward the supported side). When the wheel is spinning, the sudden lack of support at one end of its axle is equivalent to this same force. So, instead of toppling over, the wheel behaves as if a continuous force is being applied to it at 3 or 9 o'clock, depending on the direction of spin and which hand was removed. This causes the wheel to begin pivoting at the point of support while remaining upright. It should be noted that although it pivots at the point of support, it does so only because of the fact that it is supported there; the actual axis of precessional rotation is located vertically through the wheel, passing through its center of mass. Also, this explanation does not account for the effect of variation in the speed of the spinning object; it only describes how the spin axis behaves due to precession. More correctly, the object behaves according to the balance of all forces based on the magnitude of the applied force, mass and rotational speed of the object.

16.2.2 Relativistic

The special and general theories of relativity give three types of corrections to the Newtonian precession, of a gyroscope near a large mass such as Earth, described above. They are:

- Thomas precession a special relativistic correction accounting for the observer's being in a rotating non-inertial frame.

- de Sitter precession a general relativistic correction accounting for the Schwarzschild metric of curved space near a large non-rotating mass.

- Lense–Thirring precession a general relativistic correction accounting for the frame dragging by the Kerr metric of curved space near a large rotating mass.

16.3 Astronomy

In astronomy, precession refers to any of several gravity-induced, slow and continuous changes in an astronomical body's rotational axis or orbital path. Precession of the equinoxes, perihelion precession, changes in the tilt of Earth's axis to its orbit, and the eccentricity of its orbit over tens of thousands of years are all important parts of the astronomical theory of ice ages. *(See Milankovitch cycles.)*

16.3.1 Axial precession (precession of the equinoxes)

Main article: Axial precession

Axial precession is the movement of the rotational axis of an astronomical body, whereby the axis slowly traces out a cone. In the case of Earth, this type of precession is also known as the *precession of the equinoxes, lunisolar precession*, or

16.3. ASTRONOMY

precession of the equator. Earth goes through one such complete precessional cycle in a period of approximately 26,000 years or 1° every 72 years, during which the positions of stars will slowly change in both equatorial coordinates and ecliptic longitude. Over this cycle, Earth's north axial pole moves from where it is now, within 1° of Polaris, in a circle around the ecliptic pole, with an angular radius of about 23.5 degrees.

Hipparchus is the earliest known astronomer to recognize and assess the precession of the equinoxes at about 1° per century (which is not far from the actual value for antiquity, 1.38°).[4] The precession of Earth's axis was later explained by Newtonian physics. Being an oblate spheroid, Earth has a non-spherical shape, bulging outward at the equator. The gravitational tidal forces of the Moon and Sun apply torque to the equator, attempting to pull the equatorial bulge into the plane of the ecliptic, but instead causing it to precess. The torque exerted by the planets, particularly Jupiter, also plays a role.[5]

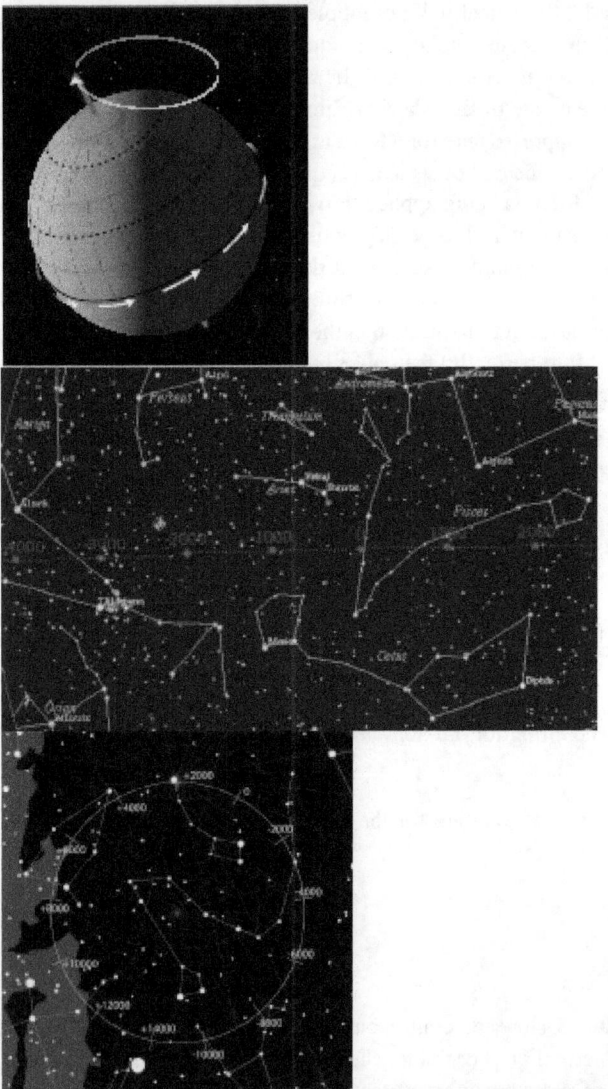

Precessional movement of the axis (left), precession of the equinox in relation to the distant stars (middle), and the path of the north celestial pole among the stars due to the precession. Vega is the bright star near the bottom (right).

16.3.2 Perihelion precession

Main article: Apsidal precession

The orbits of a planet around the Sun do not really follow an identical ellipse each time, but actually trace out a flower-petal

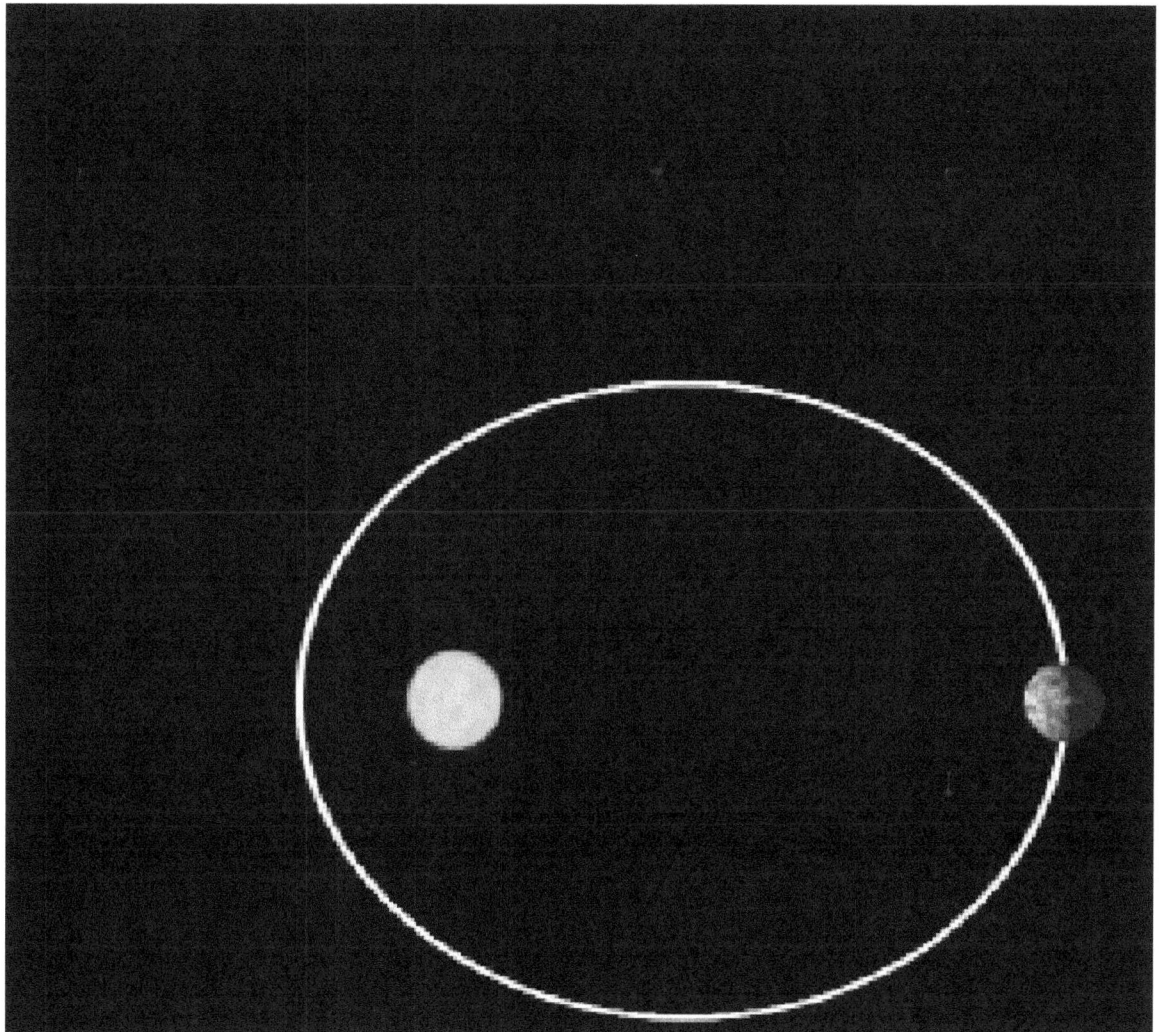

Apsidal precession—the orbit rotates gradually over time.

shape because the major axis of each planet's elliptical orbit also precesses within its orbital plane, partly in response to perturbations in the form of the changing gravitational forces exerted by other planets. This is called perihelion precession or apsidal precession.

In the adjunct image, the Earth apsidal precession is illustrated. As the Earth travels around the Sun, its elliptical orbit rotates gradually over time. The eccentricity of its ellipse and the precession rate of its orbit are exaggerated for visualization. Most orbits in the Solar System have a much smaller eccentricity and precess at a much slower rate, making them nearly circular and stationary.

Discrepancies between the observed perihelion precession rate of the planet Mercury and that predicted by classical mechanics were prominent among the forms of experimental evidence leading to the acceptance of Einstein's Theory of Relativity (in particular, his General Theory of Relativity), which accurately predicted the anomalies.[6][7] Deviating from Newton's law, Einstein's theory of gravitation predicts an extra term of A/r4, which accurately gives the observed excess turning rate of 43 arcseconds every 100 years.

The gravitational force between the Sun and moon induces the precession in Earth's orbit, which is the major cause of the widely known climate oscillation of Earth that has a period of 19,000 to 23,000 years. It follows that changes in Earth's orbital parameters (e.g., orbital inclination, the angle between Earth's rotation axis and its plane of orbit) is important to the study of Earth's climate, in particular to the study of past ice ages. *(See also nodal precession. For precession of the lunar orbit see lunar precession).*

16.4 See also

- Larmor precession
- Polar motion
- Precession (mechanical)

16.5 References

[1] Schaub, Hanspeter (2003), *Analytical Mechanics of Space Systems*, AIAA, pp. 149–150, ISBN 9781600860270, retrieved May 2014

[2] Boal, David (2001). "Lecture 26 – Torque-free rotation – body-fixed axes" (PDF). Retrieved 2008-09-17.

[3] Teodorescu, Petre P (2002). *Mechanical Systems, Classical Models*. Springer. p. 420.

[4] DIO 9.1 ‡3

[5] Bradt, Hale (2007). *Astronomy Methods*. Cambridge University Press. p. 66. ISBN 978 0 521 53551 9.

[6] Max Born (1924), *Einstein's Theory of Relativity* (The 1962 Dover edition, page 348 lists a table documenting the observed and calculated values for the precession of the perihelion of Mercury, Venus, and Earth.)

[7] An even larger value for a precession has been found, for a black hole in orbit around a much more massive black hole, amounting to 39 degrees each orbit.

16.6 External links

- Explanation and derivation of formula for precession of a top
- Precession and the Milankovich theory from educational web site From Stargazers to Starships

Chapter 17

Orbital station-keeping

In astrodynamics **orbital station-keeping** is the orbital maneuvers made by thruster burns that are needed to keep a spacecraft in a particular assigned orbit.

For many Earth satellites the effects of the non-Keplerian forces, i.e. the deviations of the gravitational force of the Earth from that of a homogeneous sphere, gravitational forces from Sun/Moon, solar radiation pressure and air-drag must be counteracted.

The deviation of Earth's gravity field from that of a homogeneous sphere and gravitational forces from Sun/Moon will in general perturb the orbital plane. For sun-synchronous orbit the precession of the orbital plane caused by the oblateness of the Earth is a desirable feature that is part of the mission design but the inclination change caused by the gravitational forces of Sun/Moon is undesirable. For geostationary spacecraft the inclination change caused by the gravitational forces of Sun/Moon must be counteracted to a rather large expense of fuel, as the inclination should be kept sufficiently small for the spacecraft to be tracked by a non-steerable antenna.

For spacecraft in low orbits the effects of atmospheric drag must often be compensated for. For some missions this is needed simply to avoid re-entry; for other missions, typically missions for which the orbit should be accurately synchronized with Earth rotation, this is necessary to avoid the orbital period shortening.

Solar radiation pressure will in general perturb the eccentricity (i.e. the eccentricity vector), see Orbital perturbation analysis (spacecraft). For some missions this must be actively counter-acted with manoeuvres. For geostationary spacecraft the eccentricity must be kept sufficiently small for a spacecraft to be tracked with a non-steerable antenna. Also for Earth observation spacecraft for which a very repetitive orbit with a fixed ground track is desirable, the eccentricity vector should be kept as fixed as possible. A large part of this compensation can be done by using a frozen orbit design, but for the fine control manoeuvres with thrusters are needed.

For spacecraft in a halo orbit around a Lagrangian point stationkeeping is even more fundamental as such an orbit is unstable; without an active control with thruster burns the smallest deviation in position/velocity would result in the spacecraft leaving the orbit completely.[1]

17.1 Station-keeping in low-earth orbit

For a spacecraft in a very low orbit the atmospheric drag is sufficiently strong to cause a re-entry before the intended end of mission if orbit raising manoeuvres are not executed from time to time.

An example of this is the International Space Station (ISS), which has an operational altitude above Earth's surface of between 330 and 410 km. Due to atmospheric drag the space station is constantly losing orbital energy. In order to compensate for this loss, which would eventually lead to a re-entry of the station, it has from time to time been re-boosted to a higher orbit. The chosen orbital altitude is a trade-off between the delta-v needed to counter-act the air drag and the delta-v needed to send payloads and people to the station. The upper limitation of orbit altitude is due to the constraints imposed by the Soyuz spacecraft. On 25 April 2008, the Automated Transfer Vehicle "Jules Verne" raised the orbit of

17.2 Station-keeping for Earth observation spacecraft

For Earth observation spacecraft typically operated in an altitude above the Earth surface of about 700 – 800 km the air-drag is very faint and a re-entry due to air-drag is not a concern. But if the orbital period should be synchronous with the Earth's rotation to maintain a fixed ground track, the faint air-drag at this high altitude must also be counter-acted by orbit raising manoeuvres in the form of thruster burns tangential to the orbit. These manoeuvres will be very small, typically in the order of a few mm/s of delta-v. If a frozen orbit design is used these very small orbit raising manoeuvres are sufficient to also control the eccentricity vector.

To maintain a fixed ground track it is also necessary to make out-of-plane manoeuvres to compensate for the inclination change caused by Sun/Moon gravitation. These are executed as thruster burns orthogonal to the orbital plane. For Sun-synchronous spacecraft having a constant geometry relative to the Sun, the inclination change due to the solar gravitation is particularly large; a delta-v in the order of 1–2 m/s per year can be needed to keep the inclination constant.

17.3 Station-keeping in geostationary orbit

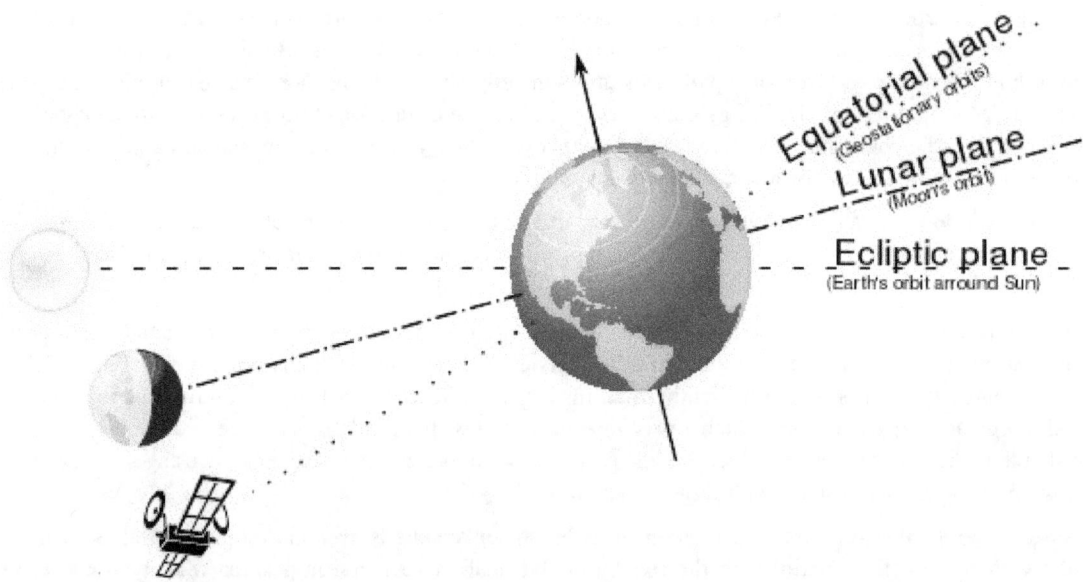

Inclined orbital planes

For geostationary spacecraft, thruster burns orthogonal to the orbital plane must be executed to compensate for the effect of the lunar/solar gravitation that perturbs the orbit pole with typically 0.85 degrees per year. The delta-v needed to compensate for this perturbation keeping the inclination to the equatorial plane amounts to in the order 45 m/s per year. This part of the GEO station-keeping is called North-South control.[2]

The East-West control is the control of the orbital period and the eccentricity vector performed by making thruster burns tangential to the orbit. These burns are then designed to keep the orbital period perfectly synchronous with the Earth rotation and to keep the eccentricity sufficiently small. Perturbation of the orbital period results from the imperfect rotational symmetry of the Earth relative the North/South axis, sometimes called the ellipticity of the Earth equator. The eccentricity (i.e. the eccentricity vector) is perturbed by the solar radiation pressure.

The fuel needed for this East-West control is much less than what is needed for the North-South control. To extend the life-time of ageing geostationary spacecraft with little fuel left one sometimes discontinues the North-South control only continuing with the East-West control. As seen from an observer on the rotating Earth the spacecraft will then move North-South with a period of 24 hours. When this North-South movement gets too large a steerable antenna is needed to track the spacecraft. An example of this is Artemis.

To save weight, it is crucial for GEO satellites to have the most fuel-efficient propulsion system. Some modern satellites are therefore employing a high specific impulse system like plasma or ion thrusters.

17.4 Station-keeping at libration points

Orbits of spacecraft are also possible around Lagrangian points—also referred to as libration points—gravity wells that exist at five points in relation to two larger solar system bodies. For example, there are five of these points in the Sun-Earth system, five in the Earth-Moon system, and so on. Small spacecraft may orbit around these gravity wells with a minimum of propellant required for station-keeping purposes. Two orbits that have been used for such purposes include halo and Lissajous orbits.[3]

Orbits around libration points are dynamically unstable, meaning small departures from equilibrium grow exponentially over time.[1] As a result, spacecraft in libration point orbits must use propulsion systems to perform orbital station-keeping.

One important libration point is Earth-Sun L_1, and three heliophysics missions have been orbiting L1 since approximately 2000. Station-keeping propellant use can be quite low, facilitating missions that can potentially last decades should other spacecraft systems remain operational. The three spacecraft—Advanced Composition Explorer (ACE), Solar Heliospheric Observatory (SOHO), and the Global Geoscience WIND satellite—each have annual station-keeping propellant requirements of approximately 1 m/s or less.[3] Earth-Sun L_2—approximately 1.5 million kilometers from Earth in the anti-sun direction—is another important Lagrangian point, and the ESA Herschel space observatory operated there in a Lissajous orbit during 2009–2013, at which time it ran out of coolant for the space telescope. Small station-keeping orbital maneuvers were executed approximately monthly to maintain the spacecraft in the station-keeping orbit.[1]

17.5 See also

- Orbital maneuvre

- Delta-v budget

- Orbital perturbation analysis (spacecraft)

17.6 References

[1] "ESA Science & Technology: Orbit/Navigation". European Space Agency. 14 June 2009. Retrieved 14 February 2015.

[2] Soop, E. M. (1994). Handbook of Geostationary Orbits. Springer. ISBN 978-0-7923-3054-7.

[3] Roberts, Craig E. (1 January 2011). "Long Term Missions at the Sun-Earth Libration Point L1: ACE, SOHO, and WIND" (pdf). *NASA Technical Reports* (NASA). 20110008638. Retrieved 14 February 2015. *Three heliophysics missions -- the Advanced Composition Explorer (ACE), Solar Heliospheric Observatory (SOHO), and the Global Geoscience WIND -- have been orbiting the Sun-Earth interior libration point L1 continuously since 1997, 1996, and 2004 ... the typical interval between burns for this trio is about three months, and the typical delta-V is much smaller than 0.5 m/sec. Typical annual stationkeeping costs have been around 1.0 m/sec for ACE and WIND, and much less than that for SOHO. All three spacecraft have ample fuel remaining; barring contingencies all three could, in principle, be maintained at L1 for decades to come.*

17.7 External links

- Station-keeping at the Encyclopedia of Astrobiology, Astronomy, and Spaceflight
- XIPS Xenon Ion Propulsion Systems
- Jules Verne boosts ISS orbit (report from the European Space Agency)

Chapter 18

Ground track

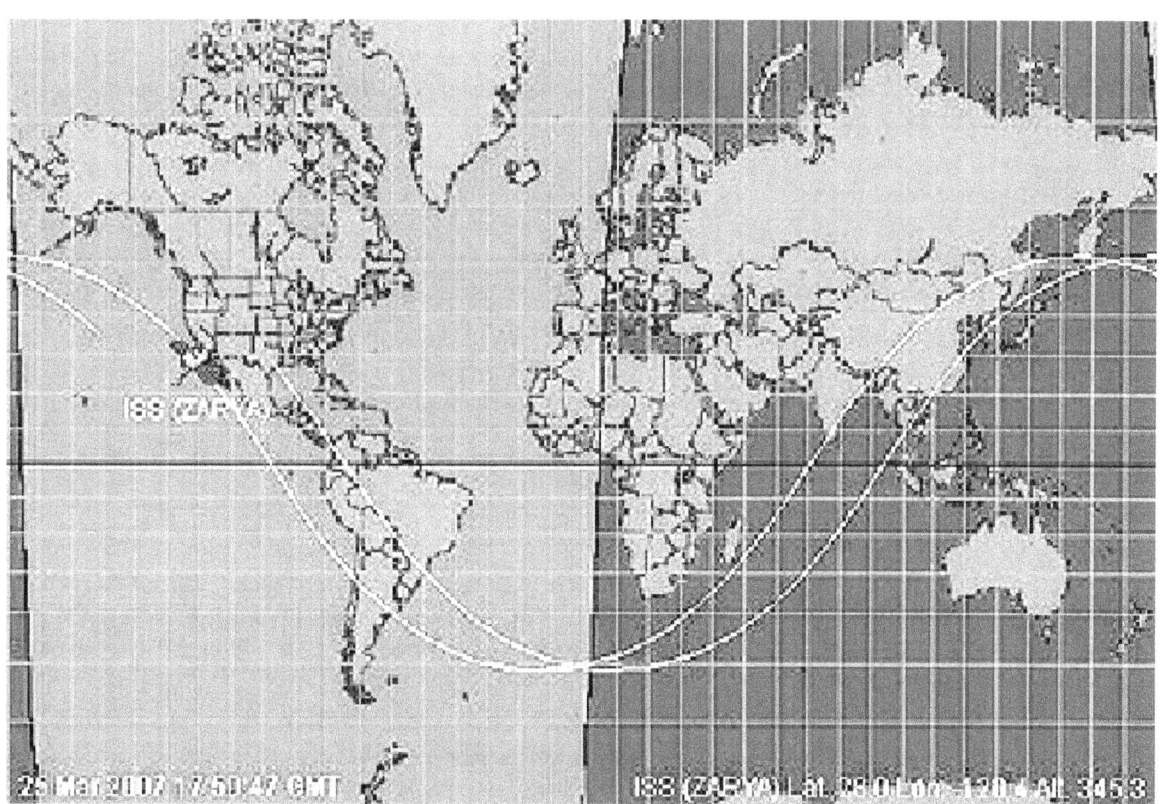

Ground track of the International Space Station for approximately two periods. The light and dark regions represent the regions of the Earth in daylight and in night, respectively.

A **ground track** or **ground trace** is the path on the surface of the Earth directly below an aircraft or satellite. In the case of a satellite, it is the projection of the satellite's orbit onto the surface of the Earth (or whatever body the satellite is orbiting).

A satellite ground track may be thought of as a path along the Earth's surface which traces the movement of an imaginary line between the satellite and the center of the Earth. In other words, the ground track is the set of points at which the satellite will pass directly overhead, or cross the zenith, in the frame of reference of a ground observer.[1]

18.1 Aircraft ground tracks

In air navigation, in order to follow a specified ground track, a pilot must adjust their heading in order to compensate for the effect of wind. Aircraft routes are planned to avoid restricted airspace and dangerous areas, and to pass near navigation beacons.

18.2 Satellite ground tracks

The ground track of a satellite can take a number of different forms, depending on the values of the orbital elements, parameters which define the size, shape, and orientation of the satellite's orbit. This article discusses closed orbits, or orbits with eccentricity less than one, and thus excludes parabolic and hyperbolic trajectories.

18.2.1 Direct and retrograde motion

Typically, satellites have a roughly sinusoidal ground track. A satellite with an orbital inclination between zero and ninety degrees is said to be in what is called a *direct* or *prograde orbit*, meaning that it orbits in the same direction as the Earth's rotation. A satellite with an orbital inclination between 90 and 180 degrees is said to be in a *retrograde orbit*.

A satellite in a direct orbit with an orbital period less than one day will tend to move from west to east along its ground track. This is called "apparent direct" motion. A satellite in a direct orbit with an orbital period *greater* than one day will tend to move from east to west along its ground track, in what is called "apparent retrograde" motion. This effect occurs because the satellite orbits more slowly than the speed at which the Earth rotates beneath it. Any satellite in a true retrograde orbit will always move from east to west along its ground track, regardless of the length of its orbital period.

Because a satellite in an eccentric orbit moves faster near perigee and slower near apogee, it is possible for a satellite to track eastward during part of its orbit and westward during another part. This phenomenon allows for ground tracks which cross over themselves, as in the geosynchronous and Molniya orbits discussed below.

18.2.2 Orbital period and ground track

A satellite whose orbital period is an integer fraction of a day (e.g., 24 hours, 12 hours, 8 hours, etc.) will follow roughly the same ground track every day. This ground track is shifted east or west depending on the longitude of the ascending node, which can vary over time due to perturbations of the orbit. If the period of the satellite is slightly longer than an integer fraction of a day, the ground track will shift west over time; if it is slightly shorter, the ground track will shift east.[1][2]

As the orbital period of a satellite increases, approaching the rotational period of the Earth (in other words, as its average orbital speed slows towards the rotational speed of the Earth), its sinusoidal ground track will become compressed longitudinally, meaning that the "nodes" (the points at which it crosses the equator) will become closer together, until at geosynchronous orbit they lie directly on top of each other. For orbital periods *longer* than the Earth's rotational period, an increase in orbital period corresponds to a longitudinal stretching out of the (apparent retrograde) ground track.

A satellite whose orbital period is *equal* to the rotational period of the Earth is said to be in a geosynchronous orbit. Its ground track will have a "figure eight" shape over a fixed location on the Earth, crossing the equator twice each day. It will track eastward when it is on the part of its orbit closest to perigee, and westward when it is closest to apogee.

A special case of the geosynchronous orbit, the geostationary orbit, has an eccentricity of zero (meaning the orbit is circular), and an inclination of zero in the Earth-Centered, Earth-Fixed coordinate system (meaning the orbital plane is not tilted relative to the Earth's equator). The "ground track" in this case consists of a single point on the Earth's equator, above which the satellite sits at all times. Note that the satellite is still orbiting the Earth — its apparent lack of motion is due to the fact the Earth is rotating about its own center of mass at the same rate as the satellite.

A geostationary orbit, as viewed from above the North Pole

18.2.3 Inclination and ground track

Orbital inclination is the angle formed between the plane of an orbit and the equatorial plane of the Earth. The geographic latitudes covered by the ground track will range from $-i$ to i, where i is the orbital inclination.[2] In other words, the greater the inclination of a satellite's orbit, the further north and south its ground track will pass. A satellite with an inclination of exactly 90° is said to be in a polar orbit, meaning it passes over the Earth's north and south poles.

Launch sites at lower latitudes are often preferred partly for the flexibility they allow in orbital inclination; the initial inclination of an orbit is constrained to be greater than or equal to the launch latitude. Vehicles launched from Cape Canaveral, for instance, must have an initial orbital inclination of at least 28°27′, the latitude of the launch site—and to achieve this minimum requires launching with a due east azimuth, which may not always be feasible given other launch constraints. At the extremes, a launch site located on the equator can launch directly into any desired inclination, while a hypothetical launch site at the north or south pole would only be able to launch (perhaps intuitively) into polar orbits. (While it is possible to perform an orbital inclination change maneuver once on orbit, such maneuvers are typically among the most costly, in terms of fuel, of all orbital maneuvers, and are typically avoided or minimized to the extent possible.)

In addition to providing for a wider range of initial orbit inclinations, low-latitude launch sites offer the benefit of requiring less energy to make orbit (at least for prograde orbits, which comprise the vast majority of launches), due to the initial velocity provided by the Earth's rotation. The desire for equatorial launch sites, coupled with geopolitical and logistical realities, has fostered the development of floating launch platforms, most notably Sea Launch.

18.2.4 Argument of perigee and ground track

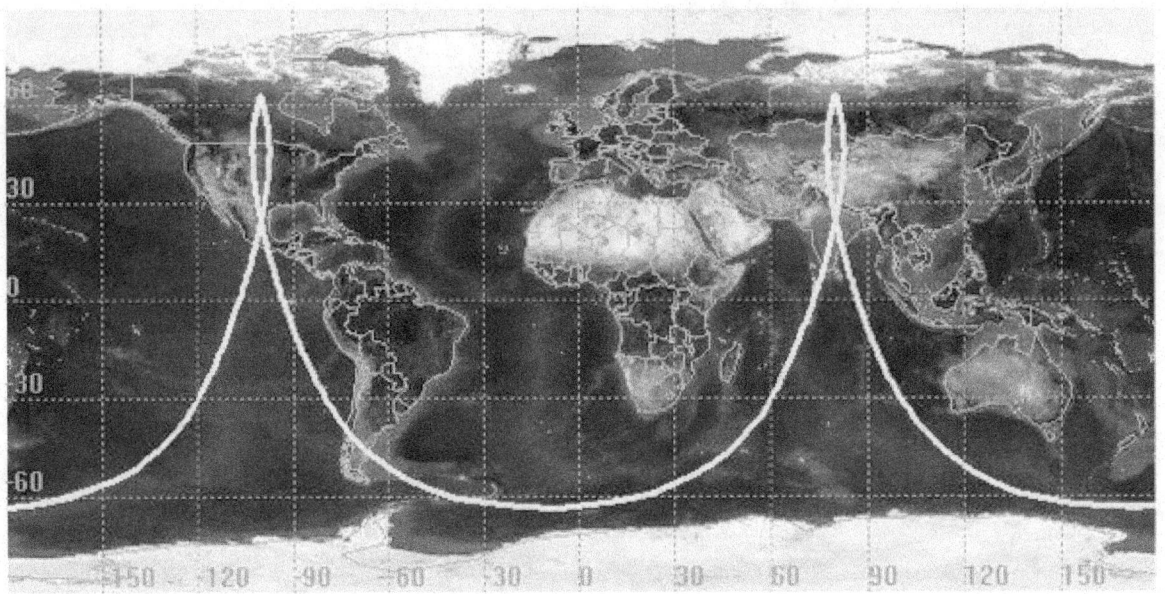

The ground track of a Molniya orbit

If the argument of perigee is zero, meaning that perigee occurs on the equatorial plane, then the ground track of the satellite will appear the same above and below the equator. (It will exhibit 180° rotational symmetry about the orbital nodes.) However, if the argument of perigee is non-zero, then the satellite will behave differently in the northern and southern hemispheres. The Molniya orbit, with an argument of perigee near 90°, is an example of such a case. In a Molniya orbit, apogee occurs at a high latitude (63°), and the orbit is highly eccentric ($e = 0.72$). This causes the satellite to "hover" over a region of the northern hemisphere for a long time, while spending very little time over the southern hemisphere. This phenomenon is known as "apogee dwell", and is desirable for communications for high latitude regions.[2]

18.3 See also

- Course (navigation)
- Terminator (solar)

18.4 References

[1] Curtis, Howard D. (2005), *Orbital Mechanics for Engineering Students* (1st ed.), Amsterdam: Elsevier Ltd., ISBN 978-0-7506-6169-0.

[2] Montenbruck, Oliver; Gill, Eberhard (2000), *Satellite Orbits* (1st ed.), The Netherlands: Springer, ISBN 3-540-67280-X.

- Lyle, S. and Capderou, Michel (2006) *Satellites: Orbits and Missions* Springer ISBN 9782287274695 pp 175-264

18.5 External links

- Satellite Tracker at eoPortal.org
- satview.org
- heavens-above.com
- satellite ground track software code at smallsats.org
- infosatellites.com
- n2yo.com

Chapter 19

Orbital maneuver

In spaceflight, an **orbital maneuver** is the use of propulsion systems to change the orbit of a spacecraft. For spacecraft far from Earth (for example those in orbits around the Sun) an orbital maneuver is called a *deep-space maneuver (DSM)*.

The rest of the flight, especially in a transfer orbit, is called **coasting**.

19.1 General

19.1.1 Rocket equation

Main article: Tsiolkovsky rocket equation

The **Tsiolkovsky rocket equation**, or **ideal rocket equation** is an equation that is useful for considering vehicles that follow the basic principle of a rocket: where a device that can apply acceleration to itself (a thrust) by expelling part of its mass with high speed and moving due to the conservation of momentum. Specifically, it is a mathematical equation that relates the delta-v (the maximum change of speed of the rocket if no other external forces act) with the effective exhaust velocity and the initial and final mass of a rocket (or other reaction engine.)

For any such maneuver (or journey involving a number of such maneuvers):

$$\Delta v = v_e \ln \frac{m_0}{m_1}$$

where:

m_0 is the initial total mass, including propellant,

m_1 is the final total mass,

v_e is the effective exhaust velocity ($v_e = I_{sp} \cdot g_0$ where I_{sp} is the specific impulse expressed as a time period and g_0 is the gravitational constant),

Δv is delta-v - the maximum change of speed of the vehicle (with no external forces acting).

19.1.2 Delta-v

Main article: Delta-v

The applied change in speed of each maneuver is referred to as delta-v ($\Delta \mathbf{v}$).

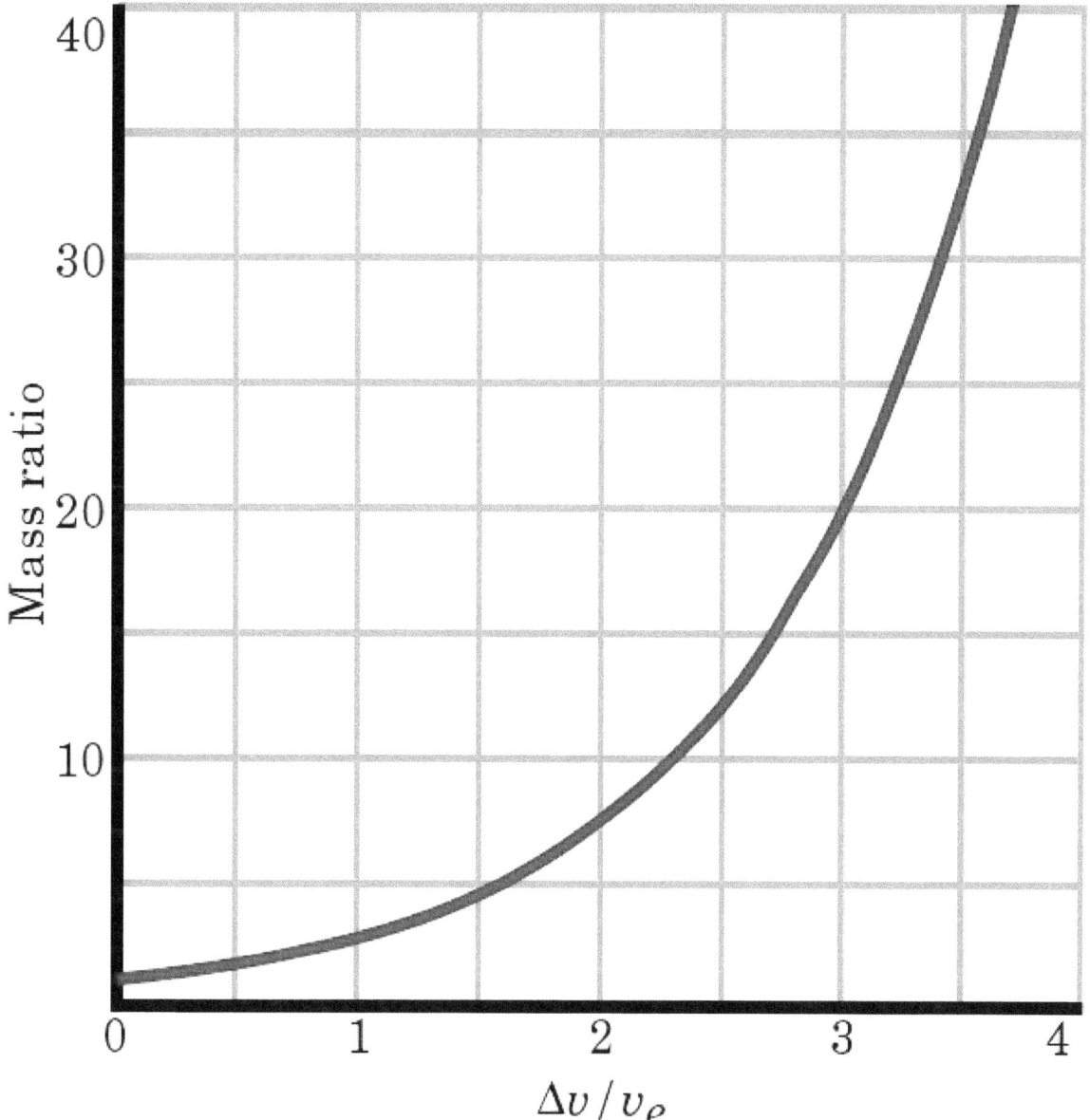

Rocket mass ratios versus final velocity calculated from the rocket equation

19.1.3 Delta-v budget

Main article: delta-v budget

The total delta-v for all and each maneuver is estimated for a mission and is called a delta-v budget. With a good approximation of the delta-v budget designers can estimate the fuel to payload requirements of the spacecraft using the rocket equation.

19.1.4 Impulsive maneuvers

An "impulsive maneuver" is the mathematical model of a maneuver as an instantaneous change in the spacecraft's velocity (magnitude and/or direction) as illustrated in figure 1. In the physical world no truly instantaneous change in velocity is

19.1. GENERAL

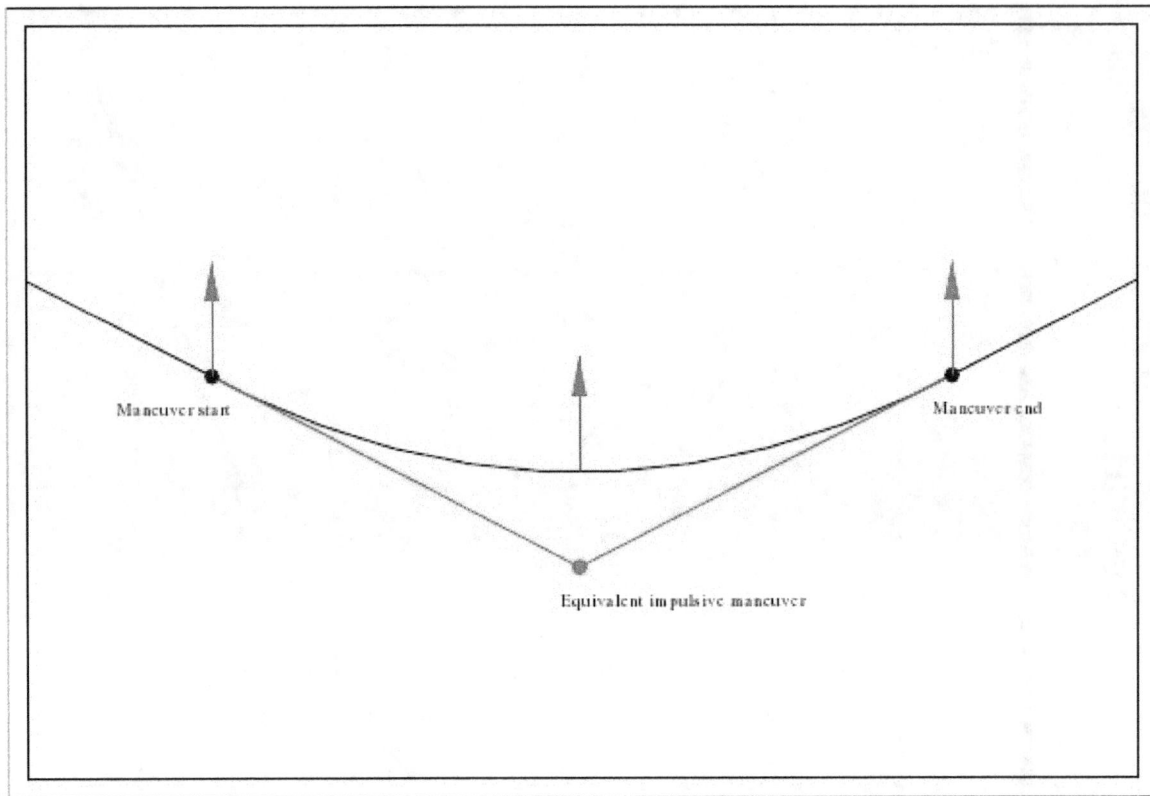

Figure 1: Approximation of a finite thrust maneuver with an impulsive change in velocity

possible as this would require an "infinite force" applied during an "infinitely short time" but as a mathematical model it in most cases describes the effect of a maneuver on the orbit very well. The off-set of the velocity vector after the end of real burn from the velocity vector at the same time resulting from the theoretical impulsive maneuver is only caused by the difference in gravitational force along the two paths (red and black in figure 1) which in general is small.

In the planning phase of space missions designers will first approximate their intended orbital changes using impulsive maneuvers that greatly reduces the complexity of finding the correct orbital transitions.

19.1.5 Applying a low thrust over a longer period of time

Applying a low thrust over a longer period of time is referred to as a **non-impulsive maneuver** (where 'non-impulsive' refers to the maneuver not being of a short time period rather than not involving *impulse*- change in momentum, which clearly must take place).

Another term is *finite burn*, where the word "finite" is used to mean "non-zero", or practically, again: over a longer period.

For a few space missions, such as those including a space rendezvous, high fidelity models of the trajectories are required to meet the mission goals. Calculating a "finite" burn requires a detailed model of the spacecraft and its thrusters. The most important of details include: mass, center of mass, moment of inertia, thruster positions, thrust vectors, thrust curves, specific impulse, thrust centroid offsets, and fuel consumption.

19.1.6 Assists

Oberth effect

Main article: Oberth effect

In astronautics, the **Oberth effect** is where the use of a rocket engine when travelling at high speed generates much more useful energy than one at low speed. Oberth effect occurs because the propellant has more usable energy (due to its kinetic energy on top of its chemical potential energy) and it turns out that the vehicle is able to employ this kinetic energy to generate more mechanical power. It is named after Hermann Oberth, the Austro-Hungarian-born, German physicist and a founder of modern rocketry, who apparently first described the effect.[1]

Oberth effect is used in a **powered flyby** or **Oberth maneuver** where the application of an impulse, typically from the use of a rocket engine, close to a gravitational body (where the gravity potential is low, and the speed is high) can give much more change in kinetic energy and final speed (i.e. higher specific energy) than the same impulse applied further from the body for the same initial orbit. For the Oberth effect to be most effective, the vehicle must be able to generate as much impulse as possible at the lowest possible altitude; thus the Oberth effect is often far less useful for low-thrust reaction engines such as ion drives, which have a low propellant flow rate.

Oberth effect also can be used to understand the behaviour of multi-stage rockets; the upper stage can generate much more usable kinetic energy than might be expected from simply considering the chemical energy of the propellants it carries.

Historically, a lack of understanding of this effect led early investigators to conclude that interplanetary travel would require completely impractical amounts of propellant, as without it, enormous amounts of energy are needed.[1]

Gravitational assist

Main article: Gravity assist

In orbital mechanics and aerospace engineering, a **gravitational slingshot**, **gravity assist maneuver**, or **swing-by** is the use of the relative movement and gravity of a planet or other celestial body to alter the path and speed of a spacecraft, typically in order to save propellant, time, and expense. Gravity assistance can be used to accelerate, decelerate and/or re-direct the path of a spacecraft.

The "assist" is provided by the motion (orbital angular momentum) of the gravitating body as it pulls on the spacecraft.[2] The technique was first proposed as a mid-course manoeuvre in 1961, and used by interplanetary probes from *Mariner 10* onwards, including the two *Voyager* probes' notable fly-bys of Jupiter and Saturn.

19.2 Transfer orbits

Orbit insertion is a general term for a maneuver that is more than a small correction. It may be used for a maneuver to change a transfer orbit or an ascent orbit into a stable one, but also to change a stable orbit into a descent: *descent orbit insertion*. Also the term **orbit injection** is used, especially for changing a stable orbit into a transfer orbit, e.g. trans-lunar injection (TLI), trans-Mars injection (TMI) and trans-Earth injection (TEI).

19.2.1 Hohmann transfer

Main article: Hohmann transfer orbit

In orbital mechanics, the **Hohmann transfer orbit** is an elliptical orbit used to transfer between two circular orbits of different altitudes, in the same plane.

The orbital maneuver to perform the Hohmann transfer uses two engine impulses which move a spacecraft onto and off the transfer orbit. This maneuver was named after Walter Hohmann, the German scientist who published a description of it

19.2. TRANSFER ORBITS

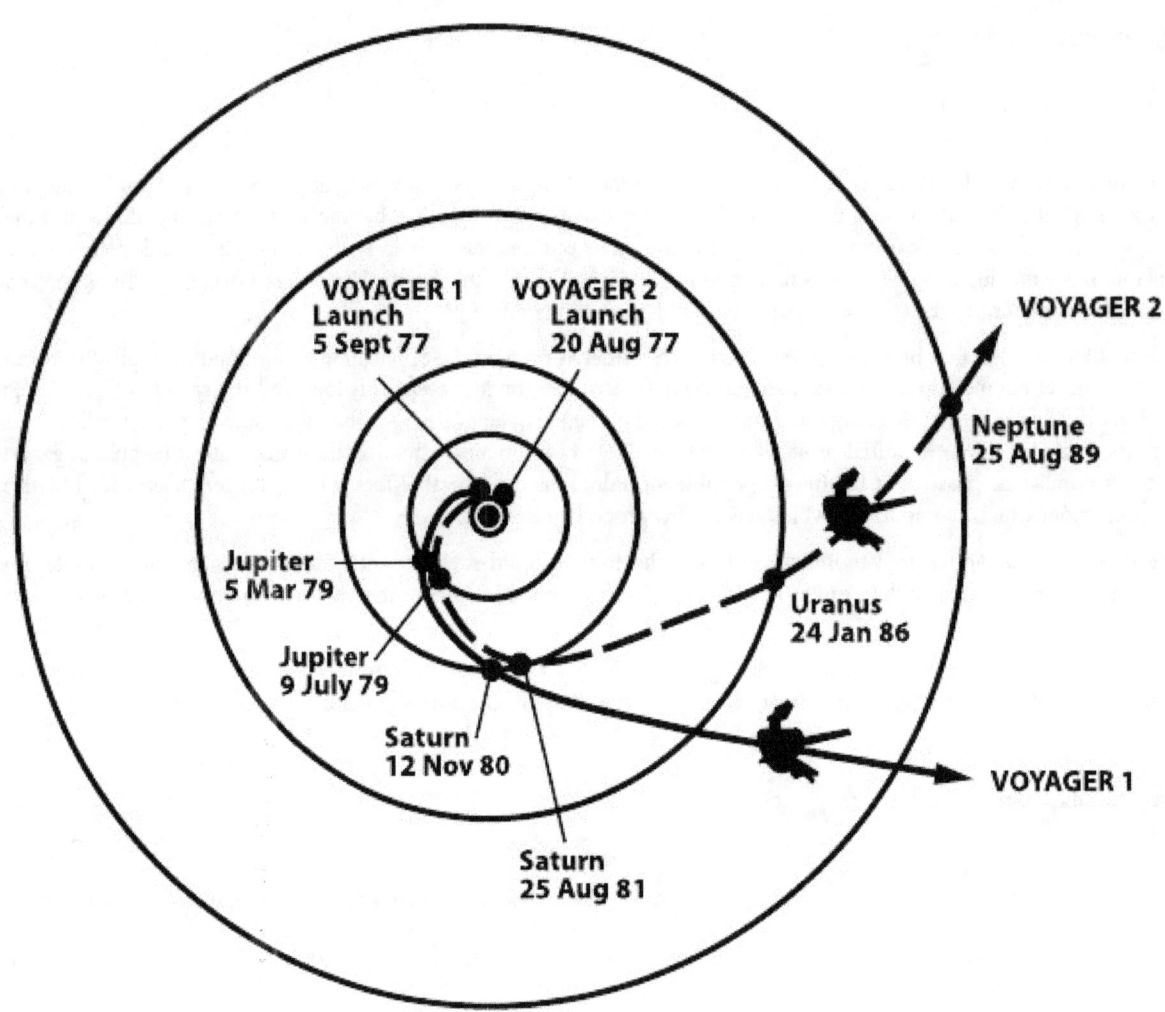

The trajectories that enabled NASA's twin Voyager spacecraft to tour the four gas giant planets and achieve velocity to escape our solar system

in his 1925 book *Die Erreichbarkeit der Himmelskörper* (*The Accessibility of Celestial Bodies*).[3] Hohmann was influenced in part by the German science fiction author Kurd Laßwitz and his 1897 book *Two Planets*.

19.2.2 Bi-elliptic transfer

Main article: Bi-elliptic transfer

In astronautics and aerospace engineering, the **bi-elliptic transfer** is an orbital maneuver that moves a spacecraft from one orbit to another and may, in certain situations, require less delta-v than a Hohmann transfer maneuver.

The bi-elliptic transfer consists of two half elliptic orbits. From the initial orbit, a delta-v is applied boosting the spacecraft into the first transfer orbit with an apoapsis at some point r_b away from the central body. At this point, a second delta-v is applied sending the spacecraft into the second elliptical orbit with periapsis at the radius of the final desired orbit, where a third delta-v is performed, injecting the spacecraft into the desired orbit.

While they require one more engine burn than a Hohmann transfer and generally requires a greater travel time, some bi-elliptic transfers require a lower amount of total delta-v than a Hohmann transfer when the ratio of final to initial semi-major axis is 11.94 or greater, depending on the intermediate semi-major axis chosen.[4]

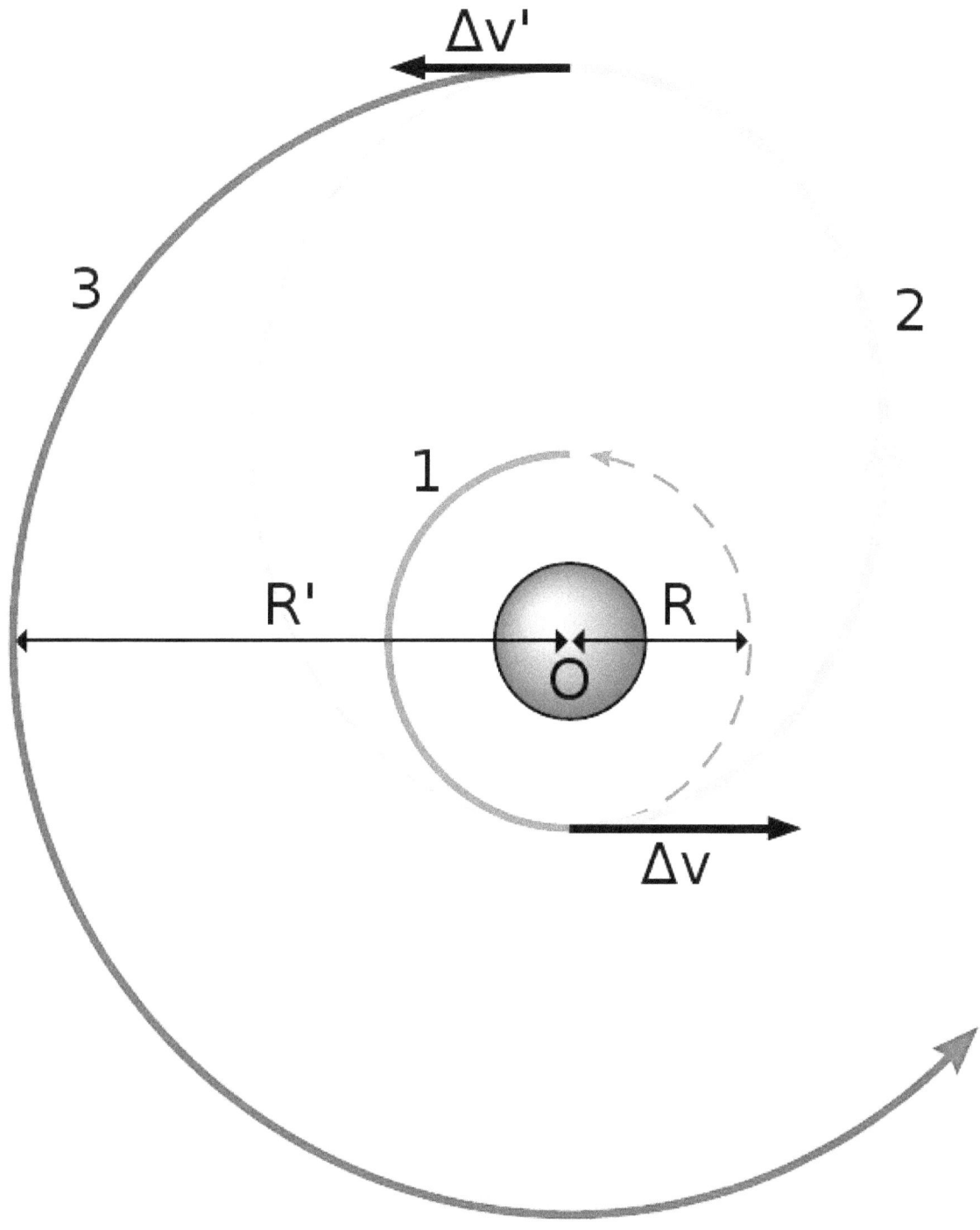

Hohmann Transfer Orbit

The idea of the bi-elliptical transfer trajectory was first published by Ary Sternfeld in 1934.[5]

19.2.3 Low energy transfer

Main article: low energy transfer

A **low energy transfer**, or low energy trajectory, is a route in space which allows spacecraft to change orbits using very little fuel.[6][7] These routes work in the Earth-Moon system and also in other systems, such as traveling between the satellites of Jupiter. The drawback of such trajectories is that they take much longer to complete than higher energy (more fuel) transfers such as Hohmann transfer orbits.

Low energy transfer are also known as weak stability boundary trajectories, or ballistic capture trajectories.

Low energy transfers follow special pathways in space, sometimes referred to as the Interplanetary Transport Network. Following these pathways allows for long distances to be traversed for little expenditure of delta-v.

19.2.4 Orbital inclination change

Main article: orbital inclination change

Orbital inclination change is an orbital maneuver aimed at changing the inclination of an orbiting body's orbit. This maneuver is also known as an orbital plane change as the plane of the orbit is tipped. This maneuver requires a change in the orbital velocity vector (delta v) at the orbital nodes (i.e. the point where the initial and desired orbits intersect, the line of orbital nodes is defined by the intersection of the two orbital planes).

In general, inclination changes can require a great deal of delta-v to perform, and most mission planners try to avoid them whenever possible to conserve fuel. This is typically achieved by launching a spacecraft directly into the desired inclination, or as close to it as possible so as to minimize any inclination change required over the duration of the spacecraft life.

Maximum efficiency of inclination change is achieved at apoapsis, (or apogee), where orbital velocity v is the lowest. In some cases, it may require less total delta v to raise the satellite into a higher orbit, change the orbit plane at the higher apogee, and then lower the satellite to its original altitude.[8]

19.2.5 Constant Thrust Trajectory

Constant-thrust and constant-acceleration trajectories involve the spacecraft firing its engine in a prolonged constant burn. In the limiting case where the vehicle acceleration is high compared to the local gravitational acceleration, the spacecraft points straight toward the target (accounting for target motion), and remains accelerating constantly under high thrust until it reaches its target. In this high-thrust case, the trajectory approaches a straight line. If it is required that the spacecraft rendezvous with the target, rather than performing a flyby, then the spacecraft must flip its orientation halfway through the journey, and decelerate the rest of the way.

In the constant-thrust trajectory,[9] the vehicle's acceleration increases during thrusting period, since the fuel use means the vehicle mass decreases. If, instead of constant thrust, the vehicle has constant acceleration, the engine thrust must decrease during the trajectory.

This trajectory requires that the spacecraft maintain a high acceleration for long durations. For interplanetary transfers, days, weeks or months of constant thrusting may be required. As a result, there are no currently available spacecraft propulsion systems capable of using this trajectory. It has been suggested that some forms of nuclear (fission or fusion based) or antimatter powered rockets would be capable of this trajectory.

19.3 Rendezvous and docking

19.3.1 Orbit phasing

Main article: Orbit phasing

In astrodynamics **orbit phasing** is the adjustment of the time-position of spacecraft along its orbit, usually described as adjusting the orbiting spacecraft's true anomaly.

19.3.2 Space rendezvous and docking

Main article: Space rendezvous

A **space rendezvous** is an orbital maneuver during which two spacecraft, one of which is often a space station, arrive at

Gemini 7 photographed from Gemini 6 in 1965

the same orbit and approach to a very close distance (e.g. within visual contact). Rendezvous requires a precise match of the orbital velocities of the two spacecraft, allowing them to remain at a constant distance through orbital station-keeping. Rendezvous may or may not be followed by docking or berthing, procedures which bring the spacecraft into physical contact and create a link between them.

19.4 See also

- Collision avoidance (spacecraft)
- In-space propulsion technologies

19.5 References

[1] NASA-TT-F-622: Ways to spaceflight p 200 - Herman Oberth

[2] http://www2.jpl.nasa.gov/basics/bsf4-1.php Basics of Space Flight, Sec. 1 Ch. 4, NASA Jet Propulsion Laboratory

[3] Walter Hohmann, *The Attainability of Heavenly Bodies* (Washington: NASA Technical Translation F-44, 1960) Internet Archive.

[4] Vallado, David Anthony (2001). *Fundamentals of Astrodynamics and Applications*. Springer. p. 317. ISBN 0-7923-6903-3.

[5] Sternfeld A., Sur les trajectoires permettant d'approcher d'un corps attractif central à partir d'une orbite keplérienne donnée. - Comptes rendus de l'Académie des sciences (Paris), vol. 198, pp. 711 - 713.

[6] Belbruno, Edward (2004). *Capture Dynamics and Chaotic Motions in Celestial Mechanics: With Applications to the Construction of Low Energy Transfers*. Princeton University Press. p. 224. ISBN 978-0-691-09480-9.

[7] Belbruno, Edward (2007). *Fly Me to the Moon: An Insider's Guide to the New Science of Space Travel*. Princeton University Press. p. 176. ISBN 978-0-691-12822-1.

[8] Braeunig, Robert A. "Basics of Space Flight: Orbital Mechanics".

[9] W. E. Moeckel, Trajectories with Constant Tangential Thrust in Central Gravitational Fields, *Technical Report R-63*, NASA Lewis Research Center, 1960 (accessed 26 March 2014)

19.6 External links

- Handbook Automated Rendezvous and Docking of Spacecraft by Wigbert Fehse

Chapter 20

Hohmann transfer orbit

In orbital mechanics, the **Hohmann transfer orbit** /ˈhoʊ.mʌn/ is an elliptical orbit used to transfer between two circular orbits of different radii in the same plane.

The orbital maneuver to perform the Hohmann transfer uses two engine impulses, one to move a spacecraft onto the transfer orbit and a second to move off it. This maneuver was named after Walter Hohmann, the German scientist who published a description of it in his 1925 book *Die Erreichbarkeit der Himmelskörper* ("The Accessibility of Celestial Bodies")[1] Hohmann was influenced in part by the German science fiction author Kurd Lasswitz and his 1897 book *Two Planets*.

20.1 Explanation

The diagram shows a Hohmann transfer orbit to bring a spacecraft from a lower circular orbit into a higher one. It is one half of an elliptic orbit that touches both the lower circular orbit the spacecraft wishes to leave (green and labeled *1* on diagram) and the higher circular orbit that it wishes to reach (red and labeled *3* on diagram). The transfer (yellow and labeled *2* on diagram) is initiated by firing the spacecraft's engine in order to accelerate it so that it will follow the elliptical orbit; this adds energy to the spacecraft's orbit. When the spacecraft has reached its destination orbit, its orbital speed (and hence its orbital energy) must be increased again in order to change the elliptic orbit to the larger circular one.

Due to the reversibility of orbits, Hohmann transfer orbits also work to bring a spacecraft from a higher orbit into a lower one; in this case, the spacecraft's engine is fired in the opposite direction to its current path, slowing the spacecraft and causing it to drop into the lower-energy elliptical transfer orbit. The engine is then fired again at the lower distance to slow the spacecraft into the lower circular orbit.

The Hohmann transfer orbit is based on two instantaneous velocity changes. Extra fuel is required to compensate for the fact that the bursts take time; this is minimized by using high thrust engines to minimize the duration of the bursts. Low thrust engines can perform an approximation of a Hohmann transfer orbit, by creating a gradual enlargement of the initial circular orbit through carefully timed engine firings. This requires a change in velocity (delta-*v*) that is up to 141% greater than the two impulse transfer orbit (see also below), and takes longer to complete.

20.2 Calculation

For a small body orbiting another very much larger body, such as a satellite orbiting the earth, the total energy of the smaller body is the sum of its kinetic energy and potential energy, and this total energy also equals half the potential at the average distance a, (the semi-major axis):

20.2. CALCULATION

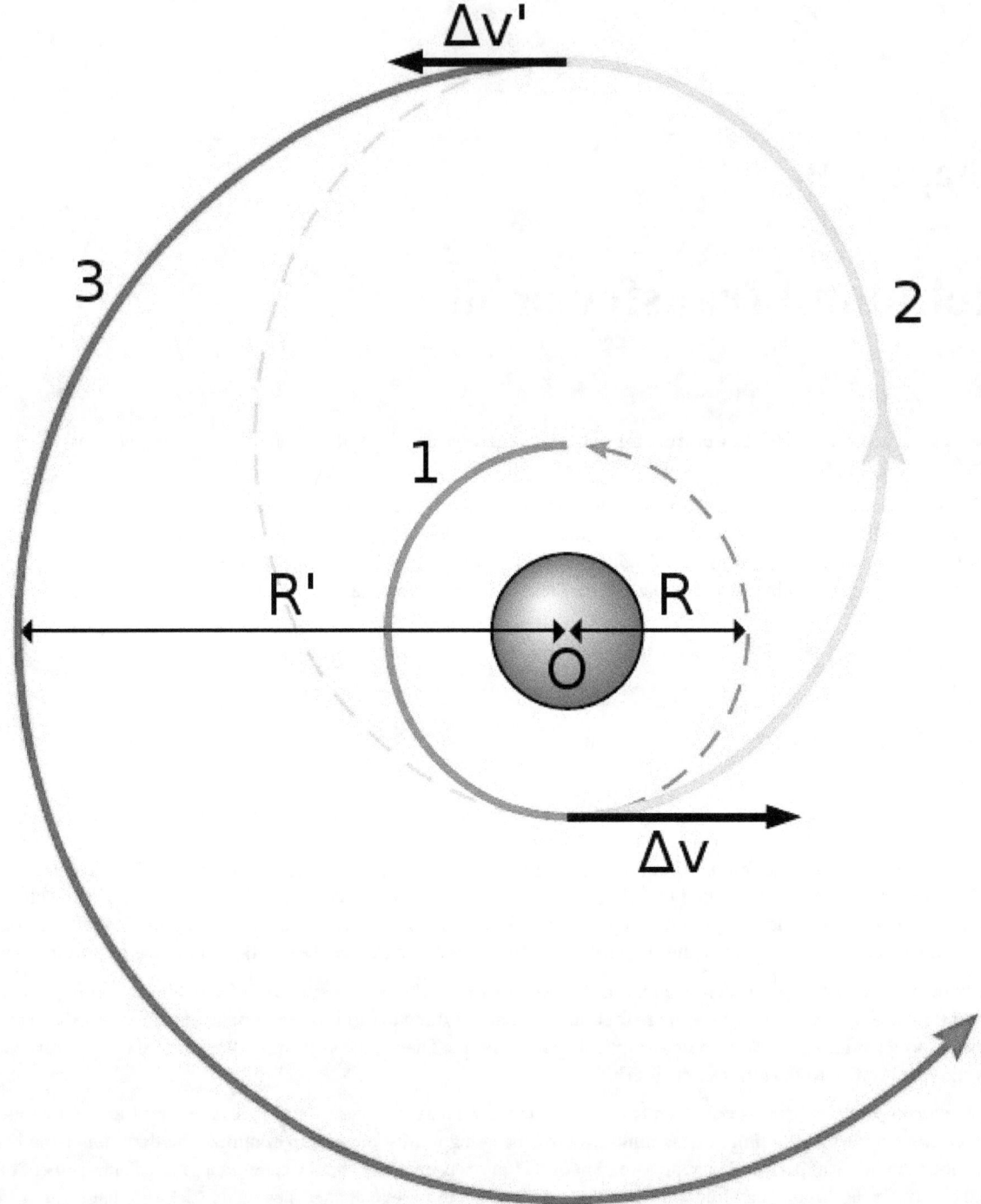

Hohmann transfer orbit, labelled 2, from a low orbit (1) to a higher orbit (3).

$$E = \tfrac{1}{2}mv^2 - \frac{GMm}{r} = \frac{-GMm}{2a}.$$

Solving this equation for velocity results in the vis-viva equation,

$$v^2 = \mu \left(\frac{2}{r} - \frac{1}{a} \right)$$

where:

- v is the speed of an orbiting body
- $\mu = GM$ is the standard gravitational parameter of the primary body, assuming $M + m$ is not significantly bigger than M (which makes $v_M \ll v$)
- r is the distance of the orbiting body from the primary focus
- a is the semi-major axis of the body's orbit.

Therefore the delta-v required for the Hohmann transfer can be computed as follows, under the assumption of instantaneous impulses:

$$\Delta v_1 = \sqrt{\frac{\mu}{r_1}} \left(\sqrt{\frac{2r_2}{r_1 + r_2}} - 1 \right),$$

to enter the elliptical orbit at $r = r_1$ from the r_1 circular orbit

$$\Delta v_2 = \sqrt{\frac{\mu}{r_2}} \left(1 - \sqrt{\frac{2r_1}{r_1 + r_2}} \right),$$

to leave the elliptical orbit at $r = r_2$ to the r_2 circular orbit where r_1 and r_2 are, respectively, the radii of the departure and arrival circular orbits; the smaller (greater) of r_1 and r_2 corresponds to the periapsis distance (apoapsis distance) of the Hohmann elliptical transfer orbit. The total Δv is then:

$$\Delta v_{total} = \Delta v_1 + \Delta v_2.$$

Whether moving into a higher or lower orbit, by Kepler's third law, the time taken to transfer between the orbits is:

$$t_H = \tfrac{1}{2} \sqrt{\frac{4\pi^2 a_H^3}{\mu}} = \pi \sqrt{\frac{(r_1 + r_2)^3}{8\mu}}$$

(one half of the orbital period for the whole ellipse), where a_H is length of semi-major axis of the Hohmann transfer orbit.

In application to traveling from one celestial body to another it is crucial to start maneuver at the time when the two bodies are properly aligned. Considering the target angular velocity being

$$\omega_2 = \sqrt{\frac{\mu}{r_2^3}}$$

angular alignment α (in radians) at the time of start between the source object and the target object shall be

$$\alpha = \pi - \omega_2 t_H = \pi \left(1 - \frac{1}{2\sqrt{2}} \sqrt{\left(\frac{r_1}{r_2} + 1 \right)^3} \right)$$

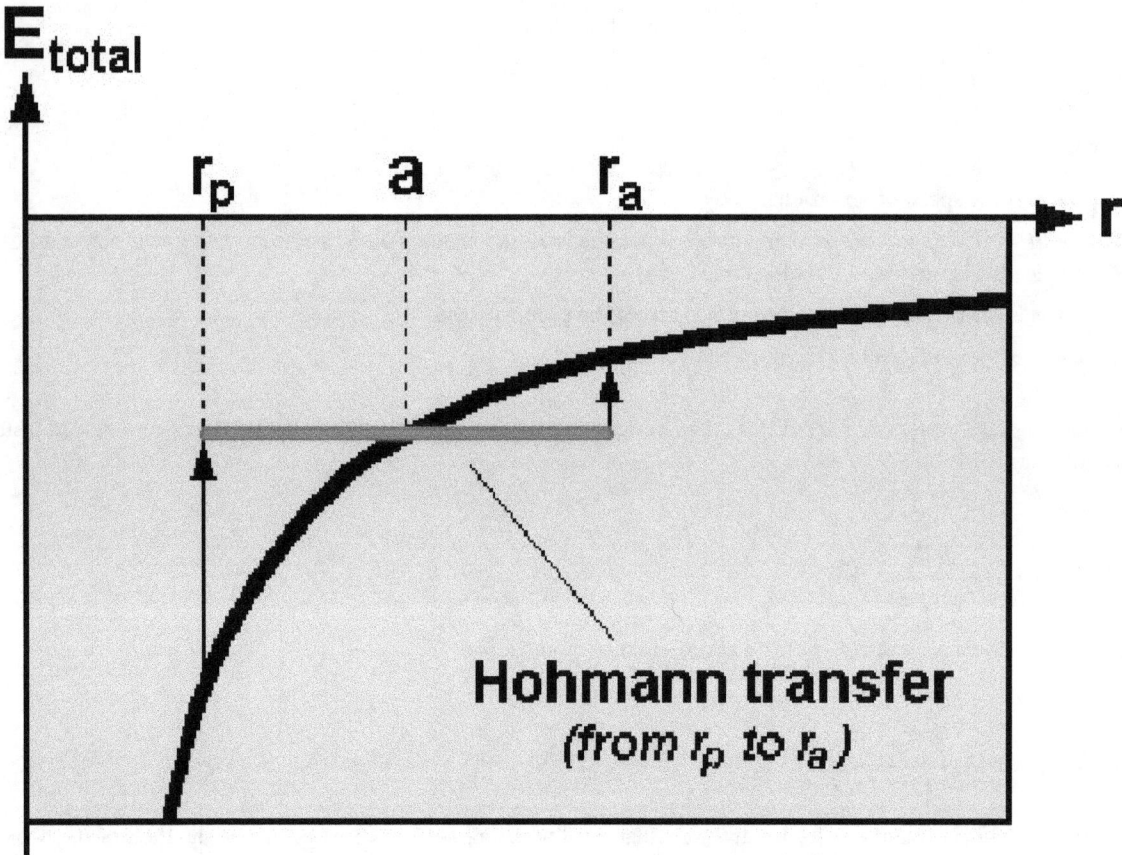

Total energy balance during a Hohmann transfer between two circular orbits with first radius r_p and second radius r_a.

20.3 Example

Consider a geostationary transfer orbit, beginning at r_1 = 6,678 km (altitude 300 km) and ending in a geostationary orbit with r_2 = 42,164 km (altitude 35,786 km).

In the smaller circular orbit the speed is 7.73 km/s; in the larger one, 3.07 km/s. In the elliptical orbit in between the speed varies from 10.15 km/s at the perigee to 1.61 km/s at the apogee.

The Δv for the two burns are thus 10.15 − 7.73 = 2.42 and 3.07 − 1.61 = 1.46 km/s, together 3.88 km/s.

It is interesting to note that this is *greater* than the Δv required for an escape orbit: 10.93 − 7.73 = 3.20 km/s. Applying a Δv at the LEO of only 0.78 km/s more (3.20−2.42) would give the rocket the escape speed, which is less than the Δv of 1.46 km/s required to circularize the geosynchronous orbit. This illustrates that at large speeds the same Δv provides more specific orbital energy, and energy increase is maximized if one spends the Δv as quickly as possible, rather than spending some, being decelerated by gravity, and then spending some more to overcome the deceleration (of course, the objective of a Hohmann transfer orbit is different).

20.4 Worst case, maximum delta-v

As the example above demonstrates, the Δv required to perform a Hohmann transfer between two circular orbits is not maximized when the destination is at infinity. Escape speed is $\sqrt{2}$ times orbital speed, so the Δv required to escape is $\sqrt{2}$ − 1 (41.4%) of the orbital speed.

The Δv required is greatest (53.0% of smaller orbital speed) if the radius of the larger orbit is 15.58 times that of the smaller orbit.[2] This number is the positive root of $x^3 - 15x^2 - 9x - 1 = 0$, which is $5 + 4\sqrt{7}\cos\left(\frac{1}{3}\tan^{-1}\frac{\sqrt{3}}{37}\right)$. For higher orbit ratios the Δv required for the second burn decreases faster than the first increases.

20.5 Application to interplanetary travel

When used to move a spacecraft from orbiting one planet to orbiting another, the situation becomes somewhat more complex, but much less delta-v is required due to the Oberth effect.

For example, consider a spacecraft travelling from the Earth to Mars. At the beginning of its journey, the spacecraft will already have a certain velocity and kinetic energy associated with its orbit around Earth. During the burn the rocket engine applies its delta-v, but the kinetic energy increases as a square law, until it is sufficient to escape the planet's gravitational potential, and then burns more so as to gain enough energy to reach the Hohmann transfer orbit (around the Sun). Because the rocket engine is able to make use of the initial kinetic energy of the propellant, far less delta-v is required over and above that needed to reach escape velocity, and the optimum situation is when the transfer burn is made at minimum altitude (low periapsis) above the planet.

At the other end, the spacecraft will need a certain velocity to orbit Mars, which will actually be less than the velocity needed to continue orbiting the Sun in the transfer orbit, let alone attempting to orbit the Sun in a Mars-like orbit. Therefore, the spacecraft will have to decelerate in order for the gravity of Mars to capture it. This capture burn should optimally be done at low altitude to also make best use of Oberth effect. Therefore, relatively small amounts of thrust at either end of the trip are needed to arrange the transfer compared to the free space situation.

However, with any Hohmann transfer, the alignment of the two planets in their orbits is crucial – the destination planet and the spacecraft must arrive at the same point in their respective orbits around the Sun at the same time. This requirement for alignment gives rise to the concept of launch windows.

The term lunar transfer orbit (LTO) is used for the moon.

20.6 Hohmann transfer versus low thrust orbits

20.6.1 Low-thrust transfer

It can be shown that going from one circular orbit to another by gradually changing the radius costs a delta-v of simply the absolute value of the difference between the two speeds. Thus for the geostationary transfer orbit 7.7 − 3.07 = 4.66 km/s, the same as, in the absence of gravity, the *deceleration* would cost. In fact, *acceleration* is applied to compensate half of the deceleration due to moving outward. Therefore the acceleration due to thrust is equal to the deceleration due to the combined effect of thrust and gravity.

Such a low-thrust maneuver requires more delta-v than a 2-burn Hohmann transfer maneuver, requiring more fuel for a given engine design. However, if only low-thrust maneuvers are required on a mission, then continuously firing a low-thrust, but very high-efficiency (high effective exhaust velocity) engine might generate this higher delta-v using less propellant mass than a high-thrust engine using an otherwise more efficient Hohmann transfer maneuver.

The amount of propellant mass used measures the efficiency of the *maneuver plus the hardware* employed for it. The total delta-v used measures the efficiency of the *maneuver only*. For electric propulsion systems, which tend to be low-thrust, the high efficiency of the propulsive system usually vastly compensates for the inability to make the more efficient Hohmann maneuver.

Transfer orbit using Electrical Propulsion or Low Thrust engines optimize the transfer time to reach the final orbit and not the delta-v as in the Hohmann transfer orbit. For geostationary orbit, the initial orbit is set to be supersynchronous and by thrusting continuously in the direction of the velocity at Apogee, the transfer orbit transforms to a circular geosynchronous one. This method however takes much longer to achieve due to the low thrust injected into the orbit.[3]

20.6.2 Interplanetary Transport Network

In 1997, a set of orbits known as the Interplanetary Transport Network was published, providing even lower propulsive delta-v (though much slower and longer) paths between different orbits than Hohmann transfer orbits.[4] The Interplanetary Transport Network is different in nature than Hohmann transfers because Hohmann transfers assume only one large body whereas the Interplanetary Transport Network does not. The Interplanetary Transport Network is able to achieve the use of less propulsive delta-v by employing gravity assist from the planets. The gravity assist provides "free" delta-v without the use of the propulsion systems.

20.7 See also

- Bi-elliptic transfer
- Delta-v budget
- Geostationary transfer orbit
- Halo orbit
- Lissajous orbit
- List of orbits
- Orbital mechanics

20.8 References

[1] Walter Hohmann, *The Attainability of Heavenly Bodies* (Washington: NASA Technical Translation F-44, 1960) Internet Archive.

[2] Vallado, David Anthony (2001). *Fundamentals of Astrodynamics and Applications*. Springer. p. 317. ISBN 0-7923-6903-3.

[3] Spitzer, Arnon (1997). *Optimal Transfer Orbit Trajectory using Electric Propulsion*. USPTO.

[4] Lo, M., S. Ross, *Surfing the Solar System: Invariant Manifolds and the Dynamics of the Solar System*, JPL IOM 312/97, 1997.

General

- Walter Hohmann (1925). *Die Erreichbarkeit der Himmelskörper*. Verlag Oldenbourg in München. ISBN 3-486-23106-5.
- Thornton, Stephen T.; Marion, Jerry B. (2003). *Classical Dynamics of Particles and Systems (5th ed.)*. Brooks Cole. ISBN 0-534-40896-6.
- Bate, R.R., Mueller, D.D., White, J.E., (1971). *Fundamentals of Astrodynamics*. Dover Publications, New York. ISBN 978-0-486-60061-1.
- Vallado, D. A. (2001). *Fundamentals of Astrodynamics and Applications, 2nd Edition*. Springer. ISBN 978-0-7923-6903-5.
- Battin, R.H. (1999). *An Introduction to the Mathematics and Methods of Astrodynamics*. American Institute of Aeronautics & Ast, Washington, DC. ISBN 978-1-56347-342-5.

20.9 External links

- ORBITAL MECHANICS (Rocket and Space Technology)
- Basics of Spaceflight - Chapter 4. Interplanetary Trajectories

Chapter 21

Delta-v

For other uses, see Delta-v (disambiguation).

Delta-v (literally "change in velocity"), symbolised as Δv and pronounced *delta-vee*, as used in spacecraft flight dynamics, is a measure of the impulse that is needed to perform a maneuver such as launch from, or landing on a planet or moon, or in-space orbital maneuver. It is a scalar that has the units of speed. As used in this context, it is *not* the same as the physical change in velocity of the vehicle.

Delta-v is produced by reaction engines, such as rocket engines and is proportional to the thrust per unit mass, and burn time, and is used to determine the mass of propellant required for the given maneuver through the Tsiolkovsky rocket equation.

For multiple maneuvers, delta-v sums linearly.

For interplanetary missions delta-v is often plotted on a porkchop plot which displays the required mission delta-v as a function of launch date.

21.1 Definition

$$\Delta v = \int_{t_0}^{t_1} \frac{|T|}{m} dt$$

where

T is the instantaneous thrust

m is the instantaneous mass

21.2 Specific cases

In the absence of external forces:

$$= \int_{t_0}^{t_1} |a|\, dt$$

where a is the coordinate acceleration.

When thrust is applied in a constant direction this simplifies to:

$$= |v_1 - v_0|$$

which is simply the magnitude of the change in velocity. However, this relation does not hold in the general case: if, for instance, a constant, unidirectional acceleration is reversed after $(t_1 - t_0)/2$ then the velocity difference is 0, but delta-v is the same as for the non-reversed thrust.

For rockets the 'absence of external forces' is taken to mean the absence of gravity, atmospheric drag as well as the absence of aerostatic back pressure on the nozzle and hence the vacuum I_{sp} is used for calculating the vehicle's delta-v capacity via the rocket equation, and the costs for the atmospheric losses are rolled into the delta-v budget when dealing with launches from a planetary surface.

21.3 Orbital maneuvers

Main article: rocket equation

Orbit maneuvers are made by firing a thruster to produce a reaction force acting on the spacecraft. The size of this force will be

$$T = v_{\text{exh}}\, \rho$$

where

V_{exh} is the velocity of the exhaust gas

ρ is the propellant flow rate to the combustion chamber

The acceleration $\dot{V}$ of the spacecraft caused by this force will be

$$\dot{v} = \frac{T}{m} = v_{\text{exh}}\, \frac{\rho}{m}$$

where m is the mass of the spacecraft

During the burn the mass of the spacecraft will decrease due to use of fuel, the time derivative of the mass being

$$\dot{m} = -\rho$$

If now the direction of the force, i.e. the direction of the nozzle, is fixed during the burn one gets the velocity increase from the thruster force of a burn starting at time t_0 and ending at t_1 as

$$\Delta v = -\int_{t_0}^{t_1} v_{\text{exh}}\, \frac{\dot{m}}{m}\, dt$$

Changing the integration variable from time t to the spacecraft mass m one gets

$$\Delta v = -\int_{m_0}^{m_1} v_{\text{exh}}\, \frac{dm}{m}$$

Assuming v_{exh} to be a constant not depending on the amount of fuel left this relation is integrated to

$$\Delta v = v_{\text{exh}}\, \ln\left(\frac{m_0}{m_1}\right)$$

which is the Tsiolkovsky rocket equation.

If for example 20% of the launch mass is fuel giving a constant V_{exh} of 2100 m/s (typical value for a hydrazine thruster) the capacity of the reaction control system is

$$\Delta V = 2100 \ \ln(\frac{1}{0.8})$$

If V_{exh} is a non-constant function of the amount of fuel left[1]

$$V_{exh} = V_{exh}(m)$$

the capacity of the reaction control system is computed by the integral (5)

The acceleration (2) caused by the thruster force is just an additional acceleration to be added to the other accelerations (force per unit mass) affecting the spacecraft and the orbit can easily be propagated with a numerical algorithm including also this thruster force.[2] But for many purposes, typically for studies or for maneuver optimization, they are approximated by impulsive maneuvers as illustrated in figure 1 with a ΔV as given by (4). Like this one can for example use a "patched conics" approach modeling the maneuver as a shift from one Kepler orbit to another by an instantaneous change of the velocity vector.

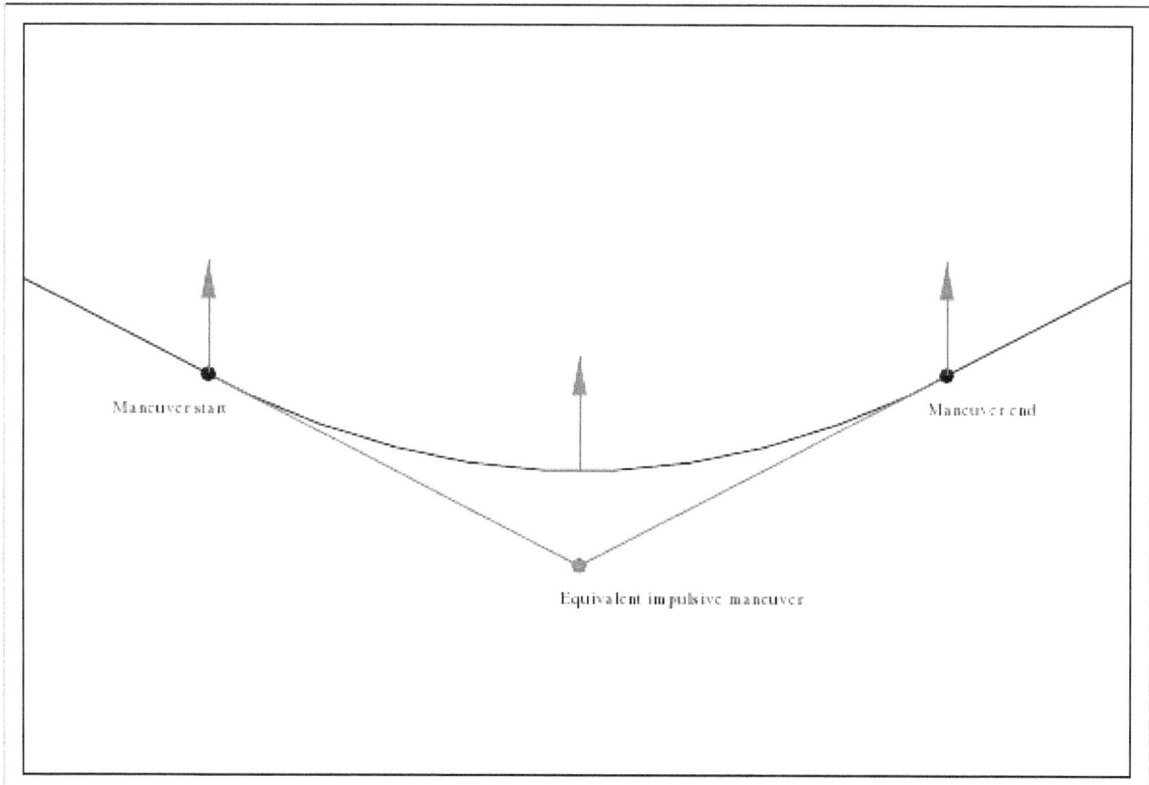

Figure 1:Approximation of a finite thrust maneuver with an impulsive change in velocity having the delta-v given by (4)

This approximation with impulsive maneuvers is in most cases very accurate, at least when chemical propulsion is used. For low thrust systems, typically electrical propulsion systems, this approximation is less accurate. But even for geostationary spacecraft using electrical propulsion for out-of-plane control with thruster burn periods extending over several hours around the nodes this approximation is fair.

21.4 Producing delta-v

Delta-v is typically provided by the thrust of a rocket engine, but can be created by other reaction engines. The time-rate of change of delta-v is the magnitude of the acceleration *caused by the engines*, i.e., the thrust per total vehicle mass. The actual acceleration vector would be found by adding thrust per mass on to the gravity vector and the vectors representing any other forces acting on the object.

The total delta-v needed is a good starting point for early design decisions since consideration of the added complexities are deferred to later times in the design process.

The rocket equation shows that the required amount of propellant dramatically increases, with increasing delta-v. Therefore in modern spacecraft propulsion systems considerable study is put into reducing the total delta-v needed for a given spaceflight, as well as designing spacecraft that are capable of producing a large delta-v.

Increasing the Delta-v provided by a propulsion system can be achieved by:

- staging
- increasing specific impulse
- improving propellant mass fraction

21.5 Multiple maneuvers

Because the mass ratios apply to any given burn, when multiple maneuvers are performed in sequence, the mass ratios multiply.

Thus it can be shown that, provided the exhaust velocity is fixed, this means that delta-v's can be added:

When M1, M2 are the mass ratios of the maneuvers, and V1, V2 are the delta-v's of the first and second maneuvers

$$M1M2$$

$$= e^{V1/V_e} e^{V2/V_e}$$
$$= e^{(V1+V2)/V_e}$$
$$= e^{V/V_e} = M$$

Where V = V1 + V2 and M = M1 M2.

Which is just the rocket equation applied to the sum of the two maneuvers.

This is convenient since it means that delta-v's can be calculated and simply added and the mass ratio calculated only for the overall vehicle for the entire mission. Thus delta-v is commonly quoted rather than mass ratios which would require multiplication.

21.6 Delta-v budgets

Main article: delta-v budget

When designing a trajectory, delta-v budget is used as a good indicator of how much propellant will be required. Propellant usage is an exponential function of delta-v in accordance with the rocket equation, it will also depend on the exhaust velocity.

It is not possible to determine delta-v requirements from conservation of energy by considering only the total energy of the vehicle in the initial and final orbits since energy is carried away in the exhaust (see also below). For example, most spacecraft are launched in an orbit with inclination fairly near to the latitude at the launch site, to take advantage of the Earth's rotational surface speed. If it is necessary, for mission-based reasons, to put the spacecraft in an orbit of different inclination, a substantial delta-v is required, though the specific kinetic and potential energies in the final orbit and the initial orbit are equal.

When rocket thrust is applied in short bursts the other sources of acceleration may be negligible, and the magnitude of the velocity change of one burst may be simply approximated by the delta-v. The total delta-v to be applied can then simply be found by addition of each of the delta-v's needed at the discrete burns, even though between bursts the magnitude and direction of the velocity changes due to gravity, e.g. in an elliptic orbit.

For examples of calculating delta-v, see Hohmann transfer orbit, gravitational slingshot, and Interplanetary Transport Network. It is also notable that large thrust can reduce gravity drag.

Delta-v is also required to keep satellites in orbit and is expended in propulsive orbital stationkeeping maneuvers. Since the propellant load on most satellites cannot be replenished, the amount of propellant initially loaded on a satellite may well determine its useful lifetime.

21.6.1 Oberth effect

Main article: Oberth effect

From power considerations, it turns out that when applying delta-v in the direction of the velocity the specific orbital energy gained per unit delta-v is equal to the instantaneous speed. This is called the Oberth effect.

For example, a satellite in an elliptical orbit is boosted more efficiently at high speed (that is, small altitude) than at low speed (that is, high altitude).

Another example is that when a vehicle is making a pass of a planet, burning the propellant at closest approach rather than further out gives significantly higher final speed, and this is even more so when the planet is a large one with a deep gravity field, such as Jupiter.

See also powered slingshots.

21.6.2 Porkchop plot

Main article: porkchop plot

Due to the relative positions of planets changing over time, different delta-vs are required at different launch dates. A diagram that shows the required delta-v plotted against time is sometimes called a *porkchop plot*. Such a diagram is useful since it enables calculation of a launch window, since launch should only occur when the mission is within the capabilities of the vehicle to be employed.[3]

21.6.3 Delta-vs around the Solar System

21.7 See also

- Delta-v budget
- Gravity drag
- Orbital maneuver
- Orbital stationkeeping
- Spacecraft propulsion

- Specific impulse
- Tsiolkovsky rocket equation
- Delta-v (physics)

21.8 References

[1] Can be the case for a "blow-down" system for which the pressure in the tank gets lower when fuel has been used and that not only the fuel rate ρ but to some lesser extent also the exhaust velocity V_{exh} decreases.

[2] The thrust force per unit mass being $\frac{f(t)}{m(t)} = V_{exh}(t)\frac{\dot{m}(t)}{m(t)}$ where $f(t)$ and $m(t)$ are given functions of time t

[3] Mars Exploration: Features

[4] Rockets and Space Transportation. See: Atomic Rocket: Missions

[5] cislunar delta-vs

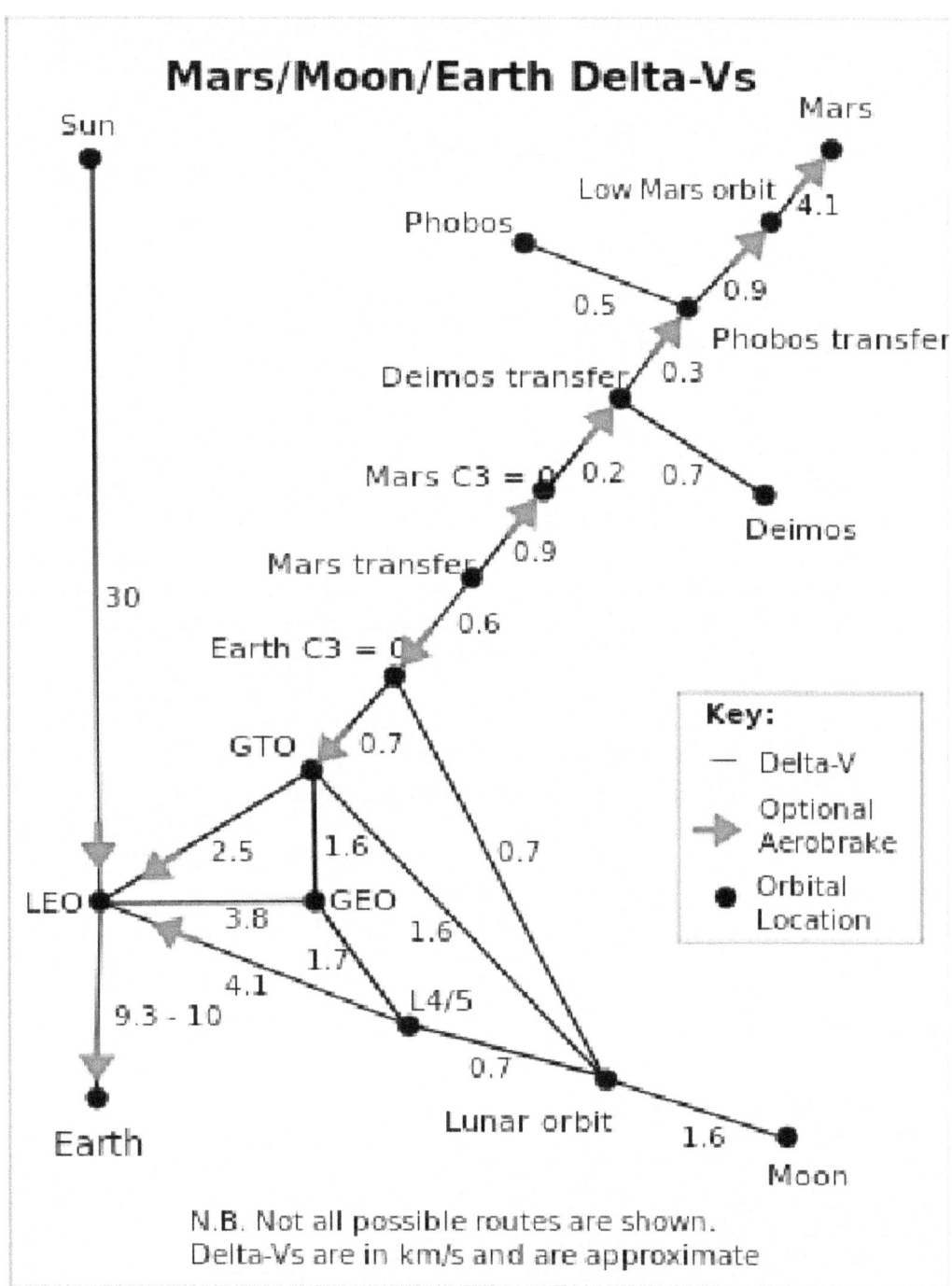

Delta-v needed for various orbital manoeuvers using conventional rockets; red arrows show where optional aerobraking can be performed in that particular direction, black numbers give delta-v in km/s that apply in either direction. Lower-delta-v transfers than shown can often be achieved, but involve rare transfer windows or take significantly longer, see: fuzzy orbital transfers. The figure 2.5 for LEO to GTO is higher than necessary[6] and the figure of 30 for LEO to the sun is also too high [7]

C3 - Escape orbit, GEO - Geosynchronous orbit, GTO - Geostationary transfer orbit, L4/5 - Earth–Moon L4L5 Lagrangian point, LEO - Low Earth orbit

Chapter 22

Bi-elliptic transfer

In astronautics and aerospace engineering, the **bi-elliptic transfer** is an orbital maneuver that moves a spacecraft from one orbit to another and may, in certain situations, require less delta-v than a Hohmann transfer maneuver.

The bi-elliptic transfer consists of two half elliptic orbits. From the initial orbit, a first burn expends delta-v to boost the spacecraft into the first transfer orbit with an apoapsis at some point r_b away from the central body. At this point a second burn sends the spacecraft into the second elliptical orbit with periapsis at the radius of the final desired orbit, where a third burn is performed, injecting the spacecraft into the desired orbit.

While they require one more engine burn than a Hohmann transfer and generally requires a greater travel time, some bi-elliptic transfers require a lower amount of total delta-v than a Hohmann transfer when the ratio of final to initial semi-major axis is 11.94 or greater, depending on the intermediate semi-major axis chosen.[1]

The idea of the bi-elliptical transfer trajectory was first published by Ary Sternfeld in 1934.[2]

22.1 Calculation

22.1.1 Delta-v

The three required changes in velocity can be obtained directly from the vis-viva equation,

$$v^2 = \mu \left(\frac{2}{r} - \frac{1}{a} \right)$$

- v is the speed of an orbiting body
- $\mu = GM$ is the standard gravitational parameter of the primary body
- r is the distance of the orbiting body from the primary
- a is the semi-major axis of the body's orbit
- r_b is the common apoapsis distance of the two transfer ellipses and is a free parameter of the maneuver.
- a_1 and a_2 are the semimajor axes of the two elliptical transfer orbits, which are given by

$$a_1 = \frac{r_0 + r_b}{2}$$

$$a_2 = \frac{r_f + r_b}{2}$$

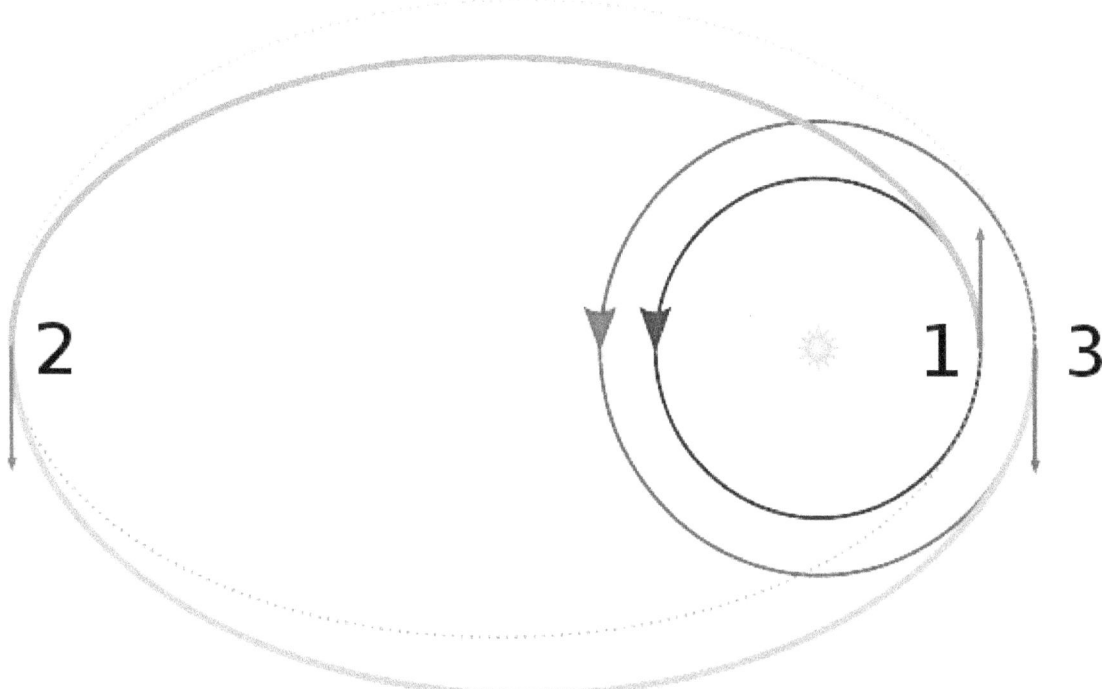

A bi-elliptic transfer from a low circular starting orbit (dark blue), to a higher circular orbit (red).

Starting from the initial circular orbit with radius r_0 (dark blue circle in the figure to the right), a prograde burn (mark 1 in the figure) puts the spacecraft on the first elliptical transfer orbit (aqua half ellipse). The magnitude of the required delta-v for this burn is:

$$\Delta v_1 = \sqrt{\frac{2\mu}{r_0} - \frac{\mu}{a_1}} - \sqrt{\frac{\mu}{r_0}}$$

When the apoapsis of the first transfer ellipse is reached at a distance r_b from the primary, a second prograde burn (mark 2) raises the periapsis to match the radius of the target circular orbit, putting the spacecraft on a second elliptic trajectory (orange half ellipse). The magnitude of the required delta-v for the second burn is:

$$\Delta v_2 = \sqrt{\frac{2\mu}{r_b} - \frac{\mu}{a_2}} - \sqrt{\frac{2\mu}{r_b} - \frac{\mu}{a_1}}$$

Lastly, when the final circular orbit with radius r_f is reached, a *retrograde* burn (mark 3) circularizes the trajectory into the final target orbit (red circle). The final retrograde burn requires a delta-v of magnitude:

$$\Delta v_3 = \sqrt{\frac{2\mu}{r_f} - \frac{\mu}{a_2}} - \sqrt{\frac{\mu}{r_f}}$$

If $r_b = r_f$, then the maneuver reduces to a Hohmann transfer (in that case Δv_3 can be verified to become zero). Thus the bi-elliptic transfer constitutes a more general class of orbital transfers, of which the Hohmann transfer is a special two-impulse case.

The maximum savings possible can be computed by assuming that $r_b = \infty$, in which case the total Δv simplifies to $(\sqrt{2} - 1)\left(\sqrt{\mu/r_0} + \sqrt{\mu/r_f}\right)$.

22.1.2 Transfer time

Like the Hohmann transfer, both transfer orbits used in the bi-elliptic transfer constitute exactly one half of an elliptic orbit. This means that the time required to execute each phase of the transfer is half the orbital period of each transfer ellipse.

Using the equation for the orbital period and the notation from above:

$$T = 2\pi \sqrt{\frac{a^3}{\mu}}$$

The total transfer time t is the sum of the time required for each half orbit. Therefore:

$$t_1 = \pi \sqrt{\frac{a_1^3}{\mu}} \quad and \quad t_2 = \pi \sqrt{\frac{a_2^3}{\mu}}$$

And finally:

$$t = t_1 + t_2$$

22.2 Example

To transfer from a circular low Earth orbit with r_0=6700 km to a new circular orbit with r_1=93800 km using a Hohmann transfer orbit requires a Δv of 2825.02+1308.70=4133.72 m/s. However, because r_1=14r_0 >11.94r_0, it is possible to do better with a bi-elliptic transfer. If the spaceship first accelerated 3061.04 m/s, thus achieving an elliptic orbit with apogee at r_2=40r_0=268000 km, then at apogee accelerated another 608.825 m/s to a new orbit with perigee at r_1=93800 km, and finally at perigee of this second transfer orbit decelerated by 447.662 m/s, entering the final circular orbit, then the total Δv would be only 4117.53 m/s, which is 16.19 m/s (0.4%) less.

The Δv saving could be further improved by increasing the intermediate apogee, at the expense of longer transfer time. For example, an apogee of 75.8r_0=507,688 km (1.3 times the distance to the Moon) would result in a 1% Δv saving over a Hohmann transfer, but a transit time of 17 days. As an impractical extreme example, an apogee of 1757r_0=11,770,000 km (30 times the distance to the Moon) would result in a 2% Δv saving over a Hohmann transfer, but the transfer would require 4.5 years (and, in practice, be perturbed by the gravitational effects of other solar system bodies). For comparison, the Hohmann transfer requires 15 hours and 34 minutes.

- Δv applied prograde
- Δv applied retrograde

Evidently, the bi-elliptic orbit spends more of its delta-v early on (in the first burn). This yields a higher contribution to the specific orbital energy and, due to the Oberth effect, is responsible for the net reduction in required delta-v.

22.3 See also

- Delta-v budget
- Oberth effect

22.4 References

[1] Vallado, David Anthony (2001). *Fundamentals of Astrodynamics and Applications*. Springer. p. 318. ISBN 0-7923-6903-3.

[2] Sternfeld, Ary J. [*sic*] (1934-02-12), "Sur les trajectoires permettant d'approcher d'un corps attractif central à partir d'une orbite képlérienne donnée" [On the allowed trajectories for approaching a central attractive body from a given Keplerian orbit], *Comptes rendus de l'Académie des sciences* (in French) (Paris) **198** (1): 711–713

Chapter 23

Gravity assist

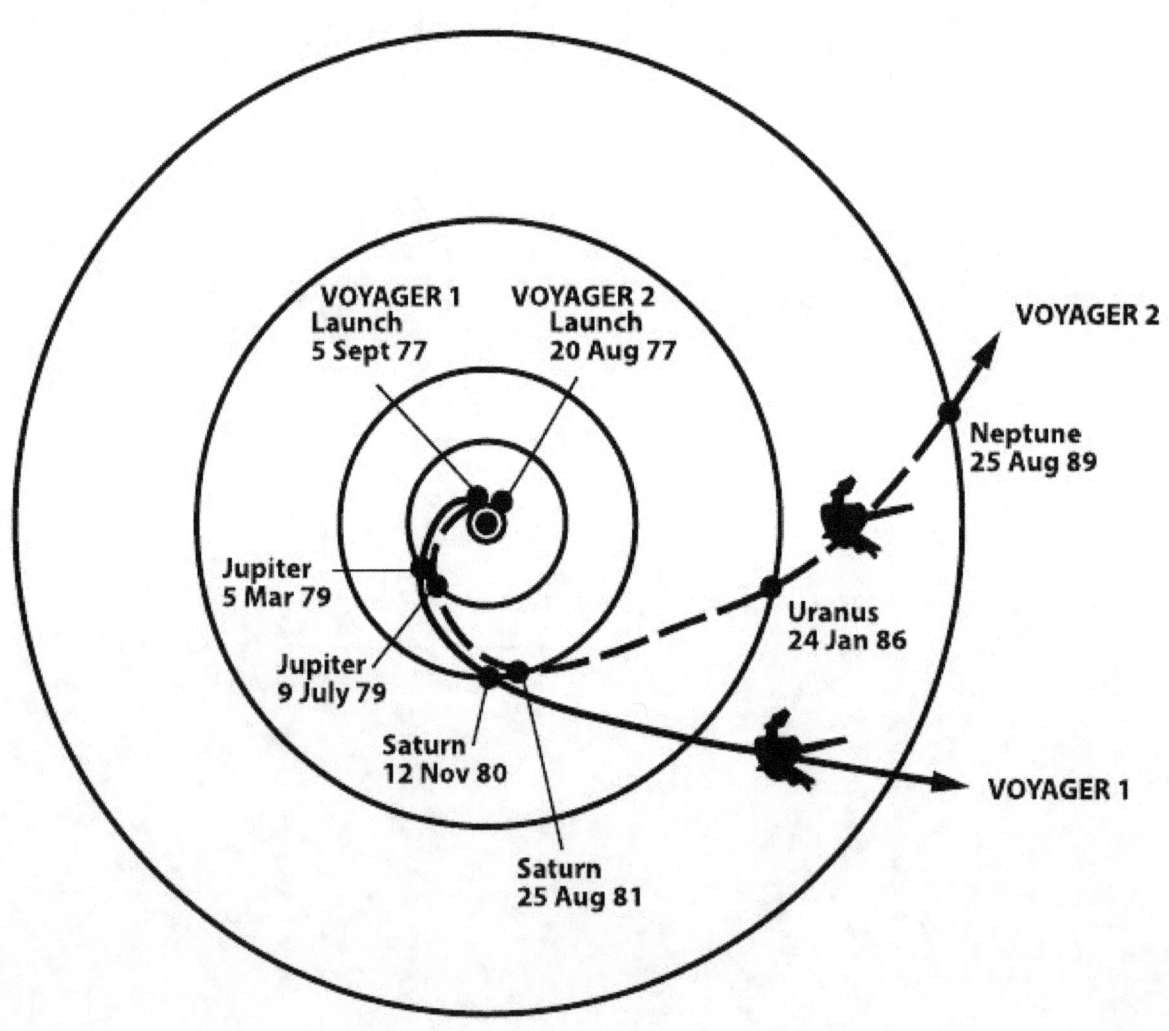

The trajectories that enabled NASA's twin Voyager spacecraft to tour the four gas giant planets and achieve velocity to escape the Solar System

For powered flybys, see Oberth maneuver.

In orbital mechanics and aerospace engineering, a **gravitational slingshot**, **gravity assist maneuver**, or **swing-by** is the use of the relative movement (e.g. orbit around the Sun) and gravity of a planet or other astronomical object to alter the path and speed of a spacecraft, typically in order to save propellant, time, and expense. Gravity assistance can be used to accelerate a spacecraft, that is, to increase or decrease its speed and/or redirect its path.

The "assist" is provided by the motion of the gravitating body as it pulls on the spacecraft.[1] The techniques were first proposed as a mid-course manoeuvre in 1961 by Michael Minovitch working on the three-body problem. It was used by interplanetary probes from *Mariner 10* onwards, including the two *Voyager* probes' notable flybys of Jupiter and Saturn.

23.1 Explanation

A gravity assist around a planet changes a spacecraft's velocity (relative to the Sun) by entering and leaving the gravitational field of a planet. The spacecraft's speed increases as it approaches the planet and decreases while escaping its gravitational pull (which is approximately the same). Because the planet orbits the sun, the spacecraft is affected by this motion during the maneuver. To increase speed, the spacecraft flies with the movement of the planet (taking a small amount of the planet's orbital energy); to decrease speed, the spacecraft flies against the movement of the planet. The sum of the kinetic energies of both bodies remains constant (see elastic collision). A slingshot maneuver can therefore be used to change the spaceship's trajectory and speed relative to the Sun.

A close terrestrial analogy is provided by a tennis ball bouncing off the front of a moving train. Imagine standing on a train platform, and throwing a ball at 30 km/h toward a train approaching at 50 km/h. The driver of the train sees the ball approaching at 80 km/h and then departing at 80 km/h after the ball bounces elastically off the front of the train. Because of the train's motion, however, that departure is at 130 km/h relative to the train platform; the ball has added twice the train's velocity to its own.

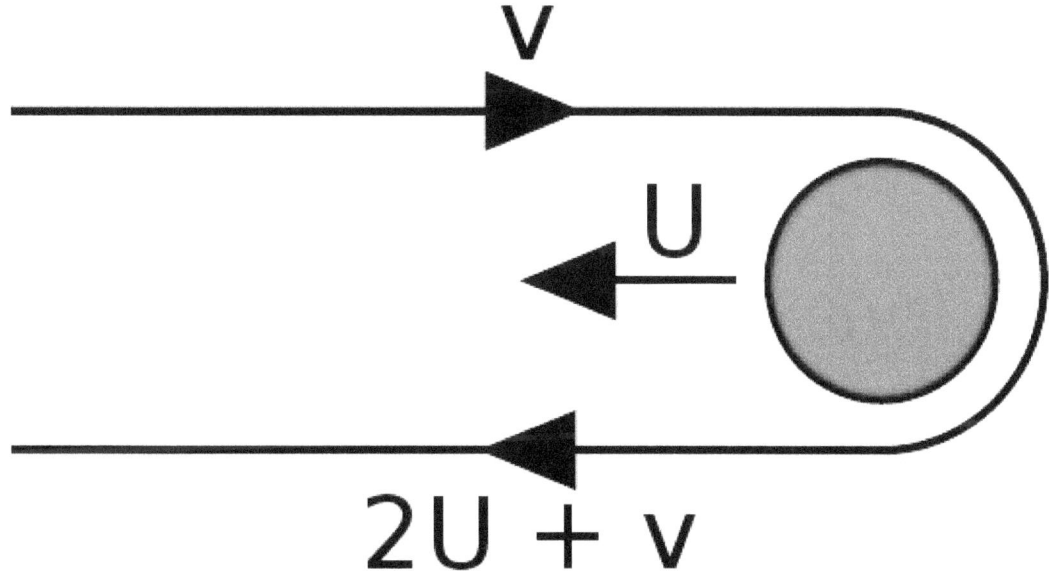

Simplified example of gravitational slingshot: the spacecraft's speed changes by up to twice the planet's speed

Translating this analogy into space, then, a "stationary" observer sees a planet moving left at speed U and a spaceship moving right at speed v. If the spaceship has the proper trajectory, it will pass close to the planet, moving at speed $U + v$ relative to the planet's surface because the planet is moving in the opposite direction at speed U. When the spaceship leaves orbit, it is moving at speed $U + v$ relative to the planet's surface but in the opposite direction (to the left). Since the planet is moving left at speed U, the total velocity of the spaceship relative to the observer will be the velocity of the

23.1. EXPLANATION

moving planet plus the velocity of the spaceship with respect to the planet. So the velocity will be $U + (U + v) = 2U + v$.

This oversimplified example is impossible to refine without additional details regarding the orbit, but if the spaceship travels in a path which forms a parabola, it can leave the planet in the opposite direction without firing its engine, and the speed gain at large distance is indeed $2U$ once it has left the gravity of the planet far behind.

This explanation might seem to violate the conservation of energy and momentum, but the spacecraft's effects on the planet have not been considered. The linear momentum gained by the spaceship is equal in magnitude to that lost by the planet, though the planet's enormous mass compared to the spacecraft makes the resulting change in its speed negligibly small. These effects on the planet are so slight (because planets are so much more massive than spacecraft) that they can be ignored in the calculation.[2]

Realistic portrayals of encounters in space require the consideration of three dimensions. The same principles apply, only adding the planet's velocity to that of the spacecraft requires vector addition, as shown below.

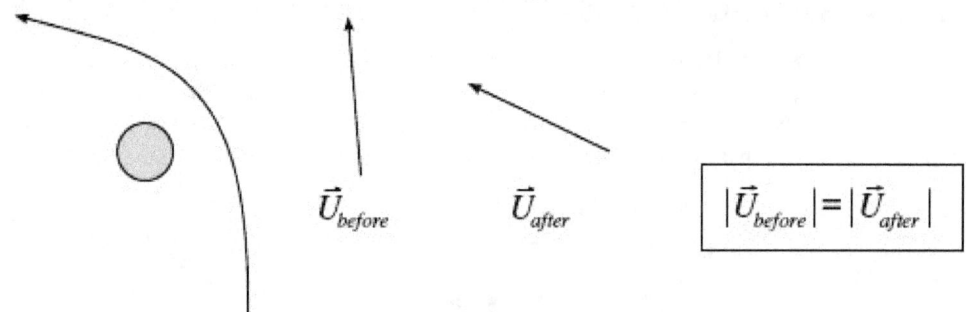

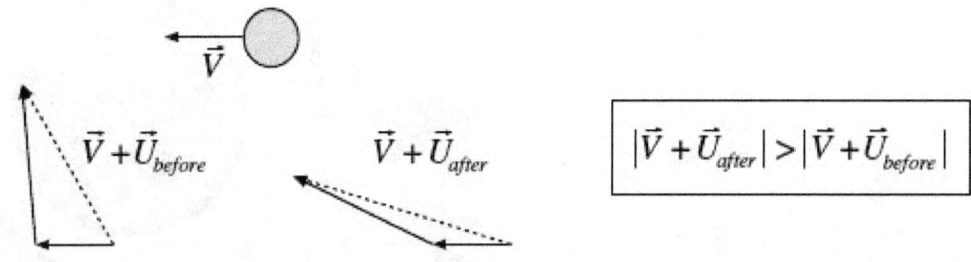

Two-dimensional schematic of gravitational slingshot. The arrows show the direction in which the spacecraft is traveling before and after the encounter. The length of the arrows shows the spacecraft's speed.

Due to the reversibility of orbits, gravitational slingshots can also be used to reduce the speed of a spacecraft. Both Mariner 10 and MESSENGER performed this maneuver to reach Mercury.

If even more speed is needed than available from gravity assist alone, the most economical way to utilize a rocket burn is to do it near the periapsis (closest approach). A given rocket burn always provides the same change in velocity (Δv), but the change in kinetic energy is proportional to the vehicle's velocity at the time of the burn (see Oberth effect). So to get the most kinetic energy from the burn, the burn must occur at the vehicle's maximum velocity, at periapsis. Powered slingshots describes this technique in more detail.

A view from MESSENGER as it uses Earth as a gravitational slingshot to decelerate to allow insertion into an orbit around Mercury.

23.2 Historical origins of the method

In his paper "Тем кто будет читать, чтобы строить" (To whoever will read [this paper] in order to build [an interplanetary rocket]),[3] published in 1938 but dated *1918–1919*,[4] Yuri Kondratyuk suggested that a spacecraft traveling between two planets could be accelerated at the beginning and end of its trajectory by using the gravity of the two planets' moons. In his 1925 paper "Проблема полета при помощи реактивных аппаратов: межпланетные полеты" [Problems of flight by jet propulsion: interplanetary flights],[5] Friedrich Zander made a similar argument.

However, neither investigator realized that gravitational assists from planets *along* a spacecraft's trajectory could propel a spacecraft and that therefore such assists could greatly reduce the amount of propellant required to travel among the planets.[6] That discovery was made by Michael Minovitch in 1961.[7]

The gravity assist maneuver was first used in 1959 when the Soviet probe Luna 3 photographed the far side of Earth's Moon. The maneuver relied on research performed under the direction of Mstislav Keldysh at the Steklov Institute of Mathematics.[8][9]

23.3 Why gravitational slingshots are used

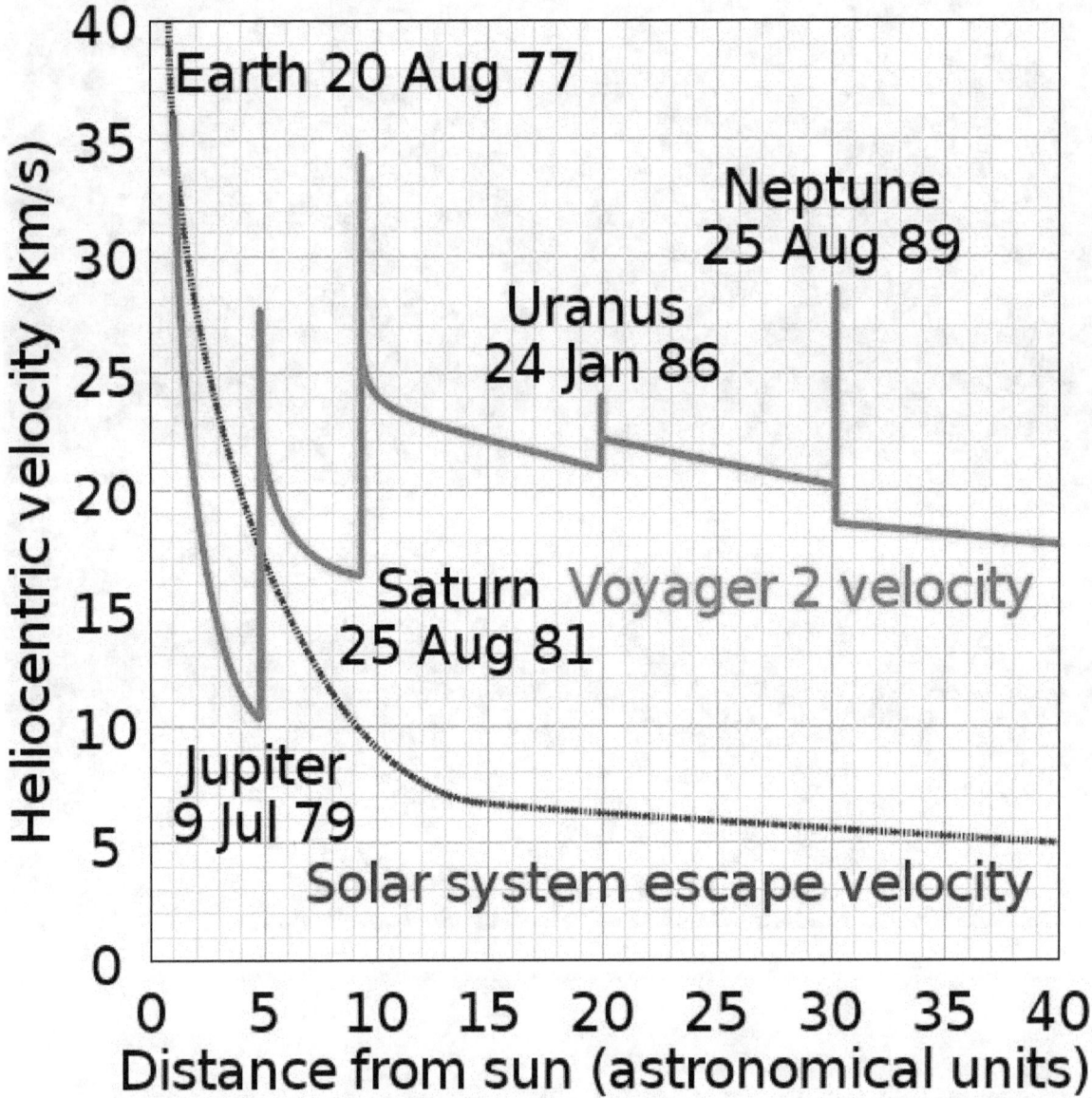

Plot of Voyager 2's heliocentric velocity against its distance from the sun, illustrating the use of gravity assist to accelerate the spacecraft by Jupiter, Saturn and Uranus. To observe Triton, Voyager 2 passed over Neptune's north pole resulting in an acceleration out of the plane of the ecliptic and reduced velocity away from the sun.[110]

A spacecraft traveling from Earth to an inner planet will increase speed because it is falling toward the Sun, and a spacecraft traveling from Earth to an outer planet will decrease speed because it is leaving the vicinity of the Sun.

Although the orbital speed of an inner planet is greater than that of the Earth, a spacecraft traveling to an inner planet, even at the minimum speed needed to reach it, is still accelerated by the Sun's gravity to a speed notably greater than the orbital speed of that destination planet. If the spacecraft's purpose is only to fly by the inner planet, then there is typically no need to slow the spacecraft. However, if the spacecraft is to be inserted into orbit about that inner planet, then there must be some way to slow down the spacecraft.

Similarly, while the orbital speed of an outer planet is less than that of the Earth, a spacecraft leaving the Earth at the minimum speed needed to travel to some outer planet is slowed by the Sun's gravity to a speed far less than the orbital speed of that outer planet. Thus, there must be some way to accelerate the spacecraft when it reaches that outer planet if it is to enter orbit about it. However, if the spacecraft is accelerated to more than the minimum required, less total propellant will be needed to enter orbit about the target planet. In addition, accelerating the spacecraft early in the flight will reduce the travel time.

Rocket engines can certainly be used to increase and decrease the speed of the spacecraft. However, rocket thrust takes propellant, propellant has mass, and even a small increment Δv (delta-v) in velocity translates to far larger requirement for propellant needed to escape Earth's gravity well. This is because not only must the primary stage engines lift that extra propellant, they must also lift more propellant still, to lift that additional propellant. Thus the liftoff mass requirement increases exponentially with an increase in the required delta-v of the spacecraft.

Since a gravity assist maneuver can change the speed of a spacecraft without expending propellant, if and when possible, combined with aerobraking, it can save significant amounts of propellant.

As an example, the Messenger mission used gravity assist maneuvering to slow the spacecraft on its way to Mercury; however, since Mercury has almost no atmosphere, aerobraking could not be used for insertion into orbit around it.

Journeys to the nearest planets, Mars and Venus, use a Hohmann transfer orbit, an elliptical path which starts as a tangent to one planet's orbit round the Sun and finishes as a tangent to the other. This method uses very nearly the smallest possible amount of fuel, but is very slow — it can take over a year to travel from Earth to Mars (fuzzy orbits use even less fuel, but are even slower).

Similarly it might take decades for a spaceship to travel to the outer planets (Jupiter, Saturn, Uranus, and Neptune) using a Hohmann transfer orbit, and it would still require far too much propellant, because the spacecraft would have to travel for 800 million km (500 million miles) or more against the force of the Sun's gravity. As gravitational assist maneuvers offer the only way to gain speed without using propellant, all missions to the outer planets have used it.

23.4 Limits to slingshot use

The main practical limit to the use of a gravity assist maneuver is that planets and other large masses are seldom in the right places to enable a voyage to a particular destination. For example the Voyager missions which started in the late 1970s were made possible by the "Grand Tour" alignment of Jupiter, Saturn, Uranus and Neptune. A similar alignment will not occur again until the middle of the 22nd century. That is an extreme case, but even for less ambitious missions there are years when the planets are scattered in unsuitable parts of their orbits.

Another limitation is the atmosphere, if any, of the available planet. The closer the spacecraft can approach, the more boost it gets, because gravity falls off with the square of distance from a planet's center. If a spacecraft gets too far into the atmosphere, the energy lost to drag can exceed that gained from the planet's gravity. On the other hand, the atmosphere can be used to accomplish aerobraking. There have also been theoretical proposals to use aerodynamic lift as the spacecraft flies through the atmosphere. This maneuver, called an aerogravity assist, could bend the trajectory through a larger angle than gravity alone, and hence increase the gain in energy.

Interplanetary slingshots using the Sun itself are not possible because the Sun is at rest relative to the Solar System as a whole. However, thrusting when near the Sun has the same effect as the powered slingshot described below. This has the potential to magnify a spacecraft's thrusting power enormously, but is limited by the spacecraft's ability to resist the heat.

An *interstellar* slingshot using the Sun is conceivable, involving for example an object coming from elsewhere in our galaxy and swinging past the Sun to boost its galactic travel. The energy and angular momentum would then come from the Sun's orbit around the Milky Way. This concept features prominently in Arthur C. Clarke's 1972 award-winning novel Rendezvous With Rama; his story concerns an interstellar spacecraft that uses the Sun to perform this sort of maneuver, and in the process alarms many nervous humans.

A rotating black hole might provide additional assistance, if its spin axis is aligned the right way. General relativity predicts that a large spinning mass-produces frame-dragging—close to the object, space itself is dragged around in the direction of the spin. Any ordinary rotating object produces this effect. Although attempts to measure frame dragging about the Sun have produced no clear evidence, experiments performed by Gravity Probe B have detected frame-dragging effects caused by Earth.[11] General relativity predicts that a spinning black hole is surrounded by a region of space, called the ergosphere, within which standing still (with respect to the black hole's spin) is impossible, because space itself is dragged

23.5. TIMELINE OF NOTABLE EXAMPLES

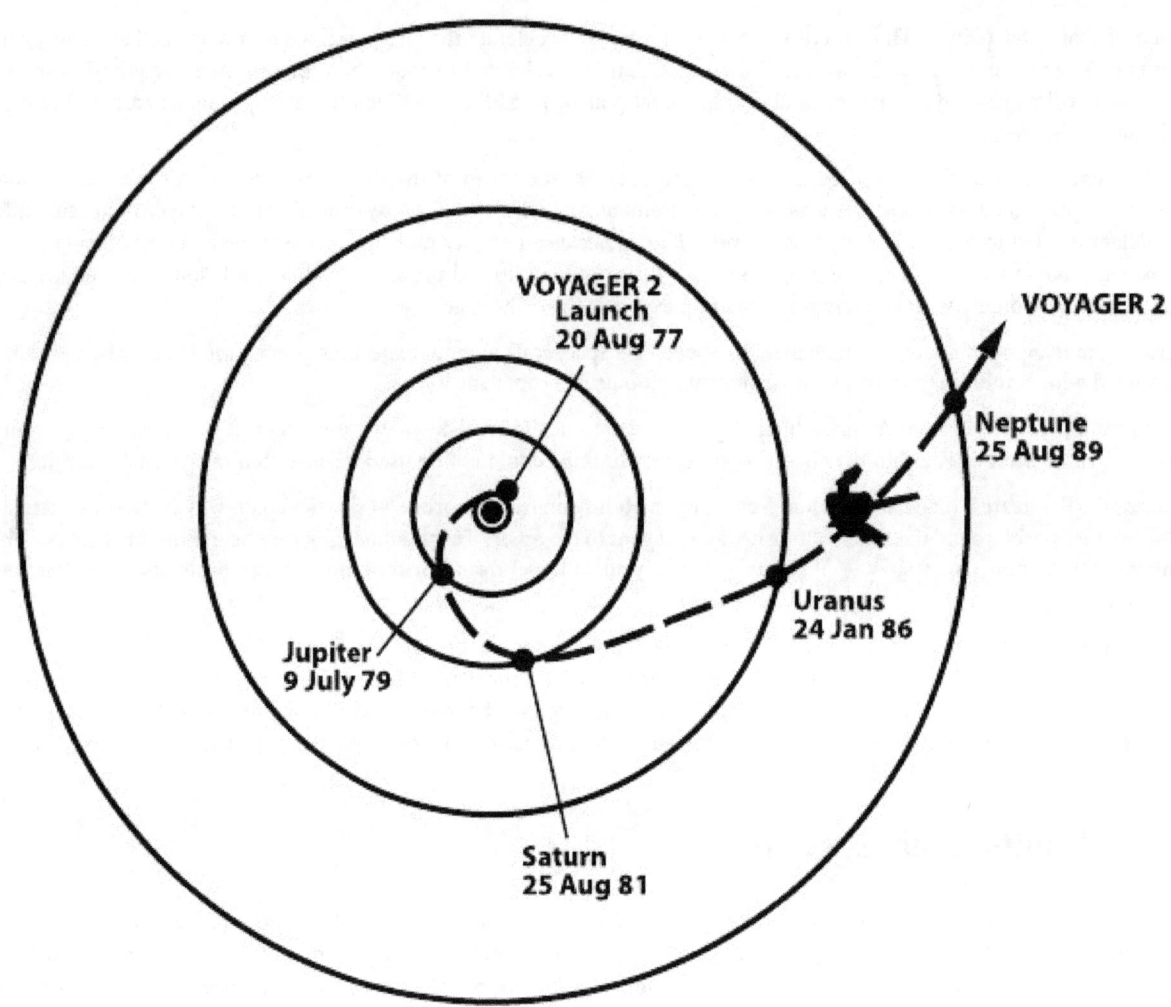

The Planetary Grand Tour trajectory of Voyager 2.

at the speed of light in the same direction as the black hole's spin. The Penrose process may offer a way to gain energy from the ergosphere, although it would require the spaceship to dump some "ballast" into the black hole, and the spaceship would have had to expend energy to carry the "ballast" to the black hole.

23.5 Timeline of notable examples

23.5.1 Mariner 10 – first use in an interplanetary trajectory

The Mariner 10 probe was the first spacecraft to use the gravitational slingshot effect to reach another planet, passing by Venus on February 5, 1974, on its way to becoming the first spacecraft to explore Mercury.

23.5.2 Voyager 1 – farthest human-made object

As of August 31, 2014, Voyager 1 is over 128.8 AU (19.2 billion km) from the Sun,[12] and is in interstellar space.[13] It gained the energy to escape the Sun's gravity completely by performing slingshot maneuvers around Jupiter and Saturn.[14][15]

23.5.3 Galileo – a change of plan

The Galileo spacecraft was launched by NASA in 1989 aboard Space Shuttle *Atlantis*. Its original mission was designed to use a direct Hohmann transfer. However, Galileo's intended booster, the cryogenically fueled (Hydrogen/Oxygen) Centaur booster rocket was prohibited as a Shuttle "cargo" for safety considerations following the loss of Space Shuttle *Challenger*. With its substituted solid rocket upperstage, the IUS, which could not provide as much delta-v, Galileo did not ascend directly to Jupiter, but flew by Venus once and Earth twice in order to reach Jupiter in December, 1995.

The Galileo engineering review speculated (but was never able to prove conclusively) that this longer flight time coupled with the stronger sunlight near Venus caused lubricant in Galileo's main antenna to fail, forcing the use of a much smaller backup antenna with a consequent lowering of data rate from the spacecraft.

Its subsequent tour of the Jovian moons also used numerous slingshot maneuvers with those moons to conserve fuel and maximize the number of encounters.

23.5.4 The Ulysses probe changed the plane of its trajectory

In 1990, NASA launched the ESA spacecraft Ulysses to study the polar regions of the Sun. All the planets orbit approximately in a plane aligned with the equator of the Sun. Thus, to enter an orbit passing over the poles of the Sun, the spacecraft would have to eliminate the 30 km/s speed it inherited from the Earth's orbit around the Sun and gain the speed needed to orbit the Sun in the pole-to-pole plane — tasks that are impossible with current spacecraft propulsion systems alone, making gravity assist maneuvers essential.

Accordingly, Ulysses was first sent toward Jupiter, aimed to arrive at a point in space just ahead and south of the planet. As it passed Jupiter, the probe fell through the planet's gravity field, exchanging momentum with the planet; this gravity assist maneuver bent the probe's trajectory northward relative to the Ecliptic Plane onto an orbit which passes over the poles of the Sun. By using this maneuver, Ulysses needed only enough propellant to send it to a point near Jupiter, which is well within current capability.

23.5.5 MESSENGER

The MESSENGER mission (launched in August 2004) made extensive use of gravity assists to slow its speed before orbiting Mercury. The MESSENGER mission included one flyby of Earth, two flybys of Venus, and three flybys of Mercury before finally arriving at Mercury in March 2011 with a velocity low enough to permit orbit insertion with available fuel. Although the flybys were primarily orbital maneuvers, each provided an opportunity for significant scientific observations.

23.5.6 The Cassini probe – multiple gravity assists

The Cassini probe passed by Venus twice, then Earth, and finally Jupiter on the way to Saturn. The 6.7-year transit was slightly longer than the six years needed for a Hohmann transfer, but cut the extra velocity (delta-v) needed to about 2 km/s, so that the large and heavy Cassini probe was able to reach Saturn, which would not have been possible in a direct transfer even with the Titan IV, the largest launch vehicle available at the time. A Hohmann transfer to Saturn would require a total of 15.7 km/s delta-v (disregarding Earth's and Saturn's own gravity wells, and disregarding aerobraking), which is not within the capabilities of current launch vehicles and spacecraft propulsion systems.

23.5.7 Solar Probe+

The NASA Solar Probe+ mission, scheduled for launch in 2018, uses multiple gravity assists at Venus to remove the Earth's angular momentum from the orbit, in order to drop down to a distance of 9.5 solar radii from the sun. This will be the closest approach to the sun of any space mission.

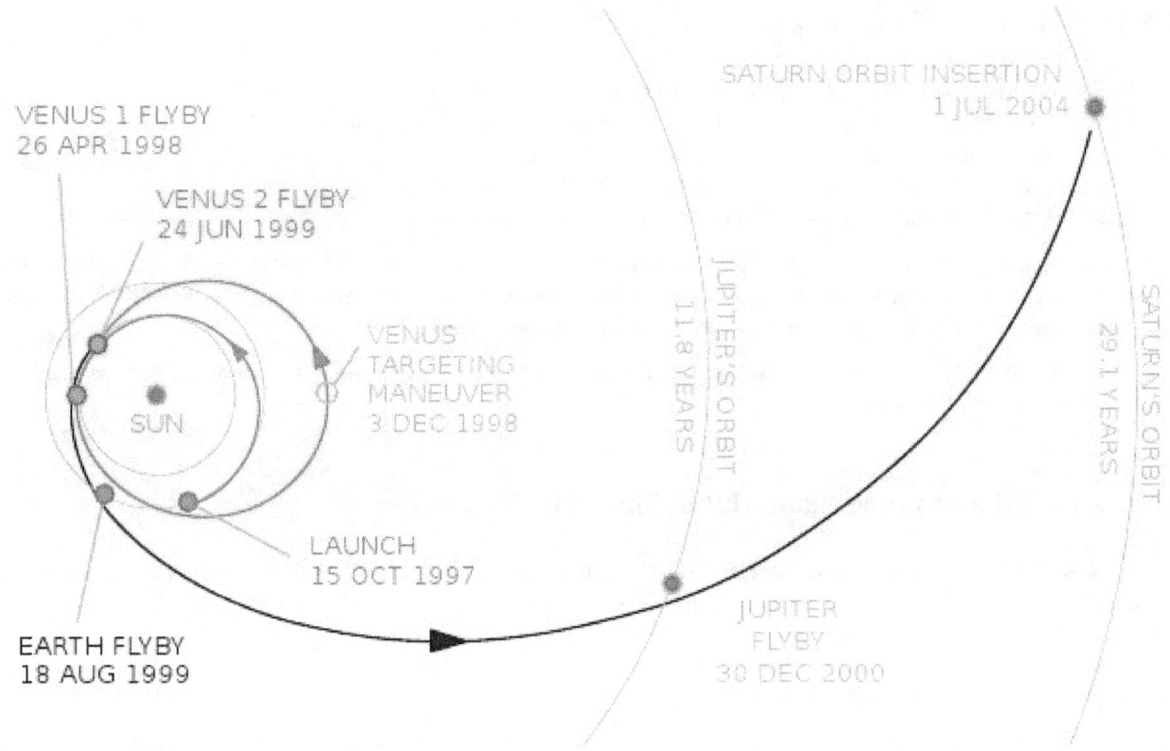

Cassini Interplanetary Trajectory

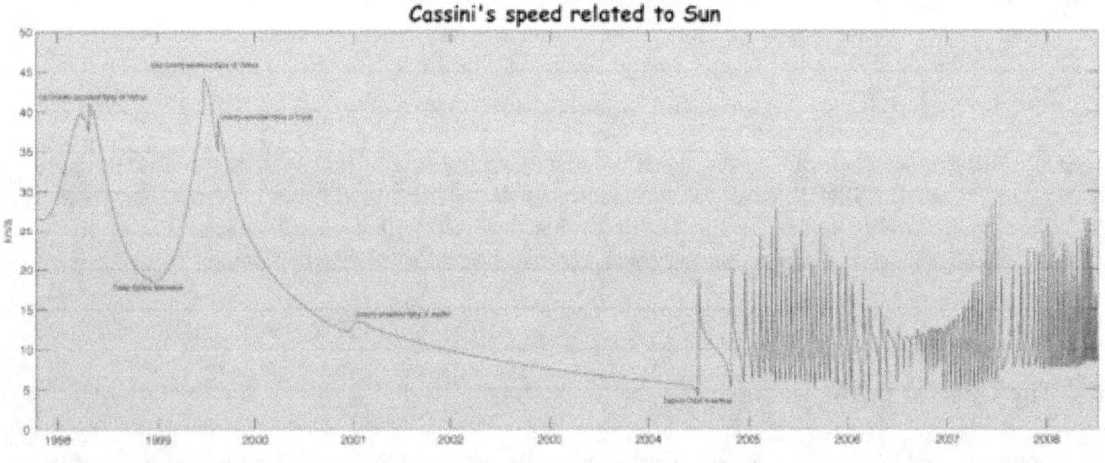

Cassini's *speed related to Sun. The various gravity assists form visible peaks on the left, while the periodic variation on the right is caused by the spacecraft's orbit around Saturn. The data was from JPL Horizons Ephemeris System. The speed above is in kilometers per second. Note also that the minimum speed achieved during Saturnian orbit is more or less equal to Saturn's own orbital velocity, which is the ~5 km/s velocity which Cassini matched to enter orbit.*

23.5.8 Rosetta – first spacecraft to match orbit with a comet

The Rosetta probe, launched in March 2004, used four gravity assist manoeuvres (including one just 250 km from the surface of Mars) to accelerate throughout the inner Solar System - enabling it to match the velocity of the 67P/Churyumov–Gerasimenko comet at their rendezvous point in August 2014.

23.6 See also

- 3753 Cruithne: an asteroid which periodically has gravitational slingshot encounters with Earth.
- Delta-v budget
- Dynamical friction
- Flyby anomaly: an anomalous delta-v increase during gravity assists
- Interplanetary Transport Network
- Gravitational keyhole
- Michael Minovitch
- n-body problem
- New Horizons: a gravity-assisted mission (flying past Jupiter) that reached Pluto on 14 July 2015.
- The Oberth effect: doing burns deep in gravity fields to gain speed
- Pioneer 10
- Pioneer 11: a gravity-assisted mission (flying past Jupiter 1974-12-03) to reach Saturn in 1979.
- Pioneer H: first Out-Of-The-Ecliptic mission (OOE) proposed, for Jupiter and solar (Sun) observations.
- Ulysses
- Voyager 1
- Voyager 2
- Messenger
- STEREO: a gravity-assisted mission which used Earth's Moon to eject two spacecraft from Earth's orbit into heliocentric orbit

23.7 References

[1] Basics of Space Flight, Sec. 1 Ch. 4, NASA Jet Propulsion Laboratory

[2] The Slingshot Effect, Durham University

[3] Kondratyuk's paper is included in the book: Mel'kumov, T. M., ed., *Pionery Raketnoy Tekhniki* [Pioneers of Rocketry: Selected Papers] (Moscow, U.S.S.R.: Institute for the History of Natural Science and Technology, Academy of Sciences of the USSR, 1964). An English translation of Kondratyuk's paper was made by NASA. See: NASA Technical Translation F-9285, pages 15-56 (Nov. 1, 1965).

[4] In 1938, when Kondratyuk submitted his manuscript "To whoever will read in order to build" for publication, he dated the manuscript *1918–1919*, although it was apparent that the manuscript had been revised at various times. See page 49 of NASA Technical Translation F-9285 (Nov. 1, 1965).

[5] Zander's 1925 paper, "Problems of flight by jet propulsion: interplanetary flights", was translated by NASA. See NASA Technical Translation F-147 (1964); specifically, Section 7: Flight Around a Planet's Satellite for Accelerating or Decelerating Spaceship, pages 290-292.

[6] See page 13 of: Dowling, Richard L.; Kosmann, William J.; Minovitch, Michael A.; and Ridenoure, Rex W., "The origin of gravity-propelled interplanetary space travel" (IAA paper no. 90-630), presented at the 41st Congress of the International Astronautical Federation, which was held 6–12 October 1990 in Dresden, G.D.R. Available on-line at: http://www.gravityassist.com/IAF1/IAF1.pdf .

[7] Minovitch, Michael, "A method for determining interplanetary free-fall reconnaissance trajectories," Jet Propulsion Laboratory Technical Memo TM-312-130, pages 38-44 (23 August 1961).

[8] T. Eneev, E. Akim. "Mstislav Keldysh. Mechanics of the space flight". *Keldysh Institute of Applied Mathematics* (in Russian).

[9] Egorov, Vsevolod Alexandrovich (1957) "Specific problems of a flight to the moon", *Physics - Uspekhi*, Vol. 63, No. 1a, pages 73–117. Egorov's work is mentioned in: Boris V. Rauschenbakh, Michael Yu. Ovchinnikov, and Susan M. P. McKenna-Lawlor, *Essential Spaceflight Dynamics and Magnetospherics* (Dordrecht, Netherlands: Kluwer Academic Publishers, 2002), pages 146–147. (The latter reference is available on-line at: Google Books.)

[10] Basics of space flight: Interplanetary Trajectories

[11] Everitt; et al. (2011). "Gravity Probe B: Final Results of a Space Experiment to Test General Relativity". *Physical Review Letters* **106** (22): 221101. arXiv:1105.3456. Bibcode:2011PhRvL.106v1101E. doi:10.1103/PhysRevLett.106.221101. PMID 21702590.

[12] http://www.heavens-above.com/SolarEscape.aspx?lat=0&lng=0&loc=Unspecified&alt=0&tz=UCT

[13] http://www.jpl.nasa.gov/interstellarvoyager/

[14] Cassini-Huygens: Operations - Gravity Assists

[15] http://www.heavens-above.com/solar-escape.asp?lat=0&lng=0&loc=Unspecified&alt=0&tz=CET

23.8 External links

- Slingshot effect
- Slingshot effect, described in terms of elastic collisions
- Animation of Cassini Huygens gravitational sling shot
- "Gravitational Slingshot" at MathPages.com.
- A Quick Gravity Assist Primer
- An artistical simulation of an unstable planetary system showing gravitational slingshots and other phenomena
- Short discussion of modifying orbits by gravity assistance part of a high school level course.

Chapter 24

Lagrangian point

"Lagrange Point" redirects here. For the video game, see Lagrange Point (video game).
This article is about three-body libration points. For two-body libration points, see Geostationary orbit#Earth orbital libration points.

In celestial mechanics, the **Lagrangian points** (/ləˈɡrɑːndʒiən/; also **Lagrange points**, **L-points**, or **libration points**) are positions in an orbital configuration of two large bodies where a small object affected only by gravity can maintain a stable position relative to the two large bodies. The Lagrange points mark positions where the combined gravitational pull of the two large masses provides precisely the centripetal force required to orbit with them. There are five such points, labeled L1 to L5, all in the orbital plane of the two large bodies. The first three are on the line connecting the two large bodies and the last two, L4 and L5, each form an equilateral triangle with the two large bodies. The two latter points are stable, which implies that objects can orbit around them in a rotating coordinate system tied to the two large bodies.

Several planets have minor planets near their L4 and L5 points (trojans) with respect to the Sun, with Jupiter in particular having more than a million of these. Artificial satellites have been placed at L1 and L2 with respect to the Sun and Earth, and Earth and the Moon for various purposes, and the Lagrangian points have been proposed for a variety of future uses in space exploration.

24.1 History

The three collinear Lagrange points (L_1, L_2, L_3) were discovered by Leonhard Euler a few years before Lagrange discovered the remaining two.[1][2]

In 1772, Joseph-Louis Lagrange published an "Essay on the three-body problem". In the first chapter he considered the general three-body problem. From that, in the second chapter, he demonstrated two special constant-pattern solutions, the collinear and the equilateral, for any three masses, with circular orbits.[3]

24.2 Lagrange points

The five Lagrangian points are labeled and defined as follows:

The L_1 point lies on the line defined by the two large masses M_1 and M_2, and between them. It is the most intuitively understood of the Lagrangian points: the one where the gravitational attraction of M_2 partially cancels M_1's gravitational attraction.

> **Explanation:** An object that orbits the Sun more closely than Earth would normally have a shorter orbital period than Earth, but that ignores the effect of Earth's own gravitational pull. If the object is directly between

24.2. LAGRANGE POINTS

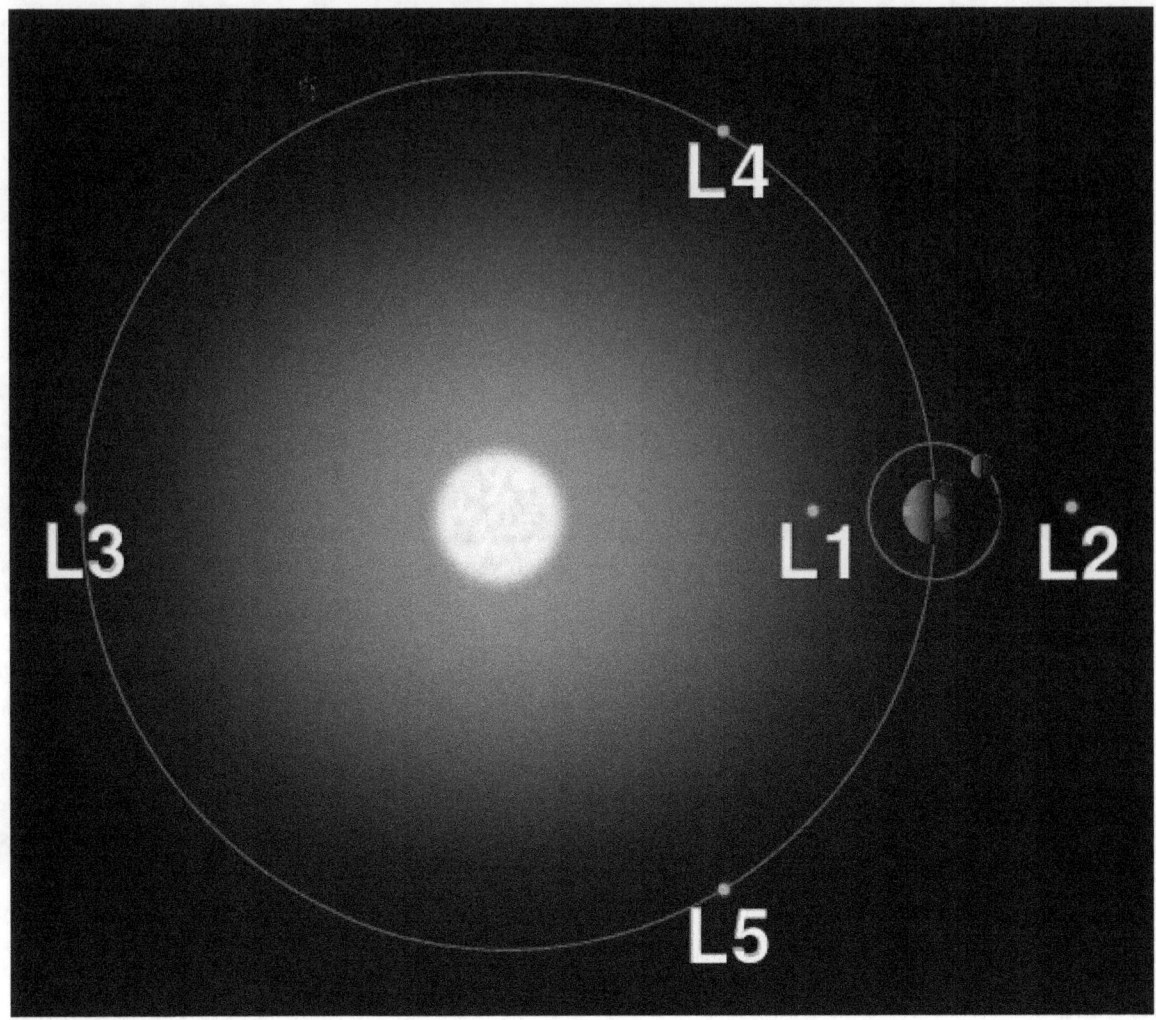

Lagrange points in the Sun–Earth system (not to scale)

Earth and the Sun, then Earth's gravity counteracts some of the Sun's pull on the object, and therefore increases the orbital period of the object. The closer to Earth the object is, the greater this effect is. At the L_1 point, the orbital period of the object becomes exactly equal to Earth's orbital period. L_1 is about 1.5 million kilometers from Earth.[4]

The **L_2** point lies on the line through the two large masses, beyond the smaller of the two. Here, the gravitational forces of the two large masses balance the centrifugal effect on a body at L_2.

Explanation: On the opposite side of Earth from the Sun, the orbital period of an object would normally be greater than that of Earth. The extra pull of Earth's gravity decreases the orbital period of the object, and at the L_2 point that orbital period becomes equal to Earth's. Like L1, L2 is about 1.5 million kilometers from Earth.

The **L_3** point lies on the line defined by the two large masses, beyond the larger of the two.

Explanation: L_3 in the Sun–Earth system exists on the opposite side of the Sun, a little outside Earth's orbit but slightly closer to the Sun than Earth is. (This apparent contradiction is because the Sun is also affected by Earth's gravity, and so orbits around the two bodies' barycenter, which is, however, well inside the body of the Sun.) At the L_3 point, the combined pull of Earth and Sun again causes the object to orbit with the same period as Earth.

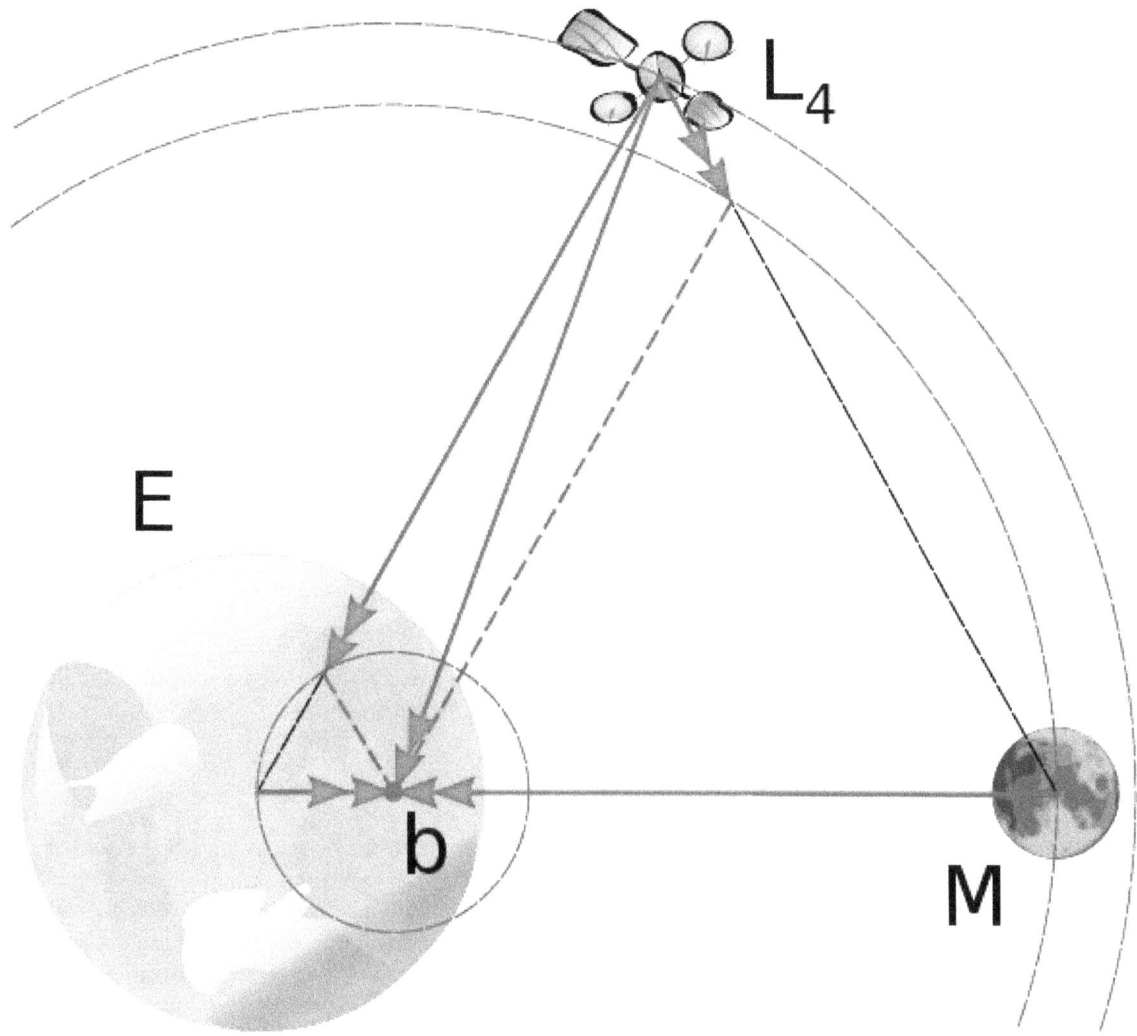

Gravitational accelerations at L_4

The L_4 and L_5 points lie at the third corners of the two equilateral triangles in the plane of orbit whose common base is the line between the centers of the two masses, such that the point lies behind (L_5) or ahead (L_4) of the smaller mass with regard to its orbit around the larger mass.

The triangular points (L_4 and L_5) are stable equilibria, provided that the ratio of M_1/M_2 is greater than 24.96.[note 1][5] This is the case for the Sun–Earth system, the Sun–Jupiter system, and, by a smaller margin, the Earth–Moon system. When a body at these points is perturbed, it moves away from the point, but the factor opposite of that which is increased or decreased by the perturbation (either gravity or angular momentum-induced speed) will also increase or decrease, bending the object's path into a stable, kidney-bean-shaped orbit around the point (as seen in the corotating frame of reference).

In contrast to L_4 and L_5, where stable equilibrium exists, the points L_1, L_2, and L_3 are positions of unstable equilibrium. Any object orbiting at one of L_1–L_3 will tend to fall out of orbit; it is therefore rare to find natural objects there, and spacecraft inhabiting these areas must employ station keeping in order to maintain their position.

24.3 Natural objects at Lagrangian points

It is common to find objects at or orbiting the L_4 and L_5 points of natural orbital systems. These are commonly called "trojans"; in the 20th century, asteroids discovered orbiting at the Sun–Jupiter L_4 and L_5 points were named after characters from Homer's *Iliad*. Asteroids at the L_4 point, which leads Jupiter, are referred to as the "Greek camp", whereas those at the L_5 point are referred to as the "Trojan camp".

Other examples of natural objects orbiting at Lagrange points:

- The Sun–Earth L_4 and L_5 points contain interplanetary dust and at least one asteroid, 2010 TK7, detected in October 2010 by Wide-field Infrared Survey Explorer (WISE) and announced during July 2011.[6][7]

- The Earth–Moon L_4 and L_5 points may contain interplanetary dust in what are called Kordylewski clouds; however, the Hiten spacecraft's Munich Dust Counter (MDC) detected no increase in dust during its passes through these points. Stability at these specific points is greatly complicated by solar gravitational influence.[8]

- Recent observations suggest that the Sun–Neptune L_4 and L_5 points, known as the Neptune trojans, may be very thickly populated,[9] containing large bodies an order of magnitude more numerous than the Jupiter trojans.

- Several asteroids also orbit near the Sun-Jupiter L_3 point, called the Hilda family.

- The Saturnian moon Tethys has two smaller moons in its L_4 and L_5 points, Telesto and Calypso. The Saturnian moon Dione also has two Lagrangian co-orbitals, Helene at its L_4 point and Polydeuces at L_5. The moons wander azimuthally about the Lagrangian points, with Polydeuces describing the largest deviations, moving up to 32 degrees away from the Saturn–Dione L_5 point. Tethys and Dione are hundreds of times more massive than their "escorts" (see the moons' articles for exact diameter figures; masses are not known in several cases), and Saturn is far more massive still, which makes the overall system stable.

- One version of the giant impact hypothesis suggests that an object named Theia formed at the Sun–Earth L_4 or L_5 points and crashed into Earth after its orbit destabilized, forming the Moon.

- Mars has four known co-orbital asteroids (5261 Eureka, 1999 UJ7, 1998 VF31 and 2007 NS2), all at its Lagrangian points.

- Earth's companion object 3753 Cruithne is in a relationship with Earth that is somewhat trojan-like, but that is different from a true trojan. Cruithne occupies one of two regular solar orbits, one of them slightly smaller and faster than Earth's, and the other slightly larger and slower. It periodically alternates between these two orbits due to close encounters with Earth. When it is in the smaller, faster orbit and approaches Earth, it gains orbital energy from Earth and moves up into the larger, slower orbit. It then falls farther and farther behind Earth, and eventually Earth approaches it from the other direction. Then Cruithne gives up orbital energy to Earth, and drops back into the smaller orbit, thus beginning the cycle anew. The cycle has no noticeable impact on the length of the year, because Earth's mass is over 20 billion (2×10^{10}) times more than that of 3753 Cruithne.

- Epimetheus and Janus, satellites of Saturn, have a similar relationship, though they are of similar masses and so actually exchange orbits with each other periodically. (Janus is roughly 4 times more massive but still light enough for its orbit to be altered.) Another similar configuration is known as orbital resonance, in which orbiting bodies tend to have periods of a simple integer ratio, due to their interaction.

- In a binary star system, the Roche lobe has its apex located at L_1; if a star overflows its Roche lobe, then it will lose matter to its companion star.

24.4 Mathematical details

Lagrangian points are the constant-pattern solutions of the restricted three-body problem. For example, given two massive bodies in orbits around their common barycenter, there are five positions in space where a third body, of comparatively negligible mass, could be placed so as to maintain its position relative to the two massive bodies. As seen in a rotating

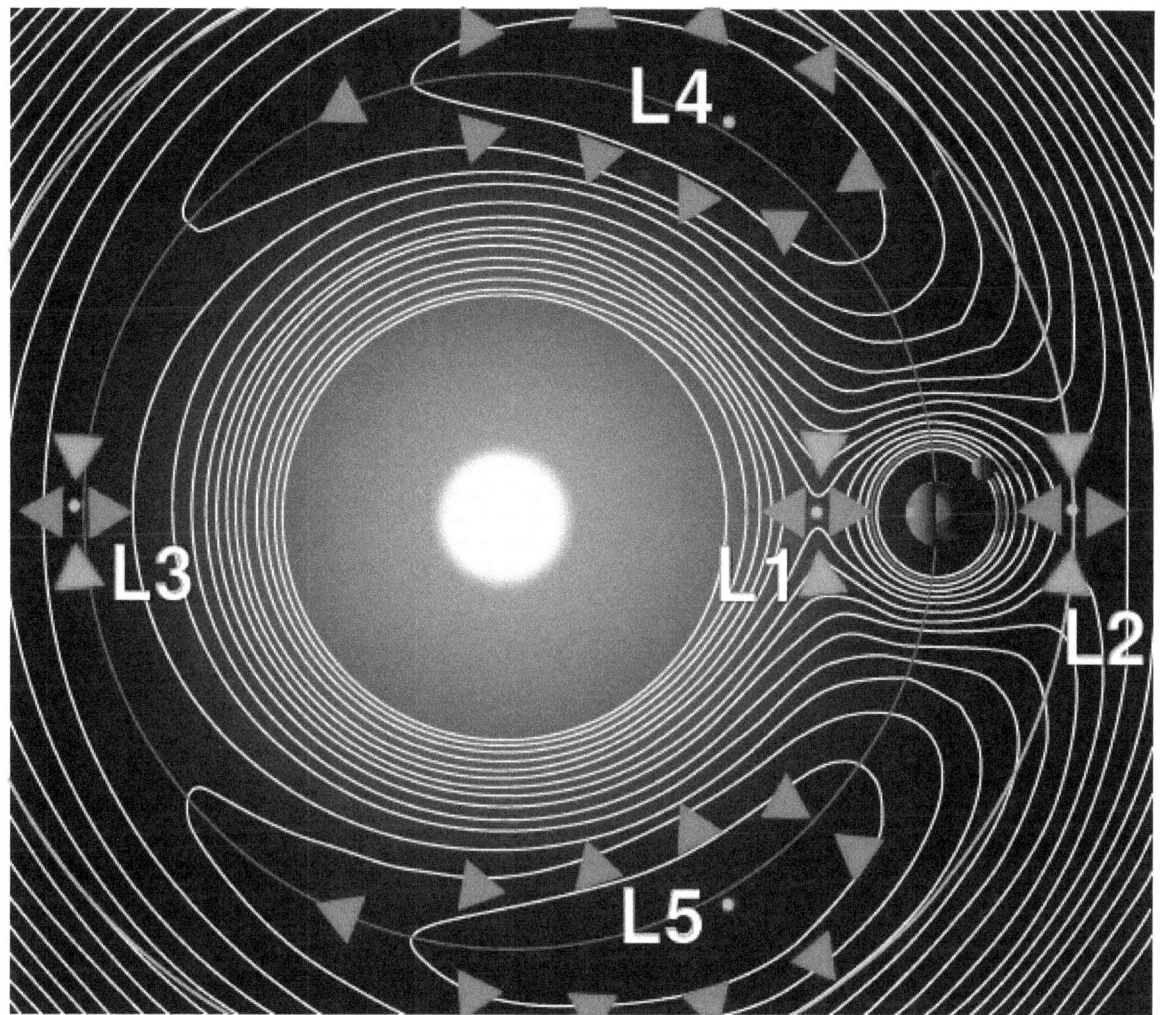

A contour plot of the effective potential due to gravity and the centrifugal force of a two-body system in a rotating frame of reference. The arrows indicate the gradients of the potential around the five Lagrange points—downhill toward them (red) or away from them (blue). Counterintuitively, the L_4 and L_5 points are the high points of the potential. At the points themselves these forces are balanced.

reference frame that matches the angular velocity of the two co-orbiting bodies, the gravitational fields of two massive bodies combined with the minor body's centrifugal force are in balance at the Lagrangian points, allowing the smaller third body to be relatively stationary with respect to the first two.[11]

24.4.1 L_1

The location of L_1 is the solution to the following equation, balancing gravitation and the centrifugal force:

$$\frac{M_1}{(R-r)^2} = \frac{M_2}{r^2} + \frac{M_1}{R^2} - \frac{r(M_1+M_2)}{R^3}$$

where r is the distance of the L_1 point from the smaller object, R is the distance between the two main objects, and M_1 and M_2 are the masses of the large and small object, respectively. (The quantity in parentheses on the right is the distance of L_1 from the center of mass.) Solving this for r involves solving a quintic function, but if the mass of the smaller object (M_2) is much smaller than the mass of the larger object (M_1) then L_1 and L_2 are at approximately equal distances r from the smaller object, equal to the radius of the Hill sphere, given by:

24.4. MATHEMATICAL DETAILS

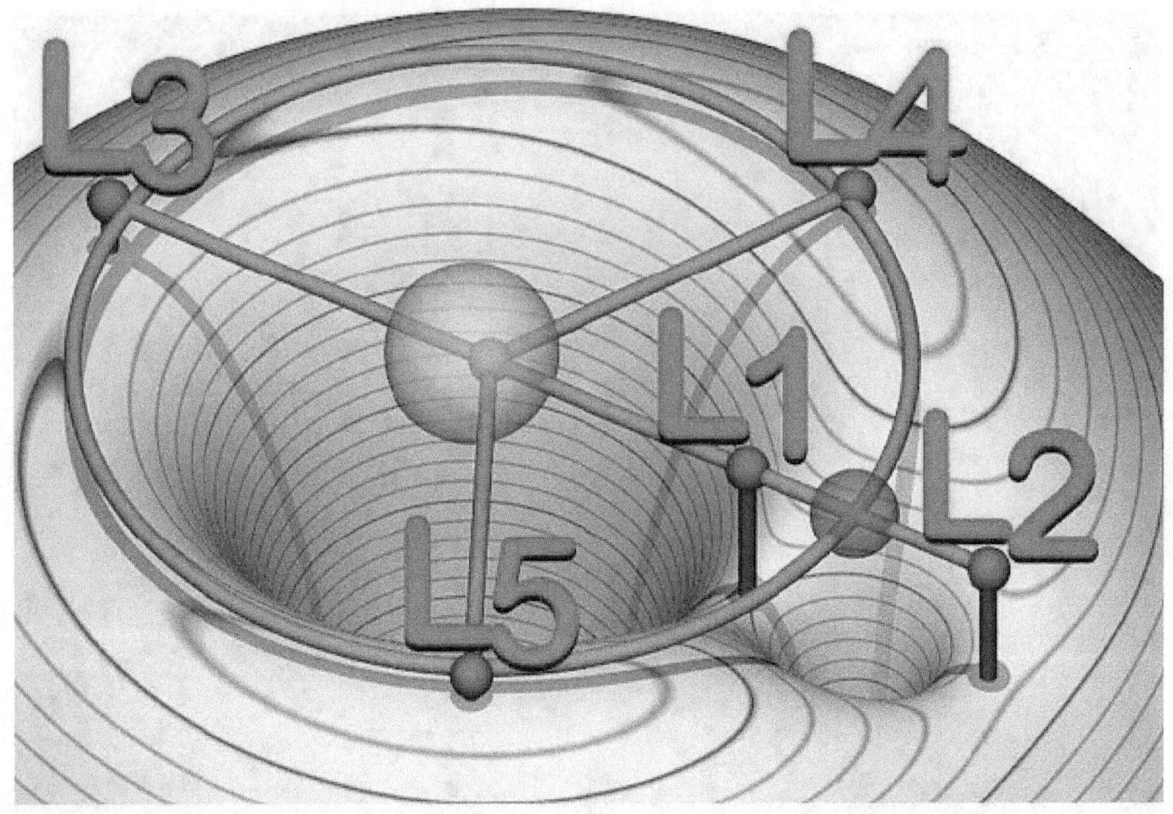

Visualisation of the relationship between the Lagrangian points (red) of a planet (blue) orbiting a star (yellow) anticlockwise, and the effective potential in the plane containing the orbit (grey rubber-sheet model with purple contours of equal potential).[10]
Click for animation.

$$r \approx R \sqrt[3]{\frac{M_2}{3M_1}}$$

This distance can be described as being such that the orbital period, corresponding to a circular orbit with this distance as radius around M_2 in the absence of M_1, is that of M_2 around M_1, divided by $\sqrt{3} \approx 1.73$:

$$T_{s,M_2}(r) = \frac{T_{M_2,M_1}(R)}{\sqrt{3}}.$$

24.4.2 L_2

The location of L_2 is the solution to the following equation, balancing gravitation and inertia:

$$\frac{M_1}{(R+r)^2} + \frac{M_2}{r^2} = \frac{M_1}{R^2} + \frac{r(M_1+M_2)}{R^3}$$

with parameters defined as for the L_1 case. Again, if the mass of the smaller object (M_2) is much smaller than the mass of the larger object (M_1) then L_2 is at approximately the radius of the Hill sphere, given by:

$$r \approx R \sqrt[3]{\frac{M_2}{3M_1}}$$

24.4.3 L₃

The location of L_3 is the solution to the following equation, balancing gravitation and the centrifugal force:

$$\frac{M_1}{(R-r)^2} + \frac{M_2}{(2R-r)^2} = \left(\frac{M_2}{M_1+M_2}R + R - r\right)\frac{M_1+M_2}{R^3}$$

with parameters defined as for the L_1 and L_2 cases except that r now indicates how much closer L_3 is to the more massive object than the smaller object. If the mass of the smaller object (M_2) is much smaller than the mass of the larger object (M_1) then:

$$r \approx R\frac{7M_2}{12M_1}$$

24.4.4 L₄ and L₅

Further information: Trojan (astronomy)

The reason these points are in balance is that, at L_4 and L_5, the distances to the two masses are equal. Accordingly, the gravitational forces from the two massive bodies are in the same ratio as the masses of the two bodies, and so the resultant force acts through the barycenter of the system; additionally, the geometry of the triangle ensures that the resultant acceleration is to the distance from the barycenter in the same ratio as for the two massive bodies. The barycenter being both the center of mass and center of rotation of the three-body system, this resultant force is exactly that required to keep the smaller body at the Lagrange point in orbital equilibrium with the other two larger bodies of system. (Indeed, the third body need not have negligible mass.) The general triangular configuration was discovered by Lagrange in work on the three-body problem.

24.5 Stability

Although the L_1, L_2, and L_3 points are nominally unstable, it turns out that it is possible to find (unstable) periodic orbits around these points, at least in the restricted three-body problem. These periodic orbits, referred to as "halo" orbits, do not exist in a full n-body dynamical system such as the Solar System. However, quasi-periodic (i.e. bounded but not precisely repeating) orbits following Lissajous-curve trajectories do exist in the n-body system. These quasi-periodic Lissajous orbits are what most of Lagrangian-point missions to date have used. Although they are not perfectly stable, a relatively modest effort at station keeping can allow a spacecraft to stay in a desired Lissajous orbit for an extended period of time. It also turns out that, at least in the case of Sun–Earth-L_1 missions, it is actually preferable to place the spacecraft in a large-amplitude (100,000–200,000 km or 62,000–124,000 mi) Lissajous orbit, instead of having it sit at the Lagrangian point, because this keeps the spacecraft off the direct line between Sun and Earth, thereby reducing the impact of solar interference on Earth–spacecraft communications. Similarly, a large-amplitude Lissajous orbit around L2 can keep a probe out of Earth's shadow and therefore ensures a better illumination of its solar panels.

24.6 Spaceflight applications

Earth–Moon L_1 allows comparatively easy access to Lunar and Earth orbits with minimal change in velocity and has this as an advantage to position a half-way manned space station intended to help transport cargo and personnel to the Moon and back.

Earth–Moon L_2 would be a good location for a communications satellite covering the Moon's far side and would be "an ideal location" for a propellant depot as part of the proposed depot-based space transportation architecture.[12]

Sun–Earth L_1 is suited for making observations of the Sun–Earth system. Objects here are never shadowed by Earth or the Moon. The first mission of this type was the International Sun Earth Explorer 3 (ISEE-3) mission used as an

24.6. SPACEFLIGHT APPLICATIONS

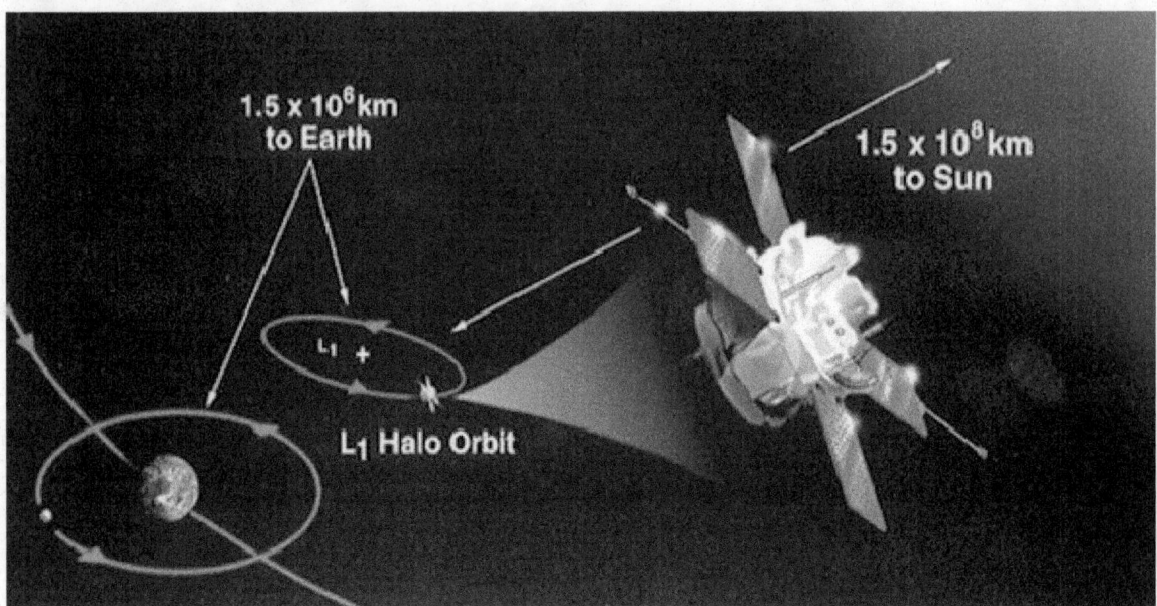

The satellite ACE in an orbit around L_1

interplanetary early warning storm monitor for solar disturbances.

Sun–Earth L_2 is a good spot for space-based observatories. Because an object around L_2 will maintain the same relative position with respect to the Sun and Earth, shielding and calibration are much simpler. It is, however, slightly beyond the reach of Earth's umbra,[13] so solar radiation is not completely blocked. From this point, the Sun, Earth and Moon are relatively closely positioned together in the sky, and hence leave a large field of view without interference – this is especially helpful for infrared astronomy.

Sun–Earth L_3 was a popular place to put a "Counter-Earth" in pulp science fiction and comic books. Once space-based observation became possible via satellites[14] and probes, it was shown to hold no such object. The Sun–Earth L_3 is unstable and could not contain an object, large or small, for very long. This is because the gravitational forces of the other planets are stronger than that of Earth (Venus, for example, comes within 0.3 AU of this L_3 every 20 months).

A spacecraft orbiting near Sun–Earth L_3 would be able to closely monitor the evolution of active sunspot regions before they rotate into a geoeffective position, so that a 7-day early warning could be issued by the NOAA Space Weather Prediction Center. Moreover, a satellite near Sun–Earth L_3 would provide very important observations not only for Earth forecasts, but also for deep space support (Mars predictions and for manned mission to near-Earth asteroids). In 2010, spacecraft transfer trajectories to Sun–Earth L_3 were studied and several designs were considered.[15]

Scientists at the B612 Foundation are planning to use Venus's L_3 point to position their planned Sentinel telescope, which aims to look back towards Earth's orbit and compile a catalogue of near-Earth asteroids.[16]

Missions to Lagrangian points generally orbit the points rather than occupy them directly.

Another interesting and useful property of the collinear Lagrangian points and their associated Lissajous orbits is that they serve as "gateways" to control the chaotic trajectories of the Interplanetary Transport Network.

24.6.1 Spacecraft at Sun–Earth L_1

International Sun Earth Explorer 3 (ISEE-3) began its mission at the Sun–Earth L_1 before leaving to intercept a comet in 1982. The Sun–Earth L_1 is also the point to which the Reboot ISEE-3 mission was attempting to return the craft as the first phase of a recovery mission (as of September 25, 2014 all efforts have failed and contact was lost).[17]

Solar and Heliospheric Observatory (SOHO) is stationed in a halo orbit at L_1, and the Advanced Composition Explorer (ACE) in a Lissajous orbit. WIND is also at L_1.

Deep Space Climate Observatory (DSCOVR), launched on 11 February 2015, began orbiting L_1 on 8 June 2015 to study the solar wind and its effects on Earth.[18] DISCOVR is unofficially known as GORESAT, because it carries a camera always oriented to Earth and capturing full-frame photos the planet similar to the Blue Marble. This concept was proposed by then-Vice President Al Gore in 1998[19] and was a centerpiece in his film An Inconvenient Truth.[20]

24.6.2 Spacecraft at Sun–Earth L_2

Spacecraft at the Sun–Earth L_2 point are in a Lissajous orbit until decommissioned, when they are sent into a heliocentric graveyard orbit.

- 1 October 2001 – October 2010—Wilkinson Microwave Anisotropy Probe[21]
- July 2009 – 29 April 2013—Herschel Space Observatory[22]
- 3 July 2009 – 21 October 2013—Space observatory Planck
- 25 August 2011 – April 2012—Chang'e 2,[23][24] from where it travelled to 4179 Toutatis and then into deep space
- January 2014 – 2018—Gaia probe
- 2018—James Webb Space Telescope will use a halo orbit
- 2020—Euclid observatory
- 2028—Advanced Telescope for High Energy Astrophysics will use a halo orbit

24.6.3 List of missions to Lagrangian points

Color key:

Past and present missions

Future and proposed missions

24.7 See also

- Co-orbital configuration
- Euler's three-body problem
- Gegenschein
- Hill sphere
- Horseshoe orbit
- L5 Society
- Lagrange point colonization
- Lagrangian mechanics
- Lissajous orbit
- List of objects at Lagrangian points
- Lunar space elevator
- Trojan wave packets

24.8 Notes

[1] Actually $\frac{25+\sqrt{621}}{2} \approx 24.9599357944$

24.9 References

[1] Koon, W. S.; M. W. Lo; J. E. Marsden; S. D. Ross (2006). *Dynamical Systems, the Three-Body Problem, and Space Mission Design*. p. 9. (16MB)

[2] Leonhard Euler, De motu rectilineo trium corporum se mutuo attrahentium (1765)

[3] Lagrange, Joseph-Louis (1867–92). "Tome 6, Chapitre II: Essai sur le problème des trois corps". *Oeuvres de Lagrange* (in French). Gauthier-Villars. pp. 229–334.

[4] Cornish, Neil J. "The Lagrangian Points" (PDF). Department of Physics, Bozeman Campus, Montana State University, USA. Retrieved 29 July 2011.

[5] *The Lagrange Points* PDF, Neil J. Cornish with input from Jeremy Goodman

[6] Space.com: *First Asteroid Companion of Earth Discovered at Last*

[7] NASA—NASA's Wise Mission Finds First Trojan Asteroid Sharing Earth's Orbit

[8] "A Search for Natural or Artificial Objects Located at the Earth–Moon Libration Points" by Robert Freitas and Francisco Valdes, *Icarus* 42, 442-447 (1980)

[9] "List Of Neptune Trojans". Minor Planet Center. Archived from the original on 2011-08-23. Retrieved 2010-10-27.

[10] ZF Seidov, "The Roche Problem: Some Analytics", The Astrophysical Journal, 603:283-284, 2004 March 1

[11] "Lagrange Points" by Enrique Zeleny, Wolfram Demonstrations Project.

[12] Zegler, Frank; Bernard Kutter (2010-09-02). "Evolving to a Depot-Based Space Transportation Architecture" (PDF). *AIAA SPACE 2010 Conference & Exposition*. AIAA. p. 4. Retrieved 2011-01-25. L2 is in deep space far away from any planetary surface and hence the thermal, micrometeoroid, and atomic oxygen environments are vastly superior to those in LEO. Thermodynamic stasis and extended hardware life are far easier to obtain without these punishing conditions seen in LEO. L2 is not just a great gateway—it is a great place to store propellants. ... L2 is an ideal location to store propellants and cargos: it is close, high energy, and cold. More importantly, it allows the continuous onward movement of propellants from LEO depots, thus suppressing their size and effectively minimizing the near-Earth boiloff penalties.

[13] Angular size of the Sun at 1 AU + 930000 miles: 31.6', angular size of Earth at 930000 miles: 29.3'

[14] STEREO mission description by NASA, http://www.nasa.gov/mission_pages/stereo/main/index.html#.UuG0NxDb-kk

[15] Tantardini, Marco; Fantino, Elena; Yuan Ren; Pierpaolo Pergola; Gerard Gómez; Josep J. Masdemont (2010). "Spacecraft trajectories to the L_3 point of the Sun–Earth three-body problem". *Celestial Mechanics and Dynamical Astronomy (Springer)*.

[16] The Sentinel Mission, B612 Foundation. Retrieved Feb 2014.

[17] "ISEE-3 is in Safe Mode". Space College. 25 September 2014. "The ground stations listening to ISEE-3 have not been able to obtain a signal since Tuesday the 16th"

[18] http://www.nesdis.noaa.gov/news_archives/DSCOVR_L1_orbit.html

[19] http://www.usatoday.com/story/tech/2015/02/07/goresat-gore-satellite-deep-space-climate/23013283/

[20] Mellow, Craig (August 2014). "Al Gore's Satellite". Air & Space/Smithsonian. Retrieved December 12, 2014.

[21] "Mission Complete! WMAP Fires Its Thrusters For The Last Time".

[22] Toobin, Adam (2013-06-19). "Herschel Space Telescope Shut Down For Good, ESA Announces". *Huffington Post*.

[23] Lakdawalla, Emily (14 June 2012). "Chang'E 2 has departed Earth's neighborhood for.....asteroid Toutatis!?". Retrieved 15 June 2012.

[24] Lakdawalla, Emily (15 June 2012). "Update on yesterday's post about Chang'E 2 going to Toutatis". Planetary Society. Retrieved 26 June 2012.

[25] "Solar System Exploration: ISEE-3/ICE". NASA. Retrieved 2010-09-28.

[26] Lakdawalla, Emily (October 3, 2008). "It's Alive!". The Planetary Science Weblog.

[27] Chang, Kenneth (August 8, 2014). "Rudderless Craft to Get Glimpse of Home Before Sinking Into Space's Depths". *The New York Times*.

[28] "ACE Mission". Caltech ACE Science Center. Retrieved 2013-03-18.

[29] Sullivan, Brian K (12 June 2015). "Space 'Buoy' May Help Protect Power Grid From Sun's Fury". *Bloomberg*.

[30] "SOHO's Orbit: An Uninterrupted View of the Sun". NASA. Retrieved 2010-09-28.

[31] "WIND Spacecraft". NASA. Retrieved 2010-09-28.

[32] "WMAP Facts". NASA. Retrieved 2013-03-18.

[33] http://map.gsfc.nasa.gov/news/events.html WMAP Ceases Communications

[34] "Herschel Factsheet". European Space Agency. 17 April 2009. Retrieved 2009-05-12.

[35] "Herschel space telescope finishes mission". *BBC news*. 29 April 2013.

[36] "Last command sent to ESA's Planck space telescope". European Space Agency. October 23, 2013. Retrieved October 23, 2013.

[37] Fox, Karen C. "First ARTEMIS Spacecraft Successfully Enters Lunar Orbit". *The Sun-Earth Connection: Heliophysics*. NASA.

[38] Hendrix, Susan. "Second ARTEMIS Spacecraft Successfully Enters Lunar Orbit". *The Sun-Earth Connection: Heliophysics*. NASA.

[39] "Worldwide launch schedule". Spaceflight Now. 27 November 2013.

[40] "ESA Gaia home". ESA. Retrieved 23 October 2013.

[41] P. E. Schmid (June 1968). "Lunar Far-Side Communication Satellites" (PDF). NASA. Retrieved 2008-07-16.

[42] O'Neill, Gerard K. (September 1974). "The Colonization of Space". *Physics Today* (American Institute of Physics) 27 (9): 32–40. Bibcode:1974PhT....27i..32O. doi:10.1063/1.3128863.

[43] "LISA Pathfinder factsheet". ESA. 11 June 2012. Retrieved 26 June 2012.

[44] http://www.business-standard.com/article/beyond-business/man-in-space-and-other-plans-114111401887_1.html

[45] "JWST factsheet". ESA. 2013-09-04. Retrieved 2013-09-07.

[46] "NASA Officially Joins ESA's 'Dark Universe' Mission". JPL/NASA. 24 January 2013. Retrieved 12 April 2013.

[47] Paul Hertz (2013-06-04), *NASA Astrophysics presentation to American Astronomical Society* (PDF), retrieved 2013-09-10

[48] Hiroshi Shibai (2014-12-31), *SPICA* (PDF), retrieved 2015-02-24

[49] NASA teams evaluating ISS-built Exploration Platform roadmap

[50] Bergin, Chris (December 2011). "Exploration Gateway Platform hosting Reusable Lunar Lander proposed". NASA Spaceflight.com. Retrieved 2011-12-05.

[51] "ESA Science & Technology: Athena to study the hot and energetic Universe". ESA. 27 June 2014. Retrieved 23 August 2014.

24.10 External links

- Joseph-Louis, Comte Lagrange, from Oeuvres Tome 6, "Essai sur le Problème des Trois Corps"—Essai (PDF); source Tome 6 (Viewer)

- "Essay on the Three-Body Problem" by J-L Lagrange, translated from the above, in http://www.merlyn.demon.co.uk/essai-3c.htm.

- Considerationes de motu corporum coelestium—Leonhard Euler—transcription and translation at http://www.merlyn.demon.co.uk/euler304.htm.

- What are Lagrange points?—European Space Agency page, with good animations

- Explanation of Lagrange points—Prof. Neil J. Cornish

- A NASA explanation—also attributed to Neil J. Cornish

- Explanation of Lagrange points—Prof. John Baez

- Geometry and calculations of Lagrange points—Dr J R Stockton

- Locations of Lagrange points, with approximations—Dr. David Peter Stern

- An online calculator to compute the precise positions of the 5 Lagrange points for any 2-body system—Tony Dunn

- Astronomy cast—Ep. 76: Lagrange Points Fraser Cain and Dr. Pamela Gay

- The Five Points of Lagrange by Neil deGrasse Tyson

- Earth, a lone Trojan discovered

Chapter 25

n-body problem

This article is about the problem in classical mechanics. For the problem in quantum mechanics, see Many-body problem. For engineering problems and simulations involving many components, see Multibody system and Multibody simulation.

In physics, the *n*-body problem is the problem of predicting the individual motions of a group of celestial objects interacting with each other gravitationally.[1] Solving this problem has been motivated by the desire to understand the motions of the Sun, Moon, planets and the visible stars. In the 20th century, understanding the dynamics of globular cluster star systems became an important *n*-body problem.[2] The *n*-body problem in general relativity is considerably more difficult to solve.

The classical physical problem can be informally stated as: *given the quasi-steady orbital properties (instantaneous position, velocity and time)*[3] *of a group of celestial bodies, predict their interactive forces; and consequently, predict their true orbital motions for all future times.*[4]

To this purpose the two-body problem has been completely solved and is discussed below; as is the famous *restricted 3-Body Problem*.[5]

25.1 History

Knowing three orbital positions of a planet's orbit – positions obtained by Sir Isaac Newton (1643-1727) from astronomer John Flamsteed[6] – Newton was able to produce an equation by straightforward analytical geometry, to predict a planet's motion; i.e., to give its orbital properties: position, orbital diameter, period and orbital velocity.[7] Having done so he and others soon discovered over the course of a few years, those equations of motion did not predict some orbits very well or even correctly.[8] Newton realized it was because gravitational interactive forces amongst all the planets was affecting all their orbits.

The above discovery goes right to the heart of the matter as to what exactly the *n*-body problem is physically: as Newton realized, it is not sufficient to just specify the initial position and velocity, or three orbital positions either, to determine a planet's true orbit: *the gravitational interactive forces have to be known too*. Thus came the awareness and rise of the n-body "problem" in the early 17th century. These gravitational attractive forces do conform to Newton's *Laws of Motion* and to his *Law of Universal Gravitation*, but the many multiple (*n*-body) interactions have historically made any exact solution intractable. Ironically, this conformity led to the wrong approach.

After Newton's time the *n*-body problem historically was not stated correctly *because it did not include a reference to those gravitational interactive forces*. Newton does not say it directly but implies in his Principia the *n*-body problem is unsolvable because of those gravitational interactive forces.[9] Newton said[10] in his Principia, paragraph 21:

> And hence it is that the attractive force is found in both bodies. The Sun attracts Jupiter and the other planets, Jupiter attracts its satellites and similarly the satellites act on one another. And although the actions

of each of a pair of planets on the other can be distinguished from each other and can be considered as two actions by which each attracts the other, yet inasmuch as they are between the same, two bodies they are not two but a simple operation between two termini. Two bodies can be drawn to each other by the contraction of rope between them. The cause of the action is twofold, namely the disposition of each of the two bodies; the action is likewise twofold, insofar as it is upon two bodies; but insofar as it is between two bodies it is single and one ...

Newton concluded via his 3rd Law that "according to this Law all bodies must attract each other." This last statement, which implies the existence of gravitational interactive forces, is key.

As shown below, the problem also conforms to Jean Le Rond D'Alembert's non-Newtonian 1st and 2nd *Principles* and to the nonlinear *n*-body problem algorithm, the latter allowing for a closed form solution for calculating those interactive forces.

The problem of finding the general solution of the *n*-body problem was considered very important and challenging. Indeed in the late 19th century King Oscar II of Sweden, advised by Gösta Mittag-Leffler, established a prize for anyone who could find the solution to the problem. The announcement was quite specific:

> Given a system of arbitrarily many mass points that attract each according to Newton's law, under the assumption that no two points ever collide, try to find a representation of the coordinates of each point as a series in a variable that is some known function of time and for all of whose values the series **converges uniformly**.

In case the problem could not be solved, any other important contribution to classical mechanics would then be considered to be prize-worthy. The prize was awarded to Poincaré, even though he did not solve the original problem. (The first version of his contribution even contained a serious error[11]). The version finally printed contained many important ideas which led to the development of chaos theory. The problem as stated originally was finally solved by Karl Fritiof Sundman for $n = 3$.

25.2 General formulation

The *n*-body problem considers N point masses $m_i, i = 1, 2, \ldots, N$ in an inertial reference frame in three dimensional space $\mathbb{R}^3$ moving under the influence of mutual gravitational attraction. Each mass m_i has a position vector $\mathbf{q}_i$. Newton's second law says that mass times acceleration $m_i d^2 \mathbf{q}_i / dt^2$ is equal to the sum of the forces on the mass. Newton's law of gravity says that the gravitational force felt on mass m_i by a single mass m_j is given by[12]

$$\mathbf{F}_{ij} = \frac{G m_i m_j (\mathbf{q}_j - \mathbf{q}_i)}{\|\mathbf{q}_j - \mathbf{q}_i\|^3},$$

where G is the gravitational constant and $\|\mathbf{q}_j - \mathbf{q}_i\|$ is the magnitude of the distance between $\mathbf{q}_i$ and $\mathbf{q}_j$.

Summing over all masses yields the *n*-body equations of motion:

$$m_i \frac{d^2 \mathbf{q}_i}{dt^2} = \sum_{j=1, j \neq i}^{N} \frac{G m_i m_j (\mathbf{q}_j - \mathbf{q}_i)}{\|\mathbf{q}_j - \mathbf{q}_i\|^3} = \frac{\partial U}{\partial \mathbf{q}_i}$$

where U is the *self-potential* energy

$$U = \sum_{1 \leq i < j \leq N} \frac{G m_i m_j}{\|\mathbf{q}_j - \mathbf{q}_i\|}.$$

Defining the momentum to be $\mathbf{p}_i = m_i d\mathbf{q}_i/dt$, Hamilton's equations of motion for the n-body problem become[13]

$$\frac{d\mathbf{q}_i}{dt} = \frac{\partial H}{\partial \mathbf{p}_i} \qquad \frac{d\mathbf{p}_i}{dt} = -\frac{\partial H}{\partial \mathbf{q}_i},$$

where the Hamiltonian function is

$$H = T + U$$

and T is the kinetic energy

$$T = \sum_{i=1}^{N} \frac{\|\mathbf{p}_i\|^2}{2m_i}.$$

Hamilton's equations show that the n-body problem is a system of $6N$ first-order differential equations, with $6N$ initial conditions as $3N$ initial position coordinates and $3N$ initial momentum values.

Symmetries in the n-body problem yield global integrals of motion that simplify the problem.[14] Translational symmetry of the problem results in the center of mass

$$\mathbf{C} = \frac{\sum_{i=1}^{N} m_i \mathbf{q}_i}{\sum_{i=1}^{N} m_i}$$

moving with constant velocity, so that $\mathbf{C} = \mathbf{L}_0 t + \mathbf{C}_0$, where $\mathbf{L}_0$ is the linear momentum and $\mathbf{C}_0$ is the initial position. The constants of motion L_0 and $\mathbf{C}_0$ represent six integrals of the motion. Rotational symmetry results in the total angular momentum being constant

$$\mathbf{A} = \sum_{i=1}^{N} \mathbf{q}_i \times \mathbf{p}_i,$$

where $\times$ is the cross product. The three components of the total angular momentum $\mathbf{A}$ yield three more constants of the motion. The last general constant of the motion is given by the conservation of energy H. Hence, every n-body problem has ten integrals of motion.

Because T and U are homogeneous functions of degree 2 and −1, respectively, the equations of motion have a scaling invariance: if $\mathbf{q}_i(t)$ is a solution, then so is $\lambda^{-2/3}\mathbf{q}_i(\lambda t)$ for any $\lambda > 0$.[15]

The moment of inertia of an n-body system is given by

$$I = \sum_{i=1}^{N} m_i \mathbf{q}_i \cdot \mathbf{q}_i = \sum_{i=1}^{N} m_i \|\mathbf{q}_i\|^2$$

and the *virial* is given by $Q = (1/2)dI/dt$. Then the *Lagrange-Jacobi formula* states that[16]

$$\frac{d^2 I}{dt^2} = 2T - U.$$

For systems in *dynamic equilibrium*, the long-term time average of $\langle d^2 I/dt^2 \rangle$ is zero. Then on average the total kinetic energy is half the total potential energy, $\langle T \rangle = \langle U \rangle / 2$, which is an example of the virial theorem for gravitational systems.[17] If M is the total mass and R a characteristic size of the system (for example, the radius containing half the mass of the system), then the critical time for a system to settle down to a dynamic equilibrium is $t_{cr} = \sqrt{GM/R^3}$.[18]

25.3 Special cases

25.3.1 Two-body problem

Main article: Two-body problem

Any discussion of planetary interactive forces has always started historically with the two-body problem. The purpose of this Section is to relate the real complexity in calculating any planetary forces. Note in this Section also, several subjects, such as gravity, barycenter, Kepler's Laws, etc.; and in the following Section too (Three-body problem) are discussed on other Wikipedia pages. Here though, these subjects are discussed from the perspective of the n-body problem.

The two-body problem ($N = 2$) was completely solved by Johann Bernoulli (1667-1748) by *classical* theory (and not by Newton) by assuming the main point-mass was *fixed*, is outlined here.[19] Consider then the motion of two bodies, say Sun-Earth, with the Sun *fixed*, then:

$$m_1 \mathbf{a}_1 = \frac{Gm_1m_2}{r_{12}^3}(\mathbf{r}_2 - \mathbf{r}_1) \quad \text{Sun-Earth}$$

$$m_2 \mathbf{a}_2 = \frac{Gm_1m_2}{r_{21}^3}(\mathbf{r}_1 - \mathbf{r}_2) \quad \text{Earth-Sun}$$

The equation describing the motion of mass m_2 relative to mass m_1 is readily obtained from the differences between these two equations and after canceling common terms gives: $\alpha + (\eta/r^3)\mathbf{r} = 0$, where

- $\mathbf{r} = \mathbf{r}_2 - \mathbf{r}_1$ is the vector position of m_2 relative to m_1 ;
- α is the *Eulerian* acceleration $d^2\mathbf{r}/dt^2$;
- and $\eta = G(m_1 + m_2)$.

The equation $\alpha + (\eta/r^3)\mathbf{r} = 0$ is the fundamental differential equation for the two-body problem Bernoulli solved in 1734. Notice for this approach forces have to be determined first, then the equation of motion resolved. This differential equation has elliptic, or parabolic or hyperbolic solutions[20], [21], [22]

It is incorrect to think of m_1 (the Sun) as fixed in space when applying Newton's *Law of Universal Gravitation*, and to do so leads to erroneous results. The fixed point for two isolated gravitationally interacting bodies is their mutual barycenter, and this two-body problem can be solved exactly, such as using Jacobi coordinates relative to the barycenter.

Dr. Clarence Cleminshaw calculated the approximate position of the Solar System's barycenter, a result achieved mainly by combining only the masses of Jupiter and the Sun. *Science Program* stated in reference to his work:

> "*The Sun contains 98 per cent of the mass in the solar system, with the superior planets beyond Mars accounting for most of the rest. On the average, the center of the mass of the Sun-Jupiter system, when the two most massive objects are considered alone, lies 462,000 miles from the Sun's center, or some 30,000 miles above the solar surface! Other large planets also influence the center of mass of the solar system, however. In 1951, for example, the systems' center of mass was not far from the Sun's center because Jupiter was on the opposite side from Saturn, Uranus and Neptune. In the late 1950s, when all four of these planets were on the same side of the Sun, the system's center of mass was more than 330,000 miles from the solar surface, Dr. C. H. Cleminshaw of Griffith Observatory in Los Angeles has calculated.*"[23]

The Sun wobbles as it rotates around the galactic center, dragging the Solar System and Earth along with it. What mathematician Kepler did in arriving at his three famous equations was curve-fit the apparent motions of the planets using Tycho Brahe's data, and *not* curve-fitting their true circular motions about the Sun (see Figure). Both Robert Hooke and Newton were well aware that Newton's *Law of Universal Gravitation* did not hold for the forces associated with elliptical orbits.[10] In fact, Newton's *Universal Law* does not account for the orbit of Mercury, the asteroid belt's gravitational

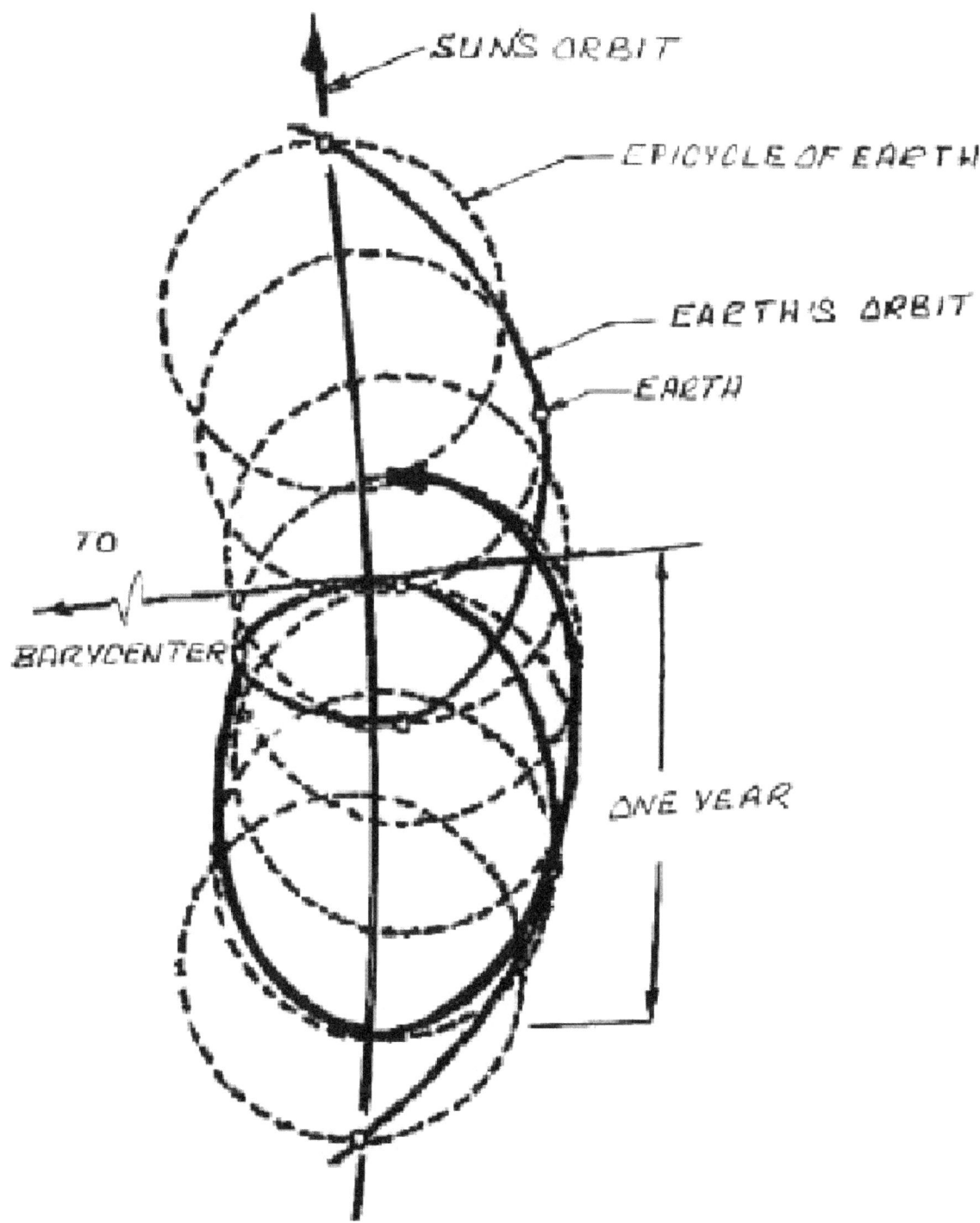

The Real Motion v.s. Kepler's Apparent Motion

behavior, or Saturn's rings.[24] Newton stated (in the 11th Section of the Principia) that the main reason, however, for failing to predict the forces for elliptical orbits was that his math model was for a body confined to a situation that hardly existed in the real world, namely, the motions of bodies attracted toward an unmoving center. Some present physics and astronomy textbooks don't emphasize the negative significance of Newton's assumption and end up teaching that his math model is in effect reality. It is to be understood that the classical two-body problem solution above is a mathematical

25.3. SPECIAL CASES

idealization. See also Kepler's first law of planetary motion.

Some modern writers have criticized Newton's *fixed* Sun as emblematic of a school of reductive thought - see Truesdell's *Essays in the History of Mechanics* referenced below. An aside: *Newtonian* physics doesn't include (among other things) relative motion and may be the root of the reason Newton "*fixed*" the Sun.[25][26]

25.3.2 Three-body problem

Main article: Three-body problem

This section relates a historically important *n*-body problem solution after simplifying assumptions were made.

In the past not much was known about the *n*-body problem for *n* equal to or greater than three.[27] The case for $n = 3$ has been the most studied. Many earlier attempts to understand the *Three-body problem* were quantitative, aiming at finding explicit solutions for special situations.

- In 1687 Isaac Newton published in the Principia the first steps in the study of the problem of the movements of three bodies subject to their mutual gravitational attractions, but his efforts resulted in verbal descriptions and geometrical sketches; see especially Book 1, Proposition 66 and its corollaries (Newton, 1687 and 1999 (transl.), see also Tisserand, 1894).

- In 1767 Euler found collinear motions, in which three bodies of any masses move proportionately along a fixed straight line. The circular restricted three-body problem is the special case in which two of the bodies are in circular orbits (approximated by the Sun-Earth-Moon system and many others).

- In 1772 Lagrange discovered two classes of periodic solution, each for three bodies of any masses. In one class, the bodies lie on a rotating straight line. In the other class, the bodies lie at the vertices of a rotating equilateral triangle. In either case, the paths of the bodies will be conic sections. Those solutions led to the study of *central configurations*, for which $\ddot{q} = kq$ for some constant $k>0$.

- A major study of the Earth-Moon-Sun system was undertaken by Charles-Eugène Delaunay, who published two volumes on the topic, each of 900 pages in length, in 1860 and 1867. Among many other accomplishments, the work already hints at chaos, and clearly demonstrates the problem of so-called "*small denominators*" in perturbation theory.

- In 1917 Forest Ray Moulton published his now classic, An Introduction to Celestial Mechanics (see references) with its plot of the *Restricted Three-body Problem* solution (see figure below).[28] An aside, see Meirovitch's book, pages 414 and 413 for his *Restricted Three-body Problem* solution.[29]

Moulton's solution may be easier to visualize (and definitely easier to solve) if one considers the more massive body (e.g., Sun) to be "*stationary*" in space, and the less massive body (e.g., Jupiter) to orbit around it, with the equilibrium points (Lagrangian points) maintaining the 60-degree spacing ahead of, and behind, the less massive body almost in its orbit (although in reality neither of the bodies are truly stationary, as they both orbit the center of mass of the whole system—about the barycenter). For sufficiently small mass ratio of the primaries, these triangular equilibrium points are stable, such that (nearly) massless particles will orbit about these points as they orbit around the larger primary (Sun). The five equilibrium points of the circular problem are known as the Lagrangian points. See figure below:

In the *Restricted 3-Body Problem* math model figure above (ref. Moulton), the Lagrangian points L_4 and L_5 are where the Trojan planetoids resided (see Lagrangian point); m_1 is the Sun and m_2 is Jupiter. L_2 is where the asteroid belt is. It has to be realized for this model, this whole Sun-Jupiter diagram is rotating about its barycenter. The *Restricted 3-Body Problem* solution predicted the Trojan planetoids before they were first seen. The *h*-circles and closed loops echo the electromagnetic fluxes issued from the Sun and Jupiter. It is conjectured, contrary to Richard H. Batin's conjecture (see References), the two h_1's are gravity sinks, in and where gravitational forces are zero, and the reason the Trojan planetoids are trapped there. The total amount of mass of the planetoids is unknown.

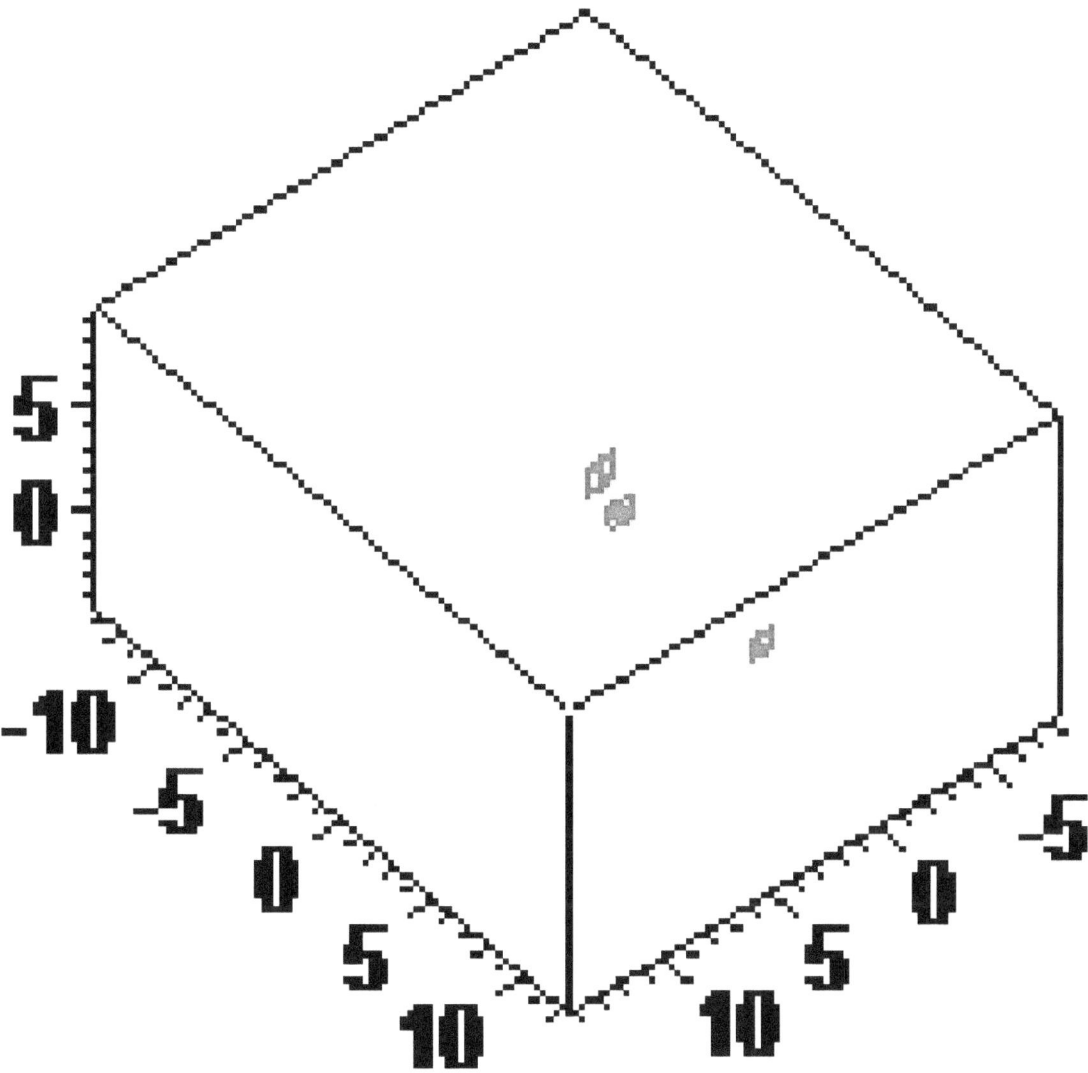

Motion of three particles under gravity, demonstrating chaotic behaviour

The *Restricted Three-body Problem* assumes the mass of one of the bodies is negligible. For a discussion of the case where the negligible body is a satellite of the body of lesser mass, see Hill sphere; for binary systems, see Roche lobe. Specific solutions to the *Three-body problem* result in chaotic motion with no obvious sign of a repetitious path.

The restricted problem (both circular and elliptical) was worked on extensively by many famous mathematicians and physicists, most notably by Poincaré at the end of the 19th century. Poincaré's work on the *Restricted Three-body Problem* was the foundation of deterministic chaos theory. In the restricted problem, there exist five equilibrium points. Three are collinear with the masses (in the rotating frame) and are unstable. The remaining two are located on the third vertex of both equilateral triangles of which the two bodies are the first and second vertices.

25.3.3 Planetary problem

The *planetary problem* is the *n*-body problem in the case that one of the masses is much larger than all the others. A prototypical example of a planetary problem is the Sun-Jupiter-Saturn system, where the mass of the Sun is about 1000 times bigger than the masses of Jupiter or Saturn.[15] An approximate solution to the problem is to decompose it

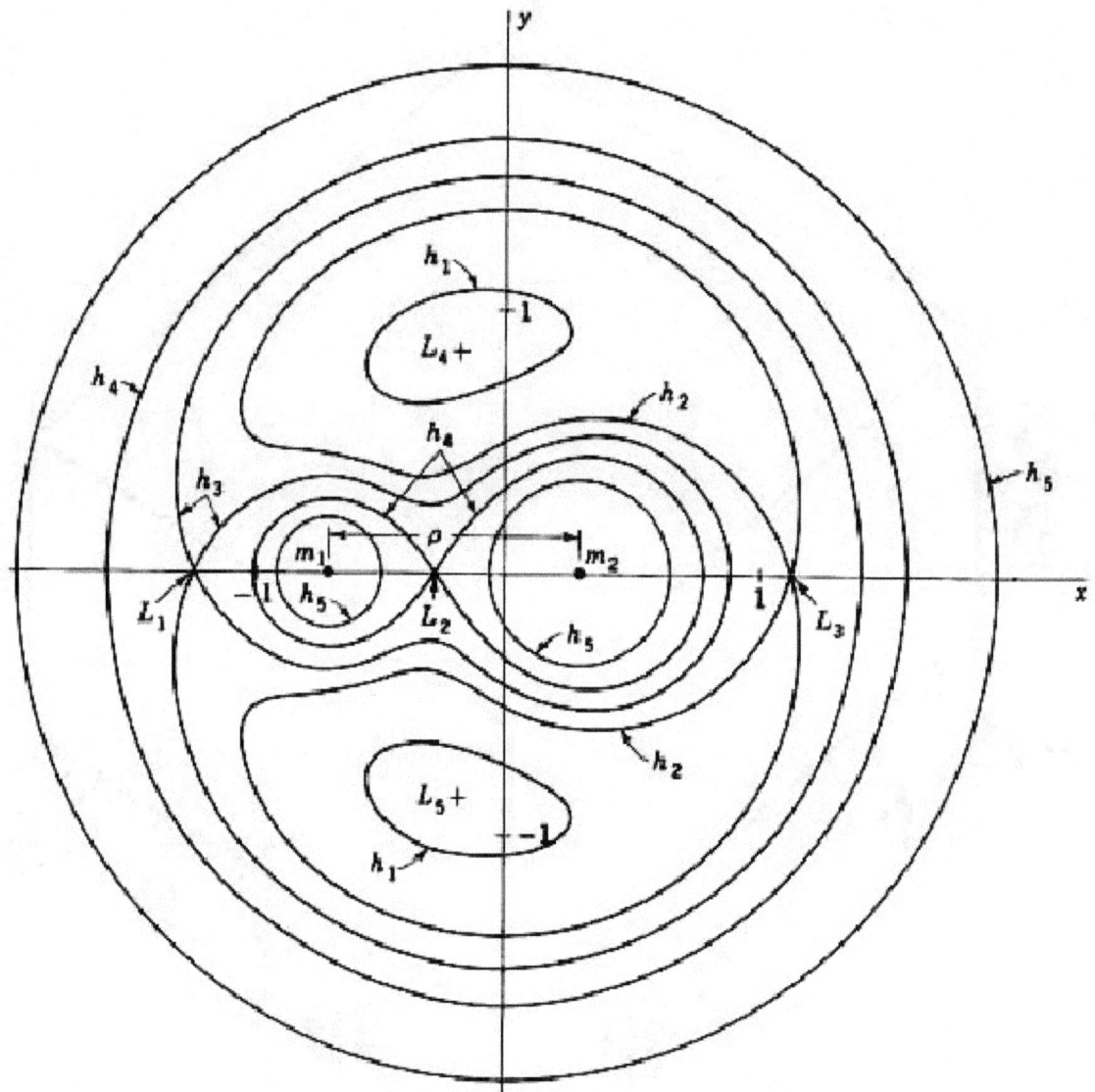

Restricted 3-Body Problem

into $N-1$ pairs of star-planet Kepler problems, treating interactions among the planets as perturbations. Perturbative approximation works well as long as there are no orbital resonances in the system, that is none of the ratios of unperturbed Kepler frequencies is a rational number. Resonances appear as small denominators in the expansion.

The existence of resonances and small denominators led to the important question of stability in the planetary problem: do planets, in nearly circular orbits around a star, remain in stable or bounded orbits over time?[15][30] In 1963, Vladimir Arnold proved using KAM theory a kind of stability of the planetary problem: there exists a set of positive measure of quasiperiodic orbits in the case of the planetary problem restricted to the plane.[30] In the KAM theory, chaotic planetary orbits would be bounded by quasiperiodic KAM tori. Arnold's result was extended to a more general theorem by Féjoz and Herman in 2004.[31]

25.3.4 Central configurations

A *central configuration* $q_1(0), \ldots, q_N(0)$ is an initial configuration such that if the particles were all released with zero velocity, they would all collapse toward the center of mass C.[30] Such a motion is called *homothetic*. Central configurations may also give rise to *homographic motions* in which all masses moves along Keplerian trajectories (elliptical, circular, parabolic, or hyperbolic), with all trajectories having the same eccentricity e. For elliptical trajectories, $e = 1$ corresponds to homothetic motion and $e = 0$ gives a *relative equilibrium motion* in which the configuration remains an isometry of the initial configuration, as if the configuration was a rigid body.[32] Central configurations have played an important role in understanding the topology of invariant manifolds created by fixing the first integrals of a system.

25.3.5 *n*-body choreography

Main article: n-body choreography

Solutions in which all masses move on the *same* curve without collisions are called choreographies.[33] A choreography for $N = 3$ was discovered by Lagrange in 1772 in which three bodies are situated at the vertices of an equilateral triangle in the rotating frame. A figure eight choreography for $N = 3$ was found numerically by C. Moore in 1993 and generalized and proven by A. Chenciner and R. Montgomery in 2000. Since then, many other choreographies have been found for $N \geq 3$.

25.4 Analytic approaches

For every solution of the problem, not only applying an isometry or a time shift but also a reversal of time (unlike in the case of friction) gives a solution as well.

In the physical literature about the n-body problem ($n \geq 3$), sometimes reference is made to *the impossibility of solving the n-body problem* (via employing the above approach). However, care must be taken when discussing the 'impossibility' of a solution, as this refers only to the method of first integrals (compare the theorems by Abel and Galois about the impossibility of solving algebraic equations of degree five or higher by means of formulas only involving roots).

25.4.1 Power series solution

One way of solving the classical *n*-body problem is "the *n*-body problem by Taylor series", which is an implementation of the Power series solution of differential equations.

We start by defining the system of differential equations:

$$\frac{d^2 x_i(t)}{dt^2} = G \sum_{k=1, k \neq i}^{n} \frac{m_k (x_k(t) - x_i(t))}{|x_k(t) - x_i(t)|^3} .$$

As xi ($t = t_0$) and $dxi(t)/dt_{=t_0}$ are given as initial conditions, every $d^2xi(t)/dt^2$ are known. Differentiating $d^2xi(t)/dt^2$ results in $d^3xi(t)/dt^3$ which at t_0 which is also known, and the Taylor series is constructed iteratively.

25.4.2 A generalized Sundman global solution

In order to generalize Sundman's result for the case $n > 3$ (or $n = 3$ and $c = 0$) one has to face two obstacles:

1. As it has been shown by Siegel, collisions which involve more than two bodies cannot be regularized analytically, hence Sundman's regularization cannot be generalized.
2. The structure of singularities is more complicated in this case: other types of singularities may occur (see below).

Lastly, Sundman's result was generalized to the case of $n > 3$ bodies by Q. Wang in the 1990s. Since the structure of singularities is more complicated, Wang had to leave out completely the questions of singularities. The central point of his approach is to transform, in an appropriate manner, the equations to a new system, such that the interval of existence for the solutions of this new system is $[0, \infty)$.

25.4.3 Singularities of the *n*-body problem

There can be two types of singularities of the *n*-body problem:

- collisions of two or more bodies, but for which $q(t)$ (the bodies' positions) remains finite. (In this mathematical sense, a "collision" means that two point-like bodies have identical positions in space.)

- singularities in which a collision does not occur, but $q(t)$ does not remain finite. In this scenario, bodies diverge to infinity in a finite time, while at the same time tending towards zero separation (an imaginary collision occurs "at infinity").

The latter ones are called Painlevé's conjecture (no-collisions singularities). Their existence has been conjectured for $n > 3$ by Painlevé (see Painlevé conjecture). Examples of this behavior for $n=5$ have been constructed by Xia[34] and a heuristic model for $n=4$ by Gerver.[35] Donald G. Saari has shown that for 4 or fewer bodies, the set of initial data giving rise to singularities has measure zero.[36]

25.5 Simulation

Main article: N-body simulation

While there are analytic solutions available for the classical (ie. non-relativistic) two-body problem and for selected configurations with $N > 2$, in general *n*-body problems must be solved or simulated using numerical methods.[18]

25.5.1 Few bodies

For a small number of bodies, an *n*-body problem can be solved using **direct methods**, also called **particle-particle methods**. These methods numerically integrate the differential equations of motion. Numerical integration for this problem can be a challenge for several reasons. First, the gravitational potential is singular; it goes to infinity as the distance between two particles goes to zero. The gravitational potential may be *softened* to remove the singularity at small distances:[18]

$$U_\epsilon = \sum_{1 \leq i < j \leq N} \frac{Gm_i m_j}{\left(\|\mathbf{q}_j - \mathbf{q}_i\|^2 + \epsilon\right)^{1/2}}.$$

Second, in general for $N > 2$, the *N*-body problem is chaotic,[37] which means that even small errors in integration may grow exponentially in time. Third, a simulation may be over large stretches of model time (e.g., millions of years) and numerical errors accumulate as integration time increases.

There are a number of techniques to reduce errors in numerical integration.[18] Local coordinate systems are used to deal with widely differing scales in some problems, for example an Earth-Moon coordinate system in the context of a solar system simulation. Variational methods and perturbation theory can yield approximate analytic trajectories upon which the numerical integration can be a correction. The use of a symplectic integrator ensures that the simulation obeys Hamilton's equations to a high degree of accuracy and in particular that energy is conserved.

25.5.2 Many bodies

Direct methods using numerical integration require on the order of $N^2/2$ computations to evaluate the potential energy over all pairs of particles, and thus have a time complexity of $O(N^2)$. For simulations with many particles, the $O(N^2)$ factor makes large-scale calculations especially time consuming.[18]

A number of approximate methods have been developed that reduce the time complexity relative to direct methods:[18]

- **Tree code methods**, such as a Barnes-Hut simulation, are *collisionless* methods used when close encounters among pairs are not important and distant particle contributions do not need to be computed to high accuracy. The potential of a distant group of particles is computed using a multipole expansion of the potential. This approximation allows for a reduction in complexity to $O(N \log(N))$.

- **Fast multipole methods** take advantage of the fact that the multipole-expanded forces from distant particles are similar for particles close to each other. It is claimed that this further approximation reduces the complexity to $O(N)$.[18]

- **Particle mesh methods** divide up simulation space into a three dimensional grid onto which the mass density of the particles is interpolated. Then calculating the potential becomes a matter of solving a Poisson equation on the grid, which can be computed in $O(N \log(N))$ time using fast Fourier transform techniques. Using adaptive mesh refinement or multigrid techniques can further reduce the complexity of the methods.

- **P3M** and **PM-Tree methods** are hybrid methods that use the particle mesh approximation for distant particles, but use more accurate methods for close particles (within a few grid intervals). P3M stands for P^3M or *Particle-Particle-Particle-Mesh* and uses direct methods with softened potentials at close range. PM-Tree methods instead use tree codes at close range. As with particle mesh methods, adaptive meshes can increase computational efficiency.

- **Mean field methods** approximate the system of particles with a time-dependent Boltzmann equation representing the mass density that is coupled to a self-consistent Poisson equation representing the potential. It is a type of smoothed-particle hydrodynamics approximation suitable for large systems.

25.5.3 Strong gravitation

In astrophysical systems with strong gravitational fields, such as those near the event horizon of a black hole, *n*-body simulations must take into account general relativity; such simulations are the domain of numerical relativity. Numerically simulating the Einstein field equations is extremely challenging[18] and a parameterized post-Newtonian formalism (PPN), such as the Einstein–Infeld–Hoffmann equations, is used if possible. The two-body problem in general relativity is analytically solvable only for the Kepler problem, in which one mass is assumed to be much larger than the other.[38]

25.6 Other *n*-body problems

Most work done on the *n*-body problem has been on the gravitational problem. But there exist other systems for which *n*-body mathematics and simulation techniques have proven useful.

In large scale electrostatics problems, such as the simulation of proteins and cellular assemblies in structural biology, the Coulomb potential has the same form as the gravitational potential, except that charges may be positive or negative, leading to repulsive as well as attractive forces.[39] *Fast Coulomb solvers* are the electrostatic counterpart to fast multipole method simulators. These are often used with periodic boundary conditions on the region simulated and Ewald summation techniques are used to speed up computations.[40]

In statistics and machine learning, some models have loss functions of a form similar to that of the gravitational potential: a sum of kernel functions over all pairs of objects, where the kernel function depends on the distance between the objects in parameter space.[41] Example problems that fit into this form include all-nearest-neighbors in manifold learning, kernel density estimation, and kernel machines. Alternative optimizations to reduce the $O(N^2)$ time complexity to $O(N)$ have been developed, such as *dual tree* algorithms, that have applicability to the gravitational *n*-body problem as well.

25.7 See also

- Celestial mechanics

- Gravitational two-body problem
- Jacobi integral
- Lunar theory
- Natural units
- Numerical model of the Solar System
- Stability of the Solar System

25.8 Notes

[1] Leimanis and Minorsky: Our interest is with Leimanis, who first discusses some history about the *n*-body problem, especially Ms. Kovalevskaya's ~1868-1888, twenty-year complex-variables approach, failure; **Section 1: The Dynamics of Rigid Bodies and Mathematical Exterior Ballistics** (Chapter 1, *the motion of a rigid body about a fixed point* (**Euler** and **Poisson** *equations*); Chapter 2, *Mathematical Exterior Ballistics*), good precursor background to the *n*-body problem; **Section 2: Celestial Mechanics** (Chapter 1, *The Uniformization of the Three-body Problem* (Restricted Three-body Problem); Chapter 2, *Capture in the Three-Body Problem*; Chapter 3, *Generalized n-body Problem*).

[2] See references cited for Heggie and Hut.

[3] *Quasi-steady* loads refers to the instantaneous inertial loads generated by instantaneous angular velocities and accelerations, as well as translational accelerations (9 variables). It is as though one took a photograph, which also recorded the instantaneous position and properties of motion. In contrast, a *steady-state* condition refers to a system's state being invariant to time; otherwise, the first derivatives and all higher derivatives are zero.

[4] R. M. Rosenberg states the *n*-body problem similarly (see References): *Each particle in a system of a finite number of particles is subjected to a Newtonian gravitational attraction from all the other particles, and to no other forces. If the initial state of the system is given, how will the particles move?* Rosenberg failed to realize, like everyone else, that it is necessary to determine the forces *first* before the motions can be determined.

[5] A general, classical solution in terms of first integrals is known to be impossible. An exact theoretical solution for arbitrary *n* can be approximated via Taylor series, but in practice such an infinite series must be truncated, giving at best only an approximate solution; and an approach now obsolete. In addition, the *n*-body problem may be solved using numerical integration, but these, too, are approximate solutions; and again obsolete. See Sverre J. Aarseth's book **Gravitational N-body Simulations** listed in the References.

[6] See David H. and Stephen P. H. Clark's **The Suppressed Scientific Discoveries of Stephen Gray and John Flamsteed, Newton's Tyranny**, W. H. Freeman and Co., 2001. A popularization of the historical events and bickering between those parties, but more importantly about the results they produced.

[7] See *"Discovery of gravitation,* **A.D. 1666**" by Sir David Brewster, in **The Great Events by Famous Historians**, Rossiter Johnson, LL.D. Editor-in-Chief, Volume XII, pp. 51-65, *The National Alumni*, 1905.

[8] Rudolf Kurth has an extensive discussion in his book (see References) on planetary perturbations. An aside: these mathematically undefined planetary perturbations (wobbles) still exist undefined even today and planetary orbits have to be constantly updated, usually yearly. See Astronomical Ephemeris and the American Ephemeris and Nautical Almanac, prepared jointly by the Nautical Almanac Offices of the United Kingdom and the United States of America.

[9] See **Principia**, Book Three, *System of the World*, "General Scholium," page 372, last paragraph. Newton was well aware his math model did not reflect physical reality. This edition referenced is from the **Great Books of the Western World**, Volume 34, which was translated by Andrew Motte and revised by Florian Cajori. This same paragraph is on page 1160 in Stephen Hawkins' huge **On the Shoulders of Giants**, 2002 edition; is a copy from Daniel Adee's 1848 addition. Cohen also has translated new editions: **Introduction to Newton's 'Principia'**, 1970; and **Isaac Newton's** *Principia*, with **Varian Readings**, 1972. Cajori also wrote a **History of Science**, which is on the Internet.

[10] See. I. Bernard Cohen's *Scientific American* article.

[11] for details of the serious error in Poincare's first submission see the article by Diacu

[12] Meyer 2009, pp. 27-28

[13] Meyer 2009, p. 28

[14] Meyer 2009, pp. 28-29

[15] Chenciner 2007

[16] Meyer 2009, p. 34

[17] "AST1100 Lecture Notes: 5 The virial theorem" (PDF). University of Oslo. Retrieved 25 March 2014.

[18] Trenti 2008

[19] See Bate, Mueller, and White: Chapter 1, *"Two-Body Orbital Mechanics,"* pp 1-49. These authors were from the *Dept. of Astronautics and Computer Science*, United States Air Force Academy. See Chapter 1. Their textbook is not filled with advanced mathematics.

[20] For the classical approach, if the common center of mass (i.e., the barycenter) of the two bodies is considered *to be at rest*, then each body travels along a conic section which has a focus at the barycenter of the system. In the case of a hyperbola it has the branch at the side of that focus. The two conics will be in the same plane. The type of conic (circle, ellipse, parabola or hyperbola) is determined by finding the sum of the combined kinetic energy of two bodies and the potential energy when the bodies are far apart. (This potential energy is always a negative value; energy of rotation of the bodies about their axes is not counted here.)

- If the sum of the energies is negative, then they both trace out ellipses.
- If the sum of both energies is zero, then they both trace out parabolas. As the distance between the bodies tends to infinity, their relative speed tends to zero.
- If the sum of both energies is positive, then they both trace out hyperbolas. As the distance between the bodies tends to infinity, their relative speed tends to some positive number.

[21] For this approach see Lindsay's **Physical Mechanics**, Chapter 3, *"Curvilinear Motion in a Plane,"* and specifically paragraph 3-9, *"Planetary Motion"*; and continue reading on to the Chapter's end, pp. 83-96. Lindsay presentation goes a long way in explaining these latter comments for the fixed *2-body problem*; i.e., when the Sun is assumed fixed.

[22] Note: The fact a parabolic orbit has zero energy arises from the assumption the gravitational potential energy goes to zero as the bodies get infinitely far apart. One could assign *any* value to the potential energy in the state of infinite separation. That state is assumed to have zero potential energy *by convention*.

[23] Science Program's **The Nature of the Universe** states Clarence Cleminshaw (1902-1985) served as Assistant Director of *Griffith Observatory* from 1938-1958 and as Director from 1958-1969. Some publications by Cleminshaw, C. H.: **Celestial Speeds**, 4 1953, equation, Kepler, orbit, comet, Saturn, Mars, velocity; Cleminshaw, C. H.: **The Coming Conjunction of Jupiter and Saturn**, 7 1960, Saturn, Jupiter, observe, conjunction; Cleminshaw, C. H.: **The Scale of The Solar System**, 7 1959, Solar system, scale, Jupiter, sun, size, light.

[24] Brush, Stephen G. Editor: **Maxwell on Saturn's Rings**, *MIT Press*, 1983.

[25] See Jacob Bronowski and Bruce Mazlish's **The Western Intellectual Tradition**, *Dorset Press*, 1986, for a discussion of this apparent lack of understanding by Newton. Also see Truesdell's **Essays in the History of Mechanics** for additional background about Newton accomplishments or lack therein.

[26] As Hufbauer points out, Newton miscalculated and published unfortunately the wrong value for the Sun's mass twice before he got it correct in his third attempt.

[27] See Leimanis and Minorsky's historical comments.

[28] See Moulton's *Restricted Three-body Problem* 's analytical and graphical solution.

[29] See Meirovitch's book: Chapters 11, *Problems in Celestial Mechanics*; 12, *Problem in Spacecraft Dynamics*; and Appendix A: *Dyadics*.

[30] Chierchia 2010

[31] Féjoz 2004

[32] See Chierchia 2010 for animations illustrating homographic motions

[33] Celletti 2008

[34] Zhihong Xia. The Existence of Noncollision Singularities in Newtonian Systems. Annals of Mathematics. Second Series, Vol. 135, No. 3 (May, 1992), pp. 411-468

[35] Joseph L. Gerver, Noncollision Singularities: Do Four Bodies Suffice?, Exp. Math. (2003), 187-198.

[36] Saari, Donald G. (1977). "A global existence theorem for the four-body problem of Newtonian mechanics". *J. Differential Equations* **26**: 80–111. Bibcode:1977JDE....26...80S. doi:10.1016/0022-0396(77)90100-0.

[37] Alligood 1996

[38] Blanchet 2001

[39] Krumscheid 2010

[40] Board 1999

[41] Ram 2010

25.9 References

- Aarseth, Sverre J.: **Gravitational N-body Simulations, Tools and Algorithms**, *Cambridge University Press*, 2003.

- Alligood, K. T., Sauer, T. D., and Yorke, J. A., **Chaos: An introduction to Dynamical systems**, Springer, pp. 46–48, 1996.

- Bate, Roger R.; Mueller, Donald D.; and White, Jerry: **Fundamentals of Astrodynamics**, *Dover*, 1971.

- Blanchet, Luc. *On the two-body problem in general relativity*, Comptes Rendus de l'Académie des Sciences-Series IV-Physics 2, no. 9 (2001): 1343-1352.

- Board, John A. Jr (1999), Humphres, Christopher W., Lambert, Christophe G., Rankin, William T., and Toukmaji, Abdulnour Y., *Ewald and Multipole Methods for Periodic N-Body Problems*, in the book **Computational Molecular Dynamics: Challenges, Methods, Ideas** by editors Deuflhard, Peter, Hermans, Jan, Leimkuhler, Benedict, Mark, Alan E., Reich, Sebastian, and Skeel, Robert D., Springer Berlin Heidelberg, pp. 459–471, doi:10.1007/978-3-642-58360-5_27, ISBN 978-3-540-63242-9.

- Bronowski, Jacob and Mazlish, Bruce: **The Western Intellectual Tradition, from Leonardo to Hegel**, *Dorsey Press*. 1986.

- Celletti, Alessandra (2008), *Computational celestial mechanics*, Scholarpedia, 3(9):4079, doi:10.4249/scholarpedia.4079

- Chenciner, Alain (2007), *Three body problem*, Scholarpedia, 2(10):2111, doi:10.4249/scholarpedia.2111

- Chierchia, Luigi and Mather, John N. (2010), *Kolmogorov-Arnold-Moser Theory*, Scholarpedia, 5(9):2123, doi:10.4249/scholarpedia.2123

- Cohen, I. Bernard: "*Newton's Discovery of Gravity*," *Scientific American*, pp. 167–179, Vol. 244, No. 3, Mar. 1980.

- Cohen, I. Bernard: **The Birth of a New Physics, Revised and Updated**, *W.W. Norton & Co.*, 1985.

- Diacu, F.: *The solution of the n-body problem*, The Mathematical Intelligencer,1996,18,p. 66–70

- Féjoz, J (2004). *Dèmonstration du 'théorème d'Arnol'd' sur la stabilité du système planétaire (d'après Herman)*, *Ergodic Theory Dynam. Systems*, vol 5 : 1521-1582.

- Heggie, Douglas and Hut, Piet: **The Gravitational Million-Body Problem, A Multidisciplinary Approach to Star Cluster Dynamics**, *Cambridge University Press*, 357 pages, 2003.

- Heggie, Douglas C.: *"Chaos in the N-body Problem of Stellar Dynamics,"* in **Predictability, Stability and Chaos in N-Body Dynamical Systems**, Ed. by Roy A. E., *Plenum Press*, 1991.

- Hufbauer, Dr. Karl C. (History of Science): **Exploring the Sun, Solar Science since Galileo**, *The Johns Hopkins University Press*, 1991. This book was sponsored by the **NASA** *History Office*.

- Krumscheid, Sebastian (2010), *Benchmark of fast Coulomb Solvers for open and periodic boundary conditions*, Jülich Supercomputing Centre, Technical Report FZJ-JSC-IB-2010-01.

- Kurth, Rudolf, **Introduction to the Mechanics of the Solar System**, *Pergamon Press*, 1959.

- Leimanis, E., and Minorsky, N.: **Dynamics and Nonlinear Mechanics**, Part I: *Some Recent Advances in the Dynamics of Rigid Bodies and Celestial Mechanics* (Leimanis), Part II: *The Theory of Oscillations* (Minorsky), *John Wiley & Sons, Inc.*, 1958.

- Lindsay, Robert Bruce: **Physical Mechanics**, 3rd Ed., *D. Van Nostrand Co., Inc.*, 1961.

- Meirovitch, Leonard: **Methods of Analytical Dynamics**, *McGraw-Hill Book Co.*, 1970.

- Meyer, Kenneth Ray and Hall, Glen R., **Introduction to Hamiltonian Dynamical Systems and the N-body Problem**, Springer Science+Business Media, LLC, 2009, doi:10.1007/978-0-387-09724-4 2.

- Mittag-Leffler, G.: *The n-body problem (Price Announcement)*, Acta Matematica, 1885/1886,7

- Moulton, Forest Ray: **An Introduction to Celestial Mechanics**, *Dover*, 1970.

- Newton, I.: *Philosophiae Naturalis Principia Mathematica*, London, 1687; also English translation of 3rd (1726) edition by I. Bernard Cohen and Anne Whitman (Berkeley, CA, 1999).

- Ram, Parikshit (2009), Dongryeol Lee, William B. March, and Alexander G. Gray. *Linear-time Algorithms for Pairwise Statistical Problems*, in **NIPS**, pp. 1527–1535. 2009.

- Rosenberg, Reinhardt M.: **Analytical Dynamics, of Discrete Systems**, Chapter 19, *About Celestial Problems*, paragraph 19.5, *The n-body Problem*, pp. 364–371, *Plenum Press*, 1977. Like Battin above, Rosenberg employs energy methods too, and to the solution of the general n-body problem but doesn't actually solve anything.

- Science Program's *The Nature of the Universe*, booklet, published by *Nelson Doubleday, Inc.*, 1968.

- Sundman, K. F.: *Memoire sur le probleme de trois corps*, Acta Mathematica 36 (1912): 105–179.

- Tisserand, F-F.: *Mecanique Celeste*, tome III (Paris, 1894), ch.III, at p. 27.

- Trenti, Michele and Hut, Piet (2008), *N-body simulations*, Scholarpedia, 3(5):3930, doi:10.4249/scholarpedia.3930

- Truesdell, Clifford: **Essays in the History of Mechanics**, *Springer-Verlay*, 1968.

- van Winter, Clasine: *The n-body problem on a Hilbert space of analytic functions*, Paper 11-29, in **Analytic Methods in Mathematical Physics**, edited by Robert P. Gilbert and Roger G. Newton, pp. 569–578, *Gordon and Breach*, 1970.

- Wang, Qiudong (1991). "The global solution of the n-body problem". *Celestial Mechanics and Dynamical Astronomy* **50** (1): 73–88. Bibcode:1991CeMDA..50...73W. doi:10.1007/BF00048987. ISSN 0923-2958. MR 1117788.

- Xia, Zhihong (1992). "The Existence of Noncollision Singularities in Newtonian Systems". *Annals Math.* **135** (3): 411–468. JSTOR 2946572.

25.10 Further reading

- Brouwer, Dirk and Clemence, Gerald M.: **Methods of Celestial Mechanics**, *Academic Press*, 1961.

- Battin, Richard H.: **An Introduction to The Mathematics and Methods of Astrodynamics**, *AIAA*, 1987. He employs energy methods rather than a Newtonian approach.

- Gelman, Harry: Part I: *The second orthogonality conditions in the theory of proper and improper rotations: Derivation of the conditions and of their main consequences*, J. Res. NBS 72B (Math. Sci.)No. 3, 1968. Part II: *The intrinsic vector*; Part III: *The Conjugacy Theorem*, J. Res. NBS 72B (Math. Sci.) No. 2, 1969. *A Note on the time dependence of the effective axis and angle of a rotation*, J. Res. NBS 72B (Math. Sci.)No. 3&4, Oct. 1971. These papers are on the Internet.

- Meriam, J. L.: **Engineering Mechanics**, Volume 1 *Statics*, Volume 2 *Dynamics*, *John Wiley & Sons*, 1978.

- Quadling, Henley: "*Gravitational N-Body Simulation: 16 bit DOS version,*" June 1994. nbody*.zip is available at the http://www.ftp.cica.indiana.edu: see external links.

- Korenev, G. V.: **The Mechanics of Guided Bodies**, *CRC Press*, 1967.

- Eisele, John A. and Mason, Robert M.: **Applied Matrix and Tensor Analysis**, *John Wiley & Sons*, 1970.

- Murray, Carl D. and Dermott, Stanley F.: **Solar System Dynamics**, *Cambridge University Press*, 606 pages, 2000.

- Crandall, Richard E.: **Topics in Advanced Scientific Computation**, Chapter 5, "*Nonlinear & Complex Systems,*" paragraph 5.1, "*N-body problems & chaos,*" pp. 215–221, *Springer-Verlag*, 1996.

- Crandall, Richard E.: **Projects in Scientific Computation**, Chapter 2, "*Exploratory Computation,*" Project 2.4.1, "*Classical Physics,*" pp. 93–97, corrected 3rd printing, *Springer-Verlag*, 1996.

- Szebehely, Victor: **Theory of Orbits**, *Academic Press*, 1967.

- Saari, D. (1990). "A visit to the Newtonian n-body problem via Elementary Complex Variables". *American Mathematical Monthly* **89**: 105–119.

- Saari, D. G.; Hulkower, N. D. (1981). "On the Manifolds of Total Collapse Orbits and of Completely Parabolic Orbits for the n-Body Problem". *Journal of Differential Equations* **41** (1): 27–43. Bibcode:1981JDE....41...27S. doi:10.1016/0022-0396(81)90051-6.

- Hagihara, Y: Celestial Mechanics. (Vol I and Vol II pt 1 and Vol II pt 2.) MIT Press, 1970.

- Boccaletti, D. and Pucacco, G.: Theory of Orbits (two volumes). Springer-Verlag, 1998.

- Havel, Karel. N-Body Gravitational Problem: Unrestricted Solution (ISBN 978-09689120-5-8). Brampton: Grevyt Press, 2008. http://www.grevytpress.com

25.11 External links

- Three-Body Problem at Scholarpedia
- More detailed information on the three-body problem
- Regular Keplerian motions in classical many-body systems
- Applet demonstrating chaos in restricted three-body problem
- Applets demonstrating many different three-body motions
- On the integration of the *n*-body equations

- Java applet simulating Solar System
- Java applet simulating a ring of bodies orbiting a large central mass
- Java applet simulating dust in the Solar System
- Java applet simulating a stable solution to the equi-mass 3-body problem
- Java applet simulating choreographies and other interesting n-body solutions
- A java applet to simulate the 3-d movement of set of particles under gravitational interaction
- Javascript Simulation of our Solar System
- The Lagrange Points - with links to the original papers of Euler and Lagrange, and to translations, with discussion
-

Chapter 26

Kepler's laws of planetary motion

For a more historical approach, see in particular the articles *Astronomia nova* and *Epitome Astronomiae Copernicanae*.

In astronomy, **Kepler's laws of planetary motion** are three scientific laws describing the motion of planets around the Sun.

1. The orbit of a planet is an ellipse with the Sun at one of the two foci.
2. A line segment joining a planet and the Sun sweeps out equal areas during equal intervals of time.[1]
3. The square of the orbital period of a planet is proportional to the cube of the semi-major axis of its orbit.

Most planetary orbits are almost circles, and careful observation and calculation is required in order to establish that they are actually ellipses. Calculations of the orbit of the planet Mars first indicated to Johannes Kepler its elliptical shape, and he inferred that other heavenly bodies, including those farther away from the Sun, also have elliptical orbits.

Kepler's work (published between 1609-1619) improved the heliocentric theory of Nicolaus Copernicus, explaining how the planets' speeds varied, and using elliptical orbits rather than circular orbits with epicycles.[2]

Isaac Newton showed in 1687 that relationships like Kepler's would apply in the solar system to a good approximation, as consequences of his own laws of motion and law of universal gravitation.

Kepler's laws are part of the foundation of modern astronomy and physics.[3]

26.1 Comparison to Copernicus

Kepler's laws improve the model of Copernicus. If the eccentricities of the planetary orbits are taken as zero, then Kepler basically agrees with Copernicus:

1. The planetary orbit is a circle
2. The Sun at the center of the orbit
3. The speed of the planet in the orbit is constant

The eccentricities of the orbits of those planets known to Copernicus and Kepler are small, so the foregoing rules give good approximations of planetary motion; but Kepler's laws fit the observations better than Copernicus's.

Kepler's corrections are not at all obvious:

1. The planetary orbit is *not* a circle, but an *ellipse*.

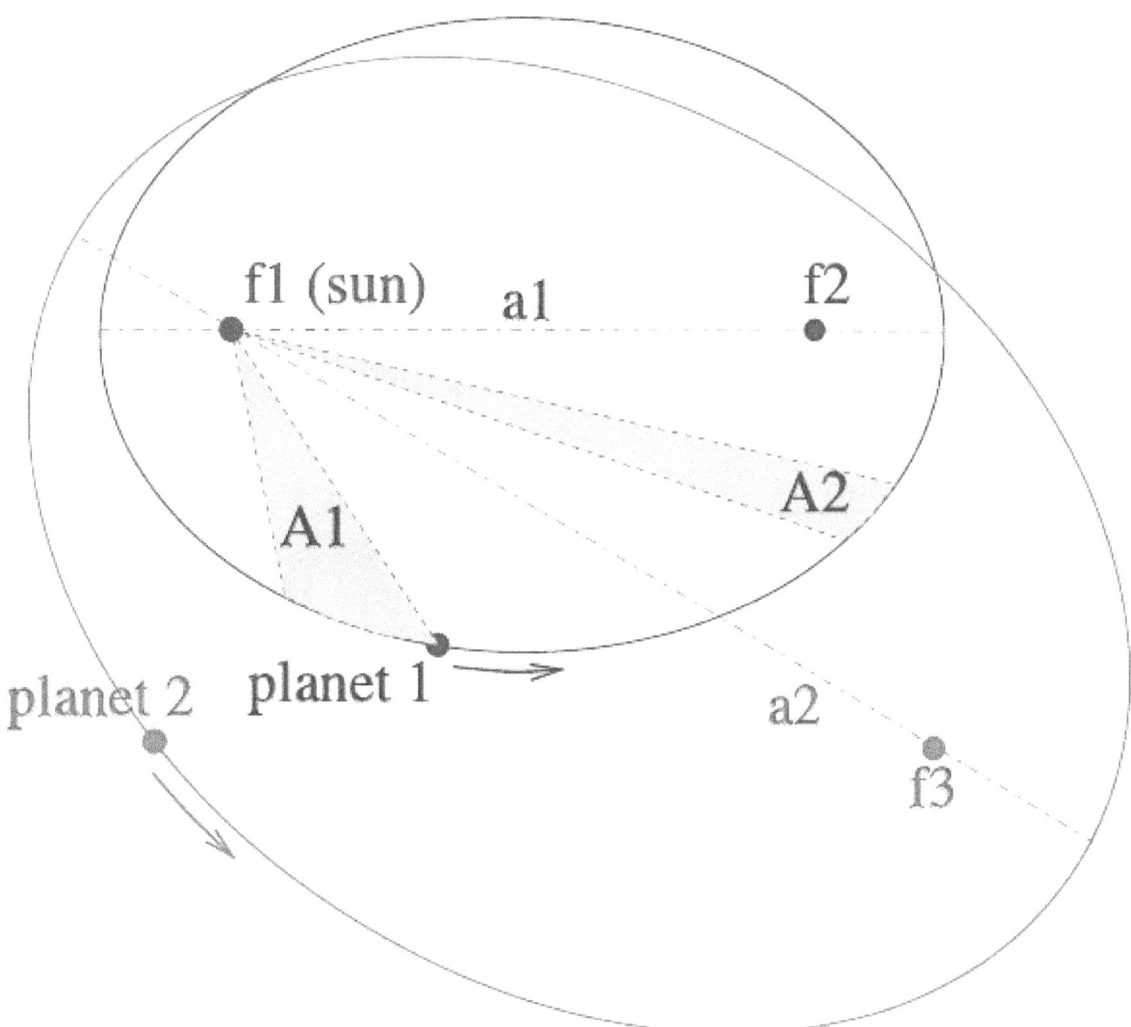

Figure 1: Illustration of Kepler's three laws with two planetary orbits.
(1) The orbits are ellipses, with focal points f_1 and f_2 for the first planet and f_1 and f_3 for the second planet. The Sun is placed in focal point f_1.
(2) The two shaded sectors A_1 and A_2 have the same surface area and the time for planet 1 to cover segment A_1 is equal to the time to cover segment A_2.
(3) The total orbit times for planet 1 and planet 2 have a ratio $a_1^{3/2} : a_2^{3/2}$.

2. The Sun is *not* at the center but at a *focal point* of the elliptical orbit.

3. Neither the linear speed nor the angular speed of the planet in the orbit is constant, but the *area speed* is constant.

The eccentricity of the orbit of the Earth makes the time from the March equinox to the September equinox, around 186 days, unequal to the time from the September equinox to the March equinox, around 179 days. A diameter would cut the orbit into equal parts, but the plane through the sun parallel to the equator of the earth cuts the orbit into two parts with areas in a 186 to 179 ratio, so the eccentricity of the orbit of the Earth is approximately

$$\varepsilon \approx \frac{\pi}{4} \frac{186 - 179}{186 + 179} \approx 0.015,$$

which is close to the correct value (0.016710219) (see Earth's orbit). The calculation is correct when perihelion, the date the Earth is closest to the Sun, falls on a solstice. The current perihelion, near January 4, is fairly close to the solstice of December 21 or 22.

26.2 Nomenclature

It took nearly two centuries for the current formulation of Kepler's work to take on its settled form. Voltaire's *Eléments de la philosophie de Newton* (*Elements of Newton's Philosophy*) of 1738 was the first publication to use the terminology of "laws".[4][5] The *Biographical Encyclopedia of Astronomers* in its article on Kepler (p. 620) states that the terminology of scientific laws for these discoveries was current at least from the time of Joseph de Lalande.[6] It was the exposition of Robert Small, in *An account of the astronomical discoveries of Kepler* (1804) that made up the set of three laws, by adding in the third.[7] Small also claimed, against the history, that these were empirical laws, based on inductive reasoning.[5][8]

Further, the current usage of "Kepler's Second Law" is something of a misnomer. Kepler had two versions of it, related in a qualitative sense, the "distance law" and the "area law". The "area law" is what became the Second Law in the set of three; but Kepler did himself not privilege it in that way.[9]

26.3 History

Johannes Kepler published his first two laws about planetary motion in 1609, having found them by analyzing the astronomical observations of Tycho Brahe.[10][2][11] Kepler's third law was published in 1619.[12][2]

Kepler in 1621 and Godefroy Wendelin in 1643 noted that Kepler's third law applies to the four brightest moons of Jupiter.[Nb 1] The second law, in the "area law" form, was contested by Nicolaus Mercator in a book from 1664; but by 1670 he was publishing in its favour in *Philosophical Transactions*, and as the century proceeded it became more widely accepted.[13] The reception in Germany changed noticeably between 1688, the year in which Newton's *Principia* was published and was taken to be basically Copernican, and 1690, by which time work of Gottfried Leibniz on Kepler had been published.[14]

Newton is credited with understanding that the second law is not special to the inverse square law of gravitation, being a consequence just of the radial nature of that law; while the other laws do depend on the inverse square form of the attraction. Carl Runge and Wilhelm Lenz much later identified a symmetry principle in the phase space of planetary motion (the orthogonal group O(4) acting) which accounts for the first and third laws in the case of Newtonian gravitation, as conservation of angular momentum does via rotational symmetry for the second law.[15]

26.4 Formulary

The mathematical model of the kinematics of a planet subject to the laws allows a large range of further calculations.

26.4.1 First law

The orbit of every planet is an ellipse with the Sun at one of the two foci.

Mathematically, an ellipse can be represented by the formula:

$$r = \frac{p}{1 + \varepsilon \cos\theta},$$

where p is the semi-latus rectum, and ε is the eccentricity of the ellipse, and r is the distance from the Sun to the planet, and θ is the angle to the planet's current position from its closest approach, as seen from the Sun. So (r, θ) are polar coordinates.

For an ellipse $0 < \varepsilon < 1$; in the limiting case $\varepsilon = 0$, the orbit is a circle with the sun at the centre (i.e. where there is no, or nil, eccentricity).

At $\theta = 0°$, perihelion, the distance is minimum

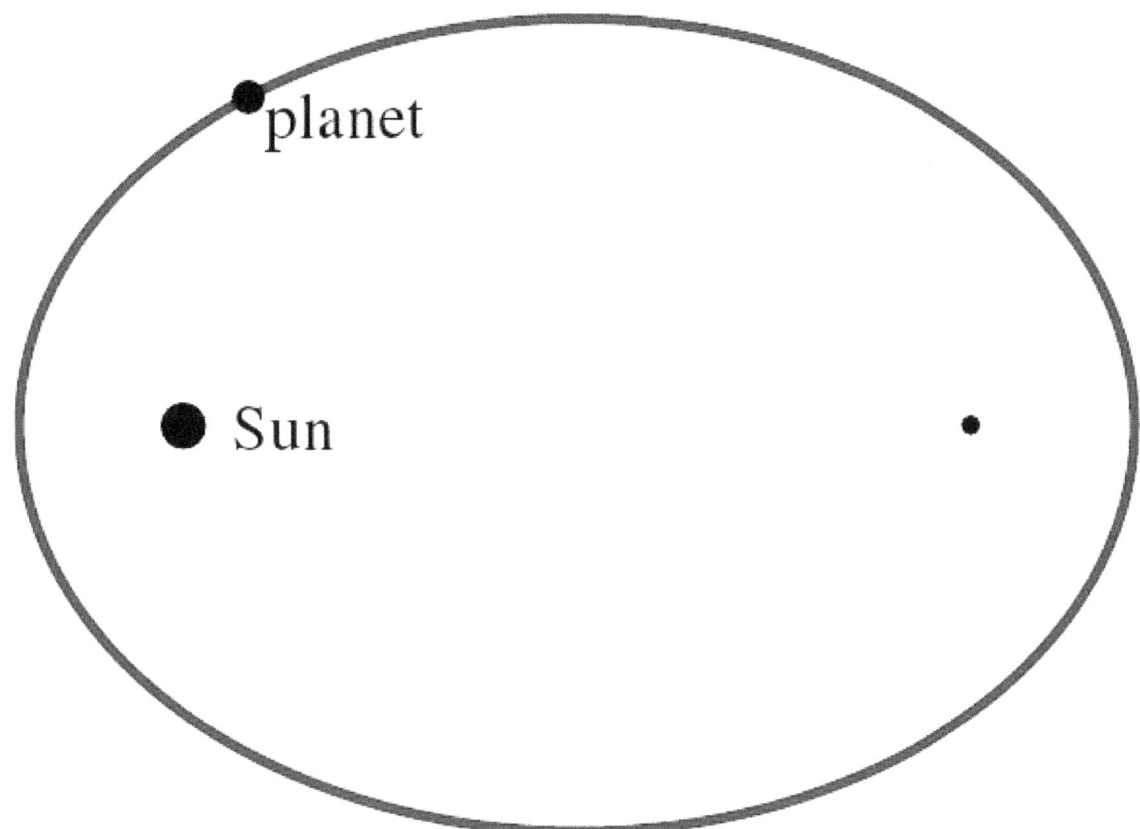

Figure 2: Kepler's first law placing the Sun at the focus of an elliptical orbit

$$r_{\min} = \frac{p}{1+\varepsilon}$$

At $\theta = 90°$ and at $\theta = 270°$ the distance is equal to p.

At $\theta = 180°$, aphelion, the distance is maximum (by definition, aphelion is - invariably - perihelion plus $180°$)

$$r_{\max} = \frac{p}{1-\varepsilon}$$

The semi-major axis a is the arithmetic mean between $r_{\min}$ and $r_{\max}$:

$$r_{\max} - a = a - r_{\min}$$

$$a = \frac{p}{1-\varepsilon^2}$$

The semi-minor axis b is the geometric mean between $r_{\min}$ and $r_{\max}$:

$$\frac{r_{\max}}{b} = \frac{b}{r_{\min}}$$

$$b = \frac{p}{\sqrt{1-\varepsilon^2}}$$

26.4. FORMULARY

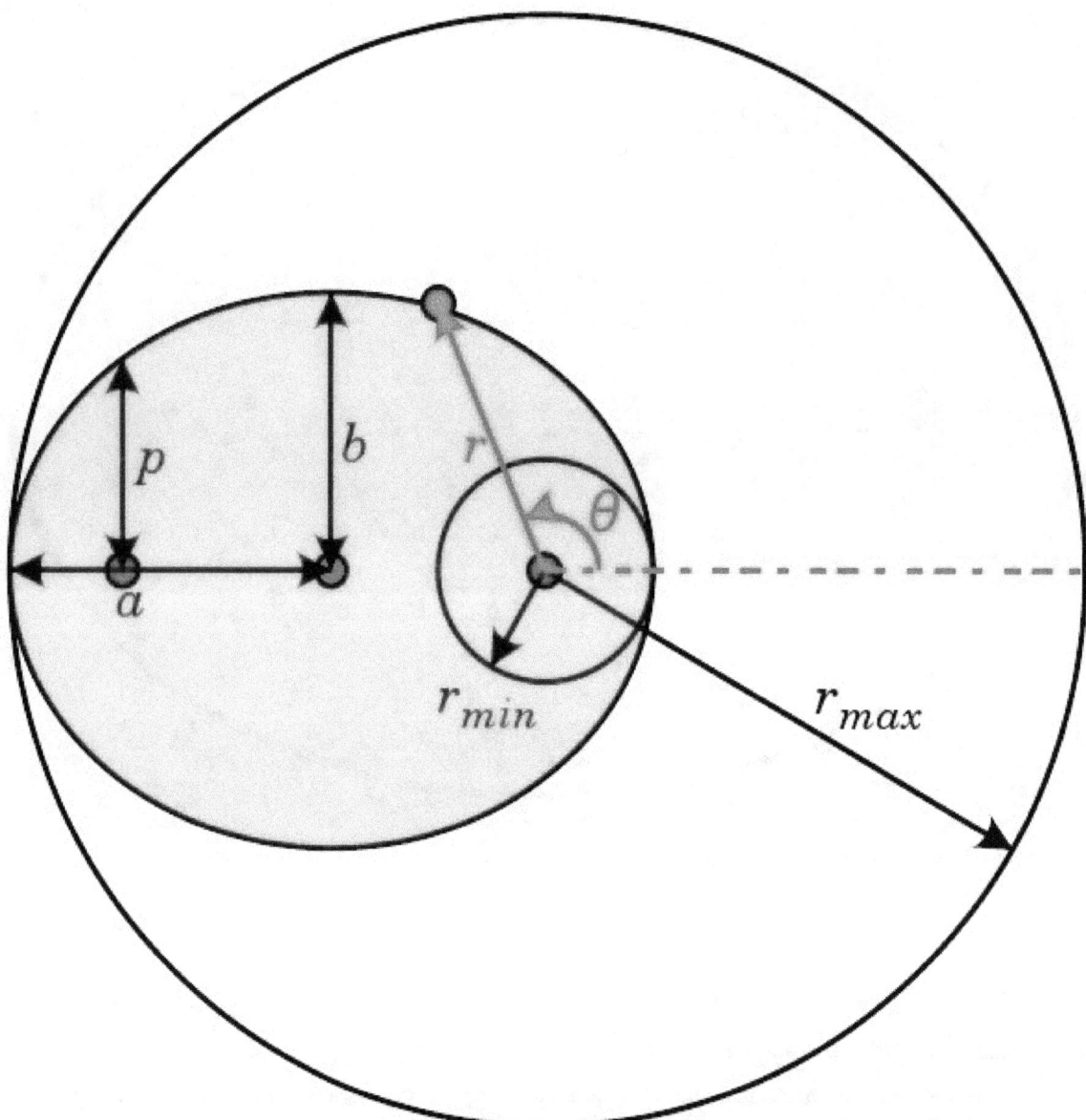

Figure 4: Heliocentric coordinate system (r, θ) for ellipse. Also shown are: semi-major axis a, semi-minor axis b and semi-latus rectum p; center of ellipse and its two foci marked by large dots. For θ = 0°, r = r_{min} and for θ = 180°, r = r_{max}.

The semi-latus rectum p is the harmonic mean between r_{min} and r_{max}:

$$\frac{1}{r_{min}} - \frac{1}{p} = \frac{1}{p} - \frac{1}{r_{max}}$$

$$pa = r_{max} r_{min} = b^2$$

The eccentricity ε is the coefficient of variation between r_{min} and r_{max}:

$$\varepsilon = \frac{r_{max} - r_{min}}{r_{max} + r_{min}}.$$

The area of the ellipse is

$A = \pi ab$.

The special case of a circle is $\varepsilon = 0$, resulting in $r = p = r_{\min} = r_{\max} = a = b$ and $A = \pi r^2$.

26.4.2 Second law

A line joining a planet and the Sun sweeps out equal areas during equal intervals of time.[1]

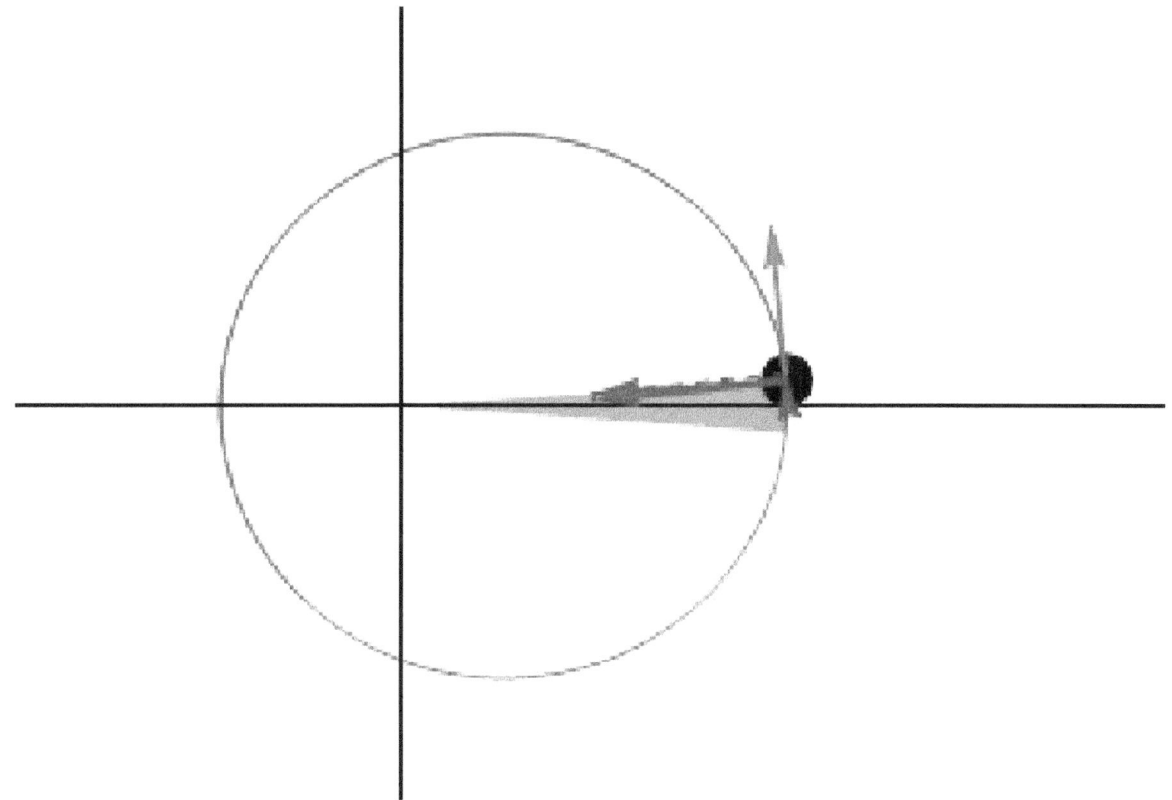

The same (blue) area is swept out in a fixed time period. The green arrow is velocity. The purple arrow directed towards the Sun is the acceleration. The other two purple arrows are acceleration components parallel and perpendicular to the velocity.

The orbital radius and angular velocity of the planet in the elliptical orbit will vary. This is shown in the animation: the planet travels faster when closer to the sun, then slower when farther from the sun. Kepler's second law states that the blue sector has constant area.

In a small time dt the planet sweeps out a small triangle having base line r and height $r\,d\theta$ and area $dA = \frac{1}{2} \cdot r \cdot r d\theta$ and so the constant areal velocity is $\frac{dA}{dt} = \frac{1}{2} r^2 \frac{d\theta}{dt}$.

The area enclosed by the elliptical orbit is πab. So the period P satisfies

$$P \cdot \tfrac{1}{2} r^2 \frac{d\theta}{dt} = \pi ab$$

and the mean motion of the planet around the Sun

$$n = 2\pi/P$$

26.5. PLANETARY ACCELERATION

satisfies

$r^2 \, d\theta = abn \, dt.$

26.4.3 Third law

> The square of the orbital period of a planet is directly proportional to the cube of the semi-major axis of its orbit.

This captures the relationship between the distance of planets from the Sun, and their orbital periods.

For a brief biography of Kepler and discussion of his third law, see: NASA: Stargaze.

Kepler enunciated in 1619 [12] this third law in a laborious attempt to determine what he viewed as the "music of the spheres" according to precise laws, and express it in terms of musical notation.[16] So it was known as the *harmonic law*.[17]

Mathematically, the law says that the expression

P^2/a^3

has the same value for all the planets in the solar system. Here P is the time taken for a planet to complete an orbit round the sun, and a is the mean value between the maximum and minimum distances between the planet and sun (i.e. the semimajor axis).

The modern formulation, with the constant evaluated, reads as:

$$\frac{T^2}{a^3} = \frac{4\pi^2}{G(M+m)}$$

where

- T is the orbital period of the orbiting body,
- M is the mass of the star,
- m is the mass of the planet,
- G is the universal gravitational constant and
- a is the semi major axis, i.e. the mean distance between bodies.

A value $M + m$ change for each planet, so the proportionality constant is not truly the same. Nevertheless, given that m is so small relative to M for planets in our solar system, the approximation $M + m \approx M$ is good.

26.5 Planetary acceleration

Isaac Newton computed in his Philosophiæ Naturalis Principia Mathematica the acceleration of a planet moving according to Kepler's first and second law.

1. The *direction* of the acceleration is towards the Sun.

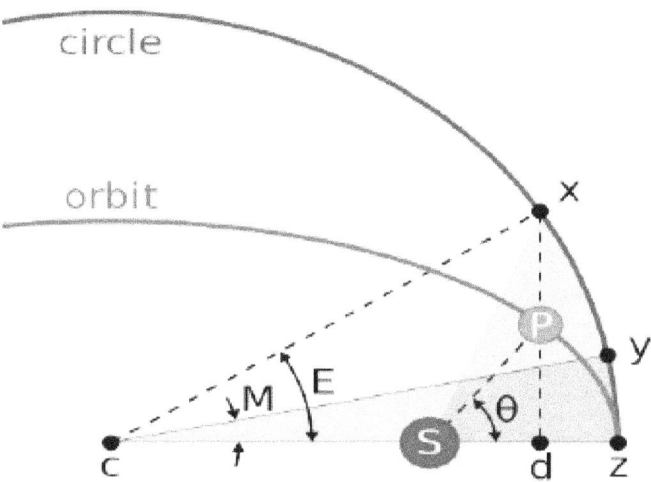

Figure 5. Geometric construction for Kepler's calculation of θ. The Sun (located at the focus) is labeled S and the planet P. The auxiliary circle is an aid to calculation. Line xd is perpendicular to the base and through the planet P. The shaded sectors are arranged to have equal areas by positioning of point y.

2. The *magnitude* of the acceleration is inversely proportional to the square of the planet's distance from the Sun (the *inverse square law*).

This implies that the Sun may be the physical cause of the acceleration of planets.

Newton defined the force acting on a planet to be the product of its mass and the acceleration (see Newton's laws of motion). So:

1. Every planet is attracted towards the Sun.

2. The force acting on a planet is in direct proportion to the mass of the planet and in inverse proportion to the square of its distance from the Sun.

The Sun plays an unsymmetrical part, which is unjustified. So he assumed, in Newton's law of universal gravitation:

1. All bodies in the solar system attract one another.

2. The force between two bodies is in direct proportion to the product of their masses and in inverse proportion to the square of the distance between them.

As the planets have small masses compared to that of the Sun, the orbits conform approximately to Kepler's laws. Newton's model improves upon Kepler's model, and fits actual observations more accurately (see two-body problem).

A deviation in the motion of a planet from Kepler's laws due to gravitational attraction by other planets is called a perturbation.

Below comes the detailed calculation of the acceleration of a planet moving according to Kepler's first and second laws.

26.5.1 Acceleration vector

See also: Polar coordinate § Vector calculus and Mechanics of planar particle motion

26.5. PLANETARY ACCELERATION

From the heliocentric point of view consider the vector to the planet $\mathbf{r} = r\hat{\mathbf{r}}$ where r is the distance to the planet and the direction $\hat{\mathbf{r}}$ is a unit vector. When the planet moves the directions change:

$$\frac{d\hat{\mathbf{r}}}{dt} = \dot{\hat{\mathbf{r}}} = \dot{\theta}\hat{\boldsymbol{\theta}}, \qquad \frac{d\hat{\boldsymbol{\theta}}}{dt} = \dot{\hat{\boldsymbol{\theta}}} = -\dot{\theta}\hat{\mathbf{r}}$$

where $\hat{\boldsymbol{\theta}}$ is the unit vector orthogonal to $\hat{\mathbf{r}}$ and pointing in the direction of rotation, and θ is the polar angle, and where a dot on top of the variable signifies differentiation with respect to time.

So differentiating the position vector twice to obtain the velocity and the acceleration vectors:

$$\dot{\mathbf{r}} = \dot{r}\hat{\mathbf{r}} + r\dot{\hat{\mathbf{r}}} = \dot{r}\hat{\mathbf{r}} + r\dot{\theta}\hat{\boldsymbol{\theta}},$$

$$\ddot{\mathbf{r}} = (\ddot{r}\hat{\mathbf{r}} + \dot{r}\dot{\hat{\mathbf{r}}}) + (\dot{r}\dot{\theta}\hat{\boldsymbol{\theta}} + r\ddot{\theta}\hat{\boldsymbol{\theta}} + r\dot{\theta}\dot{\hat{\boldsymbol{\theta}}}) = (\ddot{r} - r\dot{\theta}^2)\hat{\mathbf{r}} + (r\ddot{\theta} + 2\dot{r}\dot{\theta})\hat{\boldsymbol{\theta}}.$$

So

$$\ddot{\mathbf{r}} = a_r\hat{\mathbf{r}} + a_\theta\hat{\boldsymbol{\theta}}$$

where the **radial acceleration** is

$$a_r = \ddot{r} - r\dot{\theta}^2$$

and the **transversal acceleration** is

$$a_\theta = r\ddot{\theta} + 2\dot{r}\dot{\theta}.$$

26.5.2 The inverse square law

Kepler's laws say that

$$r^2\dot{\theta} = nab$$

is constant.

The transversal acceleration a_θ is zero:

$$\frac{d(r^2\dot{\theta})}{dt} = r(2\dot{r}\dot{\theta} + r\ddot{\theta}) = ra_\theta = 0.$$

So the acceleration of a planet obeying Kepler's laws is directed towards the sun.

The radial acceleration a_r is

$$a_r = \ddot{r} - r\dot{\theta}^2 = \ddot{r} - r\left(\frac{nab}{r^2}\right)^2 = \ddot{r} - \frac{n^2a^2b^2}{r^3}.$$

Kepler's first law states that the orbit is described by the equation:

$$\frac{p}{r} = 1 + \varepsilon \cos \theta.$$

Differentiating with respect to time

$$-\frac{p\dot{r}}{r^2} = -\varepsilon \sin \theta \, \dot{\theta}$$

or

$$p\dot{r} = nab\varepsilon \sin \theta.$$

Differentiating once more

$$p\ddot{r} = nab\varepsilon \cos \theta \, \dot{\theta} = nab\varepsilon \cos \theta \, \frac{nab}{r^2} = \frac{n^2 a^2 b^2}{r^2} \varepsilon \cos \theta.$$

The radial acceleration a_r satisfies

$$pa_r = \frac{n^2 a^2 b^2}{r^2} \varepsilon \cos \theta - p \frac{n^2 a^2 b^2}{r^3} = \frac{n^2 a^2 b^2}{r^2} \left(\varepsilon \cos \theta - \frac{p}{r} \right).$$

Substituting the equation of the ellipse gives

$$pa_r = \frac{n^2 a^2 b^2}{r^2} \left(\frac{p}{r} - 1 - \frac{p}{r} \right) = -\frac{n^2 a^2}{r^2} b^2.$$

The relation $b^2 = pa$ gives the simple final result

$$a_r = -\frac{n^2 a^3}{r^2}.$$

This means that the acceleration vector $\ddot{\mathbf{r}}$ of any planet obeying Kepler's first and second law satisfies the **inverse square law**

$$\ddot{\mathbf{r}} = -\frac{\alpha}{r^2} \hat{\mathbf{r}}$$

where

$$\alpha = n^2 a^3$$

is a constant, and $\hat{\mathbf{r}}$ is the unit vector pointing from the Sun towards the planet, and r is the distance between the planet and the Sun.

According to Kepler's third law, α has the same value for all the planets. So the inverse square law for planetary accelerations applies throughout the entire solar system.

The inverse square law is a differential equation. The solutions to this differential equation include the Keplerian motions, as shown, but they also include motions where the orbit is a hyperbola or parabola or a straight line. See Kepler orbit.

26.5.3 Newton's law of gravitation

By Newton's second law, the gravitational force that acts on the planet is:

$$\mathbf{F} = m_{\text{planet}} \ddot{\mathbf{r}} = -m_{\text{planet}} \alpha r^{-2} \hat{\mathbf{r}}$$

where m_{Planet} is the mass of the planet and α has the same value for all planets in the solar system. According to Newton's third Law, the Sun is attracted to the planet by a force of the same magnitude. Since the force is proportional to the mass of the planet, under the symmetric consideration, it should also be proportional to the mass of the Sun, m_{Sun}. So

$$\alpha = G m_{\text{Sun}}$$

where G is the gravitational constant.

The acceleration of solar system body number i is, according to Newton's laws:

$$\ddot{\mathbf{r}}_i = G \sum_{j \neq i} m_j r_{ij}^{-2} \hat{\mathbf{r}}_{ij}$$

where m_j is the mass of body j, r_{ij} is the distance between body i and body j, $\hat{\mathbf{r}}_{ij}$ is the unit vector from body i towards body j, and the vector summation is over all bodies in the world, besides i itself.

In the special case where there are only two bodies in the world, Earth and Sun, the acceleration becomes

$$\ddot{\mathbf{r}}_{\text{Earth}} = G m_{\text{Sun}} r_{\text{Earth,Sun}}^{-2} \hat{\mathbf{r}}_{\text{Earth,Sun}}$$

which is the acceleration of the Kepler motion. So this Earth moves around the Sun according to Kepler's laws.

If the two bodies in the world are Moon and Earth the acceleration of the Moon becomes

$$\ddot{\mathbf{r}}_{\text{Moon}} = G m_{\text{Earth}} r_{\text{Moon,Earth}}^{-2} \hat{\mathbf{r}}_{\text{Moon,Earth}}$$

So in this approximation the Moon moves around the Earth according to Kepler's laws.

In the three-body case the accelerations are

$$\ddot{\mathbf{r}}_{\text{Sun}} = G m_{\text{Earth}} r_{\text{Sun,Earth}}^{-2} \hat{\mathbf{r}}_{\text{Sun,Earth}} + G m_{\text{Moon}} r_{\text{Sun,Moon}}^{-2} \hat{\mathbf{r}}_{\text{Sun,Moon}}$$

$$\ddot{\mathbf{r}}_{\text{Earth}} = G m_{\text{Sun}} r_{\text{Earth,Sun}}^{-2} \hat{\mathbf{r}}_{\text{Earth,Sun}} + G m_{\text{Moon}} r_{\text{Earth,Moon}}^{-2} \hat{\mathbf{r}}_{\text{Earth,Moon}}$$

$$\ddot{\mathbf{r}}_{\text{Moon}} = G m_{\text{Sun}} r_{\text{Moon,Sun}}^{-2} \hat{\mathbf{r}}_{\text{Moon,Sun}} + G m_{\text{Earth}} r_{\text{Moon,Earth}}^{-2} \hat{\mathbf{r}}_{\text{Moon,Earth}}$$

These accelerations are not those of Kepler orbits, and the three-body problem is complicated. But Keplerian approximation is the basis for perturbation calculations. See Lunar theory.

26.6 Position as a function of time

Kepler used his two first laws to compute the position of a planet as a function of time. His method involves the solution of a transcendental equation called Kepler's equation.

The procedure for calculating the heliocentric polar coordinates (r,θ) of a planet as a function of the time t since perihelion, is the following four steps:

1. Compute the **mean anomaly** $M = nt$ where n is the mean motion.

   $n \cdot P = 2\pi$ radians where P is the period.

2. Compute the **eccentric anomaly** E by solving Kepler's equation:

$$M = E - \varepsilon \sin E$$

3. Compute the **true anomaly** θ by the equation:

$$(1 - \varepsilon)\tan^2 \frac{\theta}{2} = (1 + \varepsilon)\tan^2 \frac{E}{2}$$

4. Compute the **heliocentric distance**

$$r = a(1 - \varepsilon \cos E).$$

The important special case of circular orbit, $\varepsilon = 0$, gives $\theta = E = M$. Because the uniform circular motion was considered to be *normal*, a deviation from this motion was considered an **anomaly**.

The proof of this procedure is shown below.

26.6.1 Mean anomaly, M

The Keplerian problem assumes an elliptical orbit and the four points:

- s the Sun (at one focus of ellipse);
- z the perihelion
- c the center of the ellipse
- p the planet

and

- $a = |cz|$, distance between center and perihelion, the **semimajor axis**.
- $\varepsilon = \frac{|cs|}{a}$, the **eccentricity**.
- $b = a\sqrt{1 - \varepsilon^2}$, the **semiminor axis**.
- $r = |sp|$, the distance between Sun and planet.
- $\theta = \angle zsp$, the direction to the planet as seen from the Sun, the **true anomaly**.

The problem is to compute the polar coordinates (r, θ) of the planet from the **time since perihelion**, t.
It is solved in steps. Kepler considered the circle with the major axis as a diameter, and

- x, the projection of the planet to the auxiliary circle
- y, the point on the circle such that the sector areas $|zcy|$ and $|zsx|$ are equal.
- $M = \angle zcy$, the **mean anomaly**.

The sector areas are related by $|zsp| = \frac{b}{a} \cdot |zsx|$.

The circular sector area $|zcy| = \frac{a^2 M}{2}$.

The area swept since perihelion,

26.6. POSITION AS A FUNCTION OF TIME

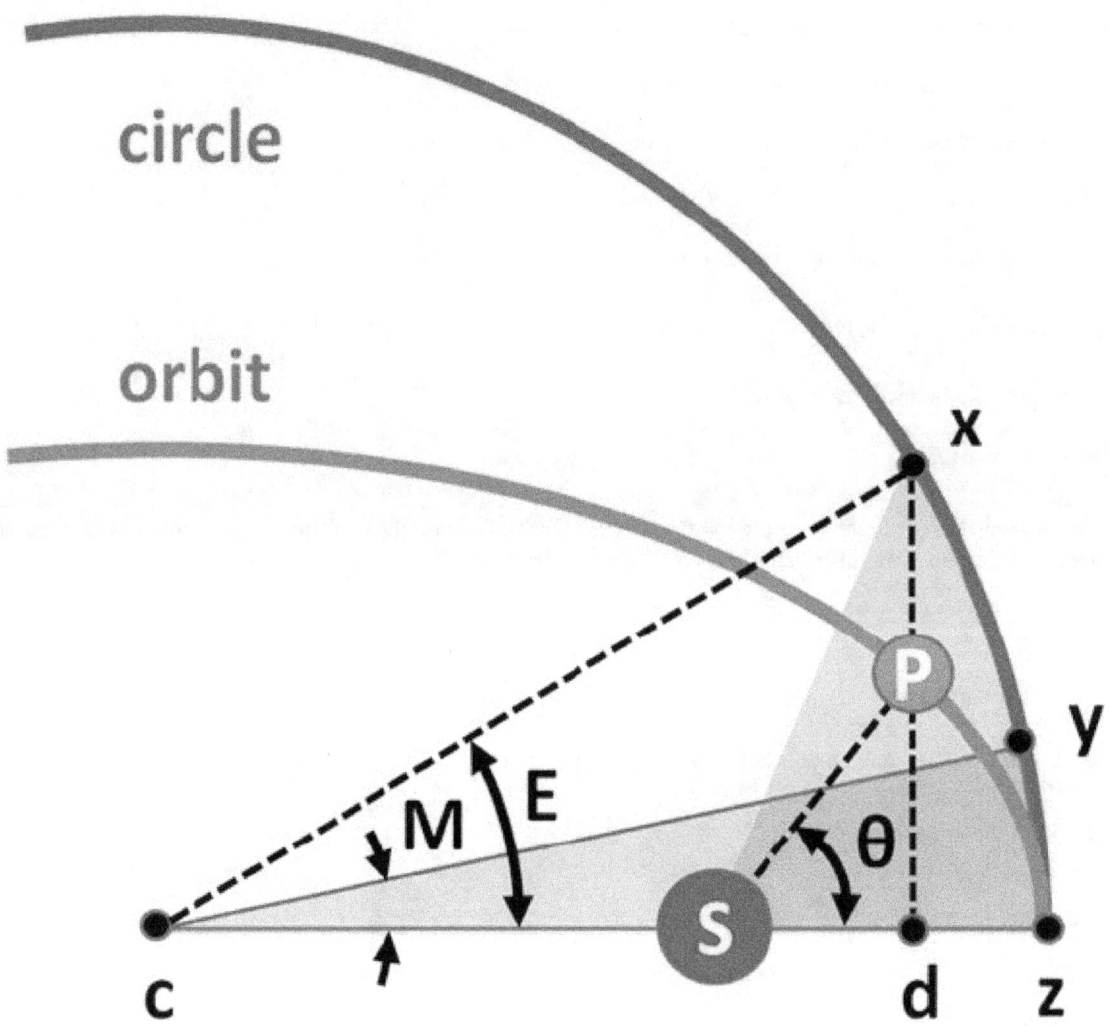

Figure 5: Geometric construction for Kepler's calculation of θ. The Sun (located at the focus) is labeled S and the planet P. The auxiliary circle is an aid to calculation. Line xd is perpendicular to the base and through the planet P. The shaded sectors are arranged to have equal areas by positioning of point y.

$$|zsp| = \frac{b}{a} \cdot |zsx| = \frac{b}{a} \cdot |zcy| = \frac{b}{a} \cdot \frac{a^2 M}{2} = \frac{abM}{2},$$

is by Kepler's second law proportional to time since perihelion. So the mean anomaly, M, is proportional to time since perihelion, t.

$$M = nt,$$

where n is the mean motion.

26.6.2 Eccentric anomaly, E

When the mean anomaly M is computed, the goal is to compute the true anomaly θ. The function $\theta = f(M)$ is, however, not elementary.[18] Kepler's solution is to use

$E = \angle zcx$, x as seen from the centre, the eccentric anomaly

as an intermediate variable, and first compute E as a function of M by solving Kepler's equation below, and then compute the true anomaly θ from the eccentric anomaly E. Here are the details.

$$|zcy| = |zsx| = |zcx| - |scx|$$

$$\frac{a^2 M}{2} = \frac{a^2 E}{2} - \frac{a\varepsilon \cdot a \sin E}{2}$$

Division by $a^2/2$ gives **Kepler's equation**

$$M = E - \varepsilon \cdot \sin E.$$

This equation gives M as a function of E. Determining E for a given M is the inverse problem. Iterative numerical algorithms are commonly used.

Having computed the eccentric anomaly E, the next step is to calculate the true anomaly θ.

26.6.3 True anomaly, θ

Note from the figure that

$$\vec{cd} = \vec{cs} + \vec{sd}$$

so that

$$a \cdot \cos E = a \cdot \varepsilon + r \cdot \cos \theta.$$

Dividing by a and inserting from Kepler's first law

$$\frac{r}{a} = \frac{1 - \varepsilon^2}{1 + \varepsilon \cdot \cos \theta}$$

to get

$$\cos E = \varepsilon + \frac{1 - \varepsilon^2}{1 + \varepsilon \cdot \cos \theta} \cdot \cos \theta = \frac{\varepsilon \cdot (1 + \varepsilon \cdot \cos \theta) + (1 - \varepsilon^2) \cdot \cos \theta}{1 + \varepsilon \cdot \cos \theta} = \frac{\varepsilon + \cos \theta}{1 + \varepsilon \cdot \cos \theta}.$$

The result is a usable relationship between the eccentric anomaly E and the true anomaly θ.

A computationally more convenient form follows by substituting into the trigonometric identity:

$$\tan^2 \frac{x}{2} = \frac{1 - \cos x}{1 + \cos x}.$$

Get

$$\tan^2 \frac{E}{2} = \frac{1 - \cos E}{1 + \cos E} = \frac{1 - \frac{\varepsilon + \cos \theta}{1 + \varepsilon \cdot \cos \theta}}{1 + \frac{\varepsilon + \cos \theta}{1 + \varepsilon \cdot \cos \theta}} = \frac{(1 + \varepsilon \cdot \cos \theta) - (\varepsilon + \cos \theta)}{(1 + \varepsilon \cdot \cos \theta) + (\varepsilon + \cos \theta)} = \frac{1 - \varepsilon}{1 + \varepsilon} \cdot \frac{1 - \cos \theta}{1 + \cos \theta} = \frac{1 - \varepsilon}{1 + \varepsilon} \cdot \tan^2 \frac{\theta}{2}.$$

Multiplying by $1+\varepsilon$ gives the result

$$(1 - \varepsilon) \cdot \tan^2 \frac{\theta}{2} = (1 + \varepsilon) \cdot \tan^2 \frac{E}{2}$$

This is the third step in the connection between time and position in the orbit.

26.6.4 Distance, r

The fourth step is to compute the heliocentric distance r from the true anomaly θ by Kepler's first law:

$$r \cdot (1 + \varepsilon \cdot \cos \theta) = a \cdot (1 - \varepsilon^2)$$

Using the relation above between θ and E the final equation for the distance r is:

$$r = a \cdot (1 - \varepsilon \cdot \cos E).$$

26.7 See also

- Circular motion
- Free-fall time
- Gravity
- Kepler orbit
- Kepler problem
- Kepler's equation
- Laplace–Runge–Lenz vector

26.8 Notes

[1] Godefroy Wendelin wrote a letter to Giovanni Battista Riccioli about the relationship between the distances of the Jovian moons from Jupiter and the periods of their orbits, showing that the periods and distances conformed to Kepler's third law. See: Joanne Baptista Riccioli, *Almagestum novum* ... (Bologna (Bononia), (Italy): Victor Benati, 1651), volume 1, page 492 Scholia III. In the margin beside the relevant paragraph is printed: *Vendelini ingeniosa speculatio circa motus & intervalla satellitum Jovis.* (Wendelin's clever speculation about the movement and distances of Jupiter's satellites.)
In 1621, Johannes Kepler had noted that Jupiter's moons obey (approximately) his third law in his *Epitome Astronomiae Copernicanae* [Epitome of Copernican Astronomy] (Linz ("Lentiis ad Danubium"), (Austria): Johann Planck, 1622), book 4, part 2, page 554.

26.9 References

[1] Bryant, Jeff; Pavlyk, Oleksandr. "Kepler's Second Law", *Wolfram Demonstrations Project*. Retrieved December 27, 2009.

[2] Holton, Gerald James; Brush, Stephen G. (2001). *Physics, the Human Adventure: From Copernicus to Einstein and Beyond* (3rd paperback ed.). Piscataway, NJ: Rutgers University Press. pp. 40–41. ISBN 0-8135-2908-5. Retrieved December 27, 2009.

[3] See also G. E. Smith, "Newton's Philosophiae Naturalis Principia Mathematica", especially the section *Historical context* ... in *The Stanford Encyclopedia of Philosophy* (Winter 2008 Edition), Edward N. Zalta (ed.).

[4] Voltaire, *[b]Eléments [/b]* de la philosophie de Newton *[Elements of Newton's Philosophy]* (London, England: 1738). See, for example:

- From p. 162: *"Par une des grandes loix de Kepler, toute Planete décrit des aires égales en temp égaux : par une autre loi non moins sûre, chaque Planete fait sa révolution autour du Soleil en telle sort, que si, sa moyenne distance au Soleil est 10. prenez le cube de ce nombre, ce qui sera 1000., & le tems de la révolution de cette Planete autour du Soleil sera proportionné à la racine quarrée de ce nombre 1000."* (By one of the great laws of Kepler, each planet describes equal areas in equal times ; by another law no less certain, each planet makes its revolution around the sun in such a way that if its mean distance from the sun is 10, take the cube of that number, which will be 1000, and the time of the revolution of that planet around the sun will be proportional to the square root of that number 1000.)

- From p. 205: *"Il est donc prouvé par la loi de Kepler & par celle de Neuton, que chaque Planete gravite vers le Soleil, ..."* (It is thus proved by the law of Kepler and by that of Newton, that each planet revolves around the sun ...)

[5] Wilson, Curtis (May 1994). "Kepler's Laws, So-Called" (PDF). *HAD News* (Washington, DC: Historical Astronomy Division, American Astronomical Society) (31): 1–2. Retrieved December 27, 2009.

[6] De la Lande, *Astronomie*, vol. 1 (Paris, France: Desaint & Saillant, 1764). See, for example:

- From page 390: *" ... mais suivant la fameuse loi de Kepler, qui sera expliquée dans le Livre suivant (892), le rapport des temps périodiques est toujours plus grand que celui des distances, une planete cinq fois plus éloignée du soleil, emploie à faire sa révolution douze fois plus de temps ou environ; ..."* (... but according to the famous law of Kepler, which will be explained in the following book [i.e., chapter] (paragraph 892), the ratio of the periods is always greater than that of the distances [so that, for example,] a planet five times farther from the sun, requires about twelve times or so more time to make its revolution [around the sun]; ...)

- From page 429: *"Les Quarrés des Temps périodiques sont comme les Cubes des Distances. 892. La plus fameuse loi du mouvement des planetes découverte par Kepler, est celle du repport qu'il y a entre les grandeurs de leurs orbites, & le temps qu'elles emploient à les parcourir; ..."* (The squares of the periods are as the cubes of the distances. 892. The most famous law of the movement of the planets discovered by Kepler is that of the relation that exists between the sizes of their orbits and the times that the [planets] require to traverse them; ...)

- From page 430: *"Les Aires sont proportionnelles au Temps. 895. Cette loi générale du mouvement des planetes devenue si importante dans l'Astronomie, sçavoir, que les aires sont proportionnelles au temps, est encore une des découvertes de Kepler; ..."* (Areas are proportional to times. 895. This general law of the movement of the planets [which has] become so important in astronomy to know, [namely] that areas are proportional to times, is one of Kepler's discoveries; ...)

- From page 435: *"On a appellé cette loi des aires proportionnelles aux temps, Loi de Kepler, aussi bien que celle de l'article 892, du nome de ce célebre Inventeur; ..."* (One called this law of areas proportional to times (the law of Kepler) as well as that of paragraph 892, by the name of that celebrated inventor; ...)

[7] Robert Small, *An account of the astronomical discoveries of Kepler* (London, England: J Mawman, 1804), pp. 298–299.

[8] Robert Small, An account of the astronomical discoveries of Kepler (London, England: J. Mawman, 1804).

[9] Bruce Stephenson (1994). *Kepler's Physical Astronomy*. Princeton University Press. p. 170. ISBN 0-691-03652-7.

[10] In his *Astronomia nova*, Kepler presented only a proof that Mars' orbit is elliptical. Evidence that the other known planets' orbits are elliptical was presented only in 1621.
See: Johannes Kepler, *Astronomia nova ...* (1609), p. 285. After having rejected circular and oval orbits, Kepler concluded that Mars' orbit must be elliptical. From the top of page 285: *"Ergo ellipsis est Planetæ iter; ..."* (Thus, an ellipse is the planet's [i.e., Mars'] path; ...) Later on the same page: " ... *ut sequenti capite patescet: ubi simul etiam demonstrabitur, nullam Planetæ relinqui figuram Orbitæ, præterquam perfecte ellipticam; ..."* (... as will be revealed in the next chapter: where it will also then be proved that any figure of the planet's orbit must be relinquished, except a perfect ellipse; ...) And then: *"Caput LIX. Demonstratio, quod orbita Martis, ... , fiat perfecta ellipsis: ..."* (Chapter 59. Proof that Mars' orbit, ... , is a perfect ellipse: ...) The geometric proof that Mars' orbit is an ellipse appears as Protheorema XI on pages 289–290.
Kepler stated that every planet travels in elliptical orbits having the Sun at one focus in: Johannes Kepler, *Epitome Astronomiae Copernicanae* [Summary of Copernican Astronomy] (Linz ("Lentiis ad Danubium"), (Austria): Johann Planck, 1622), book 5, part 1, III. De Figura Orbitæ (III. On the figure [i.e., shape] of orbits), pages 658–665. From p. 658: *"Ellipsin fieri orbitam planetæ ..."* (Of an ellipse is made a planet's orbit ...). From p. 659: " ... *Sole (Foco altero huius ellipsis) ..."* (... the Sun (the other focus of this ellipse) ...).

[11] In his *Astronomia nova ...* (1609), Kepler did not present his second law in its modern form. He did that only in his *Epitome* of 1621. Furthermore, in 1609, he presented his second law in two different forms, which scholars call the "distance law" and the "area law".

- His "distance law" is presented in: *"Caput XXXII. Virtutem quam Planetam movet in circulum attenuari cum discessu a fonte."* (Chapter 32. The force that moves a planet circularly weakens with distance from the source.) See: Johannes Kepler, *Astronomia nova* ... (1609), pp. 165–167. On page 167, Kepler states: " ... , *quanto longior est αδ quam αε, tanto diutius moratur Planeta in certo aliquo arcui excentrici apud δ, quam in æquali arcu excentrici apud ε.*" (... , as αδ is longer than αε, so much longer will a planet remain on a certain arc of the eccentric near δ than on an equal arc of the eccentric near ε.) That is, the farther a planet is from the Sun (at the point α), the slower it moves along its orbit, so a radius from the Sun to a planet passes through equal areas in equal times. However, as Kepler presented it, his argument is accurate only for circles, not ellipses.

- His "area law" is presented in: *"Caput LIX. Demonstratio, quod orbita Martis, ... , fiat perfecta ellipsis: ... "* (Chapter 59. Proof that Mars' orbit, ... , is a perfect ellipse: ...), Protheorema XIV and XV, pp. 291–295. On the top p. 294, it reads: *"Arcum ellipseos, cujus moras metitur area AKN, debere terminari in LK, ut sit AM."* (The arc of the ellipse, of which the duration is delimited [i.e., measured] by the area AKM, should be terminated in LK, so that it [i.e., the arc] is AM.) In other words, the time that Mars requires to move along an arc AM of its elliptical orbit is measured by the area of the segment AMN of the ellipse (where N is the position of the Sun), which in turn is proportional to the section AKN of the circle that encircles the ellipse and that is tangent to it. Therefore, the area that is swept out by a radius from the Sun to Mars as Mars moves along an arc of its elliptical orbit is proportional to the time that Mars requires to move along that arc. Thus, a radius from the Sun to Mars sweeps out equal areas in equal times.

In 1621, Kepler restated his second law for any planet: Johannes Kepler, *Epitome Astronomiae Copernicanae* [Summary of Copernican Astronomy] (Linz ("Lentiis ad Danubium"), (Austria): Johann Planck, 1622), book 5, page 668. From page 668: *"Dictum quidem est in superioribus, divisa orbita in particulas minutissimas æquales: accrescete iis moras planetæ per eas, in proportione intervallorum inter eas & Solem."* (It has been said above that, if the orbit of the planet is divided into the smallest equal parts, the times of the planet in them increase in the ratio of the distances between them and the sun.) That is, a planet's speed along its orbit is inversely proportional to its distance from the Sun. (The remainder of the paragraph makes clear that Kepler was referring to what is now called angular velocity.)

[12] Johannes Kepler, *Harmonices Mundi* [The Harmony of the World] (Linz, (Austria): Johann Planck, 1619), book 5, chapter 3, p. 189. From the bottom of p. 189: *"Sed res est certissima exactissimaque quod* proportio qua est inter binorum quorumcunque Planetarum tempora periodica, sit præcise sesquialtera proportionis *mediarum distantiarum,* ... " (But it is absolutely certain and exact that the *proportion between the periodic times of any two planets is precisely the sesquialternate proportion* [i.e., the ratio of 3:2] of their mean distances, ... ")

An English translation of Kepler's *Harmonices Mundi* is available as: Johannes Kepler with E.J. Aiton, A.M. Duncan, and J.V. Field, trans., *The Harmony of the World* (Philadelphia, Pennsylvania: American Philosophical Society, 1997); see especially p. 411.

[13] Wilbur Applebaum (13 June 2000). *Encyclopedia of the Scientific Revolution: From Copernicus to Newton*. Routledge. p. 603. ISBN 978-1-135-58255-5.

[14] Roy Porter (25 September 1992). *The Scientific Revolution in National Context*. Cambridge University Press. p. 102. ISBN 978-0-521-39699-8.

[15] Victor Guillemin; Shlomo Sternberg (2006). *Variations on a Theme by Kepler*. American Mathematical Soc. p. 5. ISBN 978-0-8218-4184-6.

[16] Burtt, Edwin. *The Metaphysical Foundations of Modern Physical Science*. p. 52.

[17] Gerald James Holton, Stephen G. Brush (2001). *Physics, the Human Adventure*. Rutgers University Press. p. 45. ISBN 0-8135-2908-5.

[18] MÜLLER, M (1995). "EQUATION OF TIME -- PROBLEM IN ASTRONOMY". Acta Physica Polonica A. Retrieved 23 February 2013.

26.10 Bibliography

- Kepler's life is summarized on pages 523–627 and Book Five of his *magnum opus*, *Harmonice Mundi* (*harmonies of the world*), is reprinted on pages 635–732 of *On the Shoulders of Giants*: The Great Works of Physics and Astronomy (works by Copernicus, Kepler, Galileo, Newton, and Einstein). Stephen Hawking, ed. 2002 ISBN 0-7624-1348-4

- A derivation of Kepler's third law of planetary motion is a standard topic in engineering mechanics classes. See, for example, pages 161–164 of Meriam, J. L. (1971) [1966]. "Dynamics, 2nd ed". New York: John Wiley. ISBN 0-471-59601-9..

- Murray and Dermott, Solar System Dynamics, Cambridge University Press 1999, ISBN 0-521-57597-4

- V.I. Arnold, Mathematical Methods of Classical Mechanics, Chapter 2. Springer 1989, ISBN 0-387-96890-3

26.11 External links

- B.Surendranath Reddy; animation of Kepler's laws: applet

- "Derivation of Kepler's Laws" (from Newton's laws) at *Physics Stack Exchange*.

- Crowell, Benjamin, *Conservation Laws*, http://www.lightandmatter.com/area1book2.html, an online book that gives a proof of the first law without the use of calculus. (see section 5.2, p. 112)

- David McNamara and Gianfranco Vidali, *Kepler's Second Law - Java Interactive Tutorial*, http://www.phy.syr.edu/courses/java/mc_html/kepler.html, an interactive Java applet that aids in the understanding of Kepler's Second Law.

- Audio - Cain/Gay (2010) Astronomy Cast Johannes Kepler and His Laws of Planetary Motion

- University of Tennessee's Dept. Physics & Astronomy: Astronomy 161 page on Johannes Kepler: The Laws of Planetary Motion

- Equant compared to Kepler: interactive model

- Kepler's Third Law:interactive model

- Solar System Simulator (Interactive Applet)

- Kepler and His Laws, educational web pages by David P. Stern

Chapter 27

Tsiolkovsky rocket equation

The **Tsiolkovsky rocket equation**, or **ideal rocket equation**, describes the motion of vehicles that follow the basic principle of a rocket: a device that can apply acceleration to itself (a thrust) by expelling part of its mass with high speed and thereby move due to the conservation of momentum. The equation relates the delta-v (the maximum change of velocity of the rocket if no other external forces act) with the effective exhaust velocity and the initial and final mass of a rocket (or other reaction engine).

For any such maneuver (or journey involving a number of such maneuvers):

$$\Delta v = v_e \ln \frac{m_0}{m_1}$$

where:

m_0 Is the initial total mass, including propellant, The mass measurements can be made in any unit form (kg, lb, tonnes, etc). This is because the ratios will still be the same.

m_1 Is the final total mass without propellant, also known as dry mass.

v_e Is the effective exhaust velocity,

Δv Is delta-v - the maximum change of velocity of the vehicle (with no external forces acting),

ln refers to the natural logarithm function.

(The equation can also be written using the specific impulse instead of the effective exhaust velocity by applying the formula $v_e = I_{sp} \cdot g_0$ where I_{sp} is the specific impulse expressed as a time period and g_0 is standard gravity ≈9.8 m/s^2.)

The equation is named after Russian scientist Konstantin Tsiolkovsky who independently derived it and published it in his 1903 work.[1] The equation had been derived earlier by the British mathematician William Moore in 1813.[2]

27.1 History

This equation was independently derived by Konstantin Tsiolkovsky towards the end of the 19th century and is sometimes known under his name, but more often simply referred to as 'the rocket equation' (or sometimes the 'ideal rocket equation'). However, a recently discovered pamphlet *"A Treatise on the Motion of Rockets"* by William Moore[2] shows that the earliest known derivation of this kind of equation was in fact at the Royal Military Academy at Woolwich in England in 1813,[3] and was used for weapons research.

While the derivation of the rocket equation is a straightforward calculus exercise, Tsiolkovsky is honored as being the first to apply it to the question of whether rockets could achieve speeds necessary for space travel.

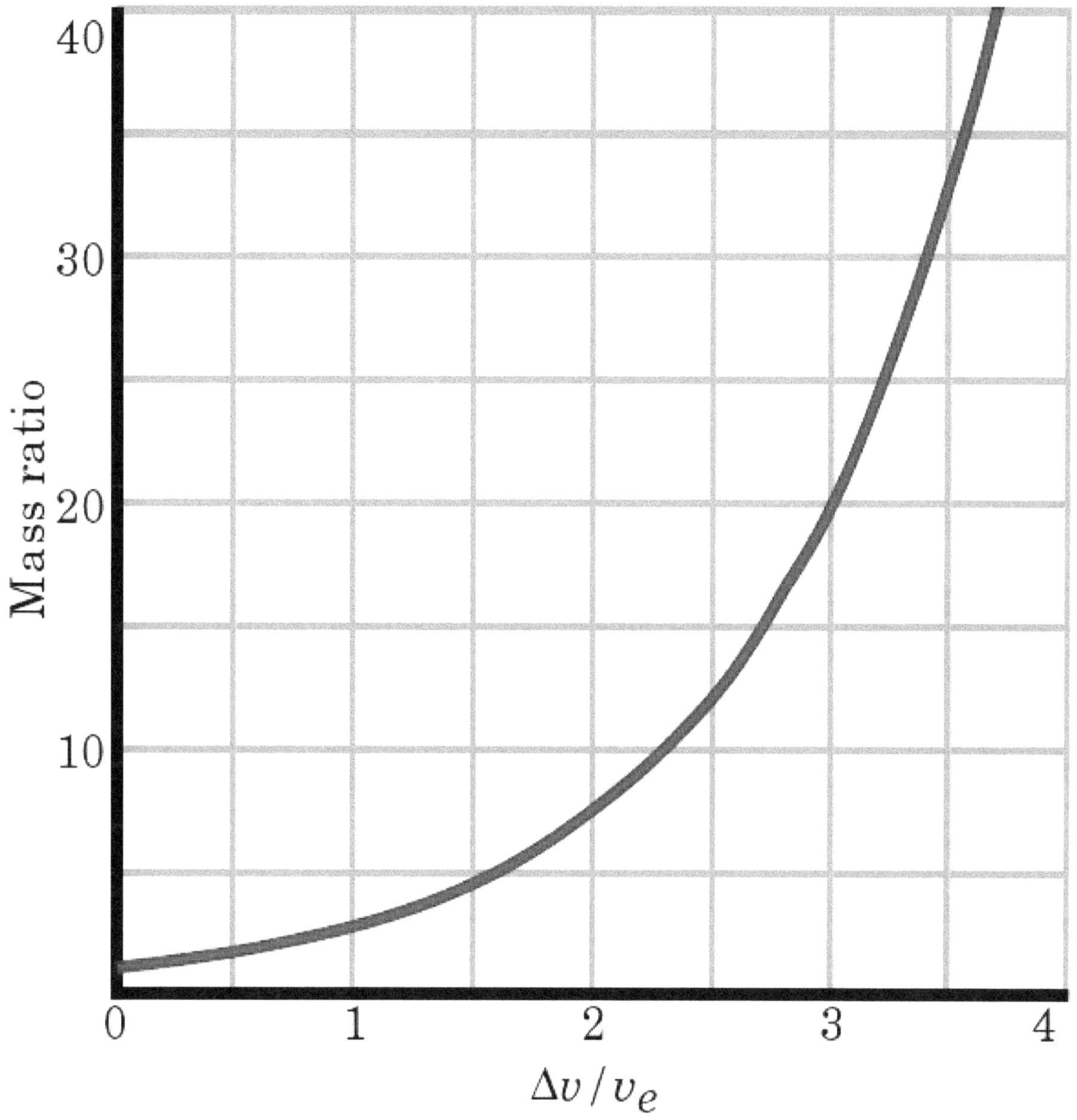

Rocket mass ratios versus final velocity calculated from the rocket equation.

27.2 Derivation

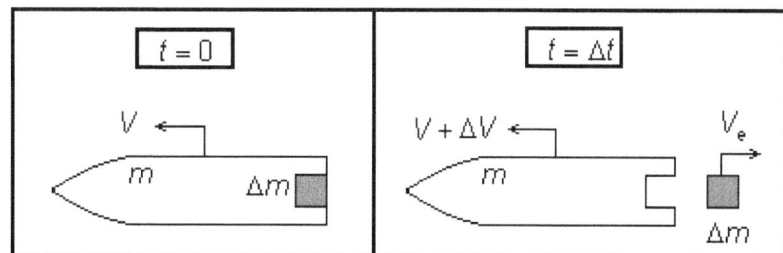

Consider the following system:

In the following derivation, "the rocket" is taken to mean "the rocket and all of its unburned propellant".

Newton's second law of motion relates external forces (F_i) to the change in linear momentum of the whole system

27.2. DERIVATION

(including rocket and exhaust) as follows:

$$\sum F_i = \lim_{\Delta t \to 0} \frac{P_2 - P_1}{\Delta t}$$

where P_1 is the momentum of the rocket at time $t=0$:

$$P_1 = (m + \Delta m) V$$

and P_2 is the momentum of the rocket and exhausted mass at time $t = \Delta t$:

$$P_2 = m(V + \Delta V) + \Delta m V_e$$

and where, with respect to the observer:

The velocity of the exhaust V_e in the observer frame is related to the velocity of the exhaust in the rocket frame v_e by (since exhaust velocity is in the negative direction)

$$V_e = V - v_e$$

Solving yields:

$$P_2 - P_1 = m \Delta V - v_e \Delta m$$

and, using $dm = -\Delta m$, since ejecting a positive Δm results in a decrease in mass,

$$\sum F_i = m \frac{dV}{dt} + v_e \frac{dm}{dt}$$

If there are no external forces then $\sum F_i = 0$ (conservation of linear momentum) and

$$m \frac{dV}{dt} = -v_e \frac{dm}{dt}$$

Assuming v_e is constant, this may be integrated to yield:

$$\Delta V = v_e \ln \frac{m_0}{m_1}$$

or equivalently

$$m_1 = m_0 e^{-\Delta V / v_e} \text{ or } m_0 = m_1 e^{\Delta V / v_e} \text{ or } m_0 - m_1 = m_1(e^{\Delta V / v_e} - 1)$$

where m_0 is the initial total mass including propellant, m_1 the final total mass, and v_e the velocity of the rocket exhaust with respect to the rocket (the specific impulse, or, if measured in time, that multiplied by gravity-on-Earth acceleration).

The value $m_0 - m_1$ is the total mass of propellant expended, and hence:

$$M_f = 1 - \frac{m_1}{m_0} = 1 - e^{-\Delta V/v_e}$$

where M_f is the propellant mass fraction (the part of the initial total mass that is spent as working mass).

ΔV (delta v) is the integration over time of the magnitude of the acceleration produced by using the rocket engine (what would be the actual acceleration if external forces were absent). In free space, for the case of acceleration in the direction of the velocity, this is the increase of the speed. In the case of an acceleration in opposite direction (deceleration) it is the decrease of the speed. Of course gravity and drag also accelerate the vehicle, and they can add or subtract to the change in velocity experienced by the vehicle. Hence delta-v is not usually the actual change in speed or velocity of the vehicle.

If special relativity is taken into account, the following equation can be derived for a relativistic rocket,[4] with Δv again standing for the rocket's final velocity (after burning off all its fuel and being reduced to a rest mass of m_1) in the inertial frame of reference where the rocket started at rest (with the rest mass including fuel being m_0 initially), and c standing for the speed of light in a vacuum:

$$\frac{m_0}{m_1} = \left[\frac{1+\frac{\Delta v}{c}}{1-\frac{\Delta v}{c}}\right]^{\frac{c}{2v_e}}$$

Writing $\frac{m_0}{m_1}$ as R, a little algebra allows this equation to be rearranged as

$$\frac{\Delta v}{c} = \frac{R^{\frac{2v_e}{c}} - 1}{R^{\frac{2v_e}{c}} + 1}$$

Then, using the identity $R^{\frac{2v_e}{c}} = \exp\left[\frac{2v_e}{c} \ln R\right]$ (here "exp" denotes the exponential function; *see also* Natural logarithm as well as the "power" identity at Logarithmic identities) and the identity $\tanh x = \frac{e^{2x}-1}{e^{2x}+1}$ (*see* Hyperbolic function), this is equivalent to

$$\Delta v = c \cdot \tanh\left(\frac{v_e}{c} \ln \frac{m_0}{m_1}\right)$$

27.3 Terms of the equation

27.3.1 Delta-*v*

Main article: Delta-v

Delta-*v* (literally "change in velocity"), symbolised as Δv and pronounced *delta-vee*, as used in spacecraft flight dynamics, is a measure of the impulse that is needed to perform a maneuver such as launch from, or landing on a planet or moon, or in-space orbital maneuver. It is a scalar that has the units of speed. As used in this context, it is *not* the same as the physical change in velocity of the vehicle.

Delta-*v* is produced by reaction engines, such as rocket engines and is proportional to the thrust per unit mass, and burn time, and is used to determine the mass of propellant required for the given manoeuvre through the rocket equation.

For multiple manoeuvres, delta-*v* sums linearly.

For interplanetary missions delta-*v* is often plotted on a porkchop plot which displays the required mission delta-*v* as a function of launch date.

27.3.2 Mass fraction

Main article: mass fraction

In aerospace engineering, the propellant mass fraction is the portion of a vehicle's mass which does not reach the destination, usually used as a measure of the vehicle's performance. In other words, the propellant mass fraction is the ratio between the propellant mass and the initial mass of the vehicle. In a spacecraft, the destination is usually an orbit, while for aircraft it is their landing location. A higher mass fraction represents less weight in a design. Another related measure is the payload fraction, which is the fraction of initial weight that is payload.

27.3.3 Effective exhaust velocity

Main article: effective exhaust velocity

The effective exhaust velocity is often specified as a specific impulse and they are related to each other by:

$$v_e = g_0 I_{sp},$$

where

I_{sp}

- v_e is the specific impulse measured in m/s, which is the same as the effective exhaust velocity measured in m/s (or ft/s if g is in ft/s^2),

- g_0 is the acceleration due to gravity at the Earth's surface, 9.81 m/s^2 (in Imperial units 32.2 ft/s^2).

27.4 Applicability

The rocket equation captures the essentials of rocket flight physics in a single short equation. It also holds true for rocket-like reaction vehicles whenever the effective exhaust velocity is constant, and can be summed or integrated when the effective exhaust velocity varies. The rocket equation only accounts for the reaction force from the rocket engine; it does not include other forces that may act on a rocket, such as aerodynamic or gravitational forces. As such, when using it to calculate the propellant requirement for launch from (or powered descent to) a planet with an atmosphere, the effects of these forces must be included in the delta-V requirement (see Examples below). The equation does not apply to non-rocket systems such as aerobraking, gun launches, space elevators, launch loops, or tether propulsion.

The rocket equation can be applied to orbital maneuvers in order to determine how much propellant is needed to change to a particular new orbit, or to find the new orbit as the result of a particular propellant burn. When applying to orbital maneuvers, one assumes an impulsive maneuver, in which the propellant is discharged and delta-v applied instantaneously. This assumption is relatively accurate for short-duration burns such as for mid-course corrections and orbital insertion maneuvers. As the burn duration increases, the result is less accurate due to the effect of gravity on the vehicle over the duration of the maneuver. For low-thrust, long duration propulsion, such as electric propulsion, more complicated analysis based on the propagation of the spacecraft's state vector and the integration of thrust are used to predict orbital motion.

27.5 Examples

Assume an exhaust velocity of 4,500 meters per second (15,000 ft/s) and a Δv of 9,700 meters per second (32,000 ft/s) (Earth to LEO, including Δv to overcome gravity and aerodynamic drag).

- Single-stage-to-orbit rocket: $1 - e^{-9.7/4.5} = 0.884$, therefore 88.4% of the initial total mass has to be propellant. The remaining 11.6% is for the engines, the tank, and the payload. In the case of a space shuttle, it would also include the orbiter.

- Two-stage-to-orbit: suppose that the first stage should provide a Δv of 5,000 meters per second (16,000 ft/s); $1 - e^{-5.0/4.5} = 0.671$, therefore 67.1% of the initial total mass has to be propellant to the first stage. The remaining mass is 32.9%. After disposing of the first stage, a mass remains equal to this 32.9%, minus the mass of the tank and engines of the first stage. Assume that this is 8% of the initial total mass, then 24.9% remains. The second stage should provide a Δv of 4,700 meters per second (15,000 ft/s); $1 - e^{-4.7/4.5} = 0.648$, therefore 64.8% of the remaining mass has to be propellant, which is 16.2%, and 8.7% remains for the tank and engines of the second stage, the payload, and in the case of a space shuttle, also the orbiter. Thus together 16.7% is available for all engines, the tanks, the payload, and the possible orbiter.

27.6 Stages

In the case of sequentially thrusting rocket stages, the equation applies for each stage, where for each stage the initial mass in the equation is the total mass of the rocket after discarding the previous stage, and the final mass in the equation is the total mass of the rocket just before discarding the stage concerned. For each stage the specific impulse may be different.

For example, if 80% of the mass of a rocket is the fuel of the first stage, and 10% is the dry mass of the first stage, and 10% is the remaining rocket, then

$$\Delta v = v_e \ln \frac{100}{100 - 80}$$
$$= v_e \ln 5$$
$$= 1.61 v_e.$$

With three similar, subsequently smaller stages with the same v_e for each stage, we have

$$\Delta v = 3 v_e \ln 5 = 4.83 v_e$$

and the payload is 10%*10%*10% = 0.1% of the initial mass.

A comparable SSTO rocket, also with a 0.1% payload, could have a mass of 11.1% for fuel tanks and engines, and 88.8% for fuel. This would give

$$\Delta v = v_e \ln(100/11.2) = 2.19 v_e.$$

If the motor of a new stage is ignited before the previous stage has been discarded and the simultaneously working motors have a different specific impulse (as is often the case with solid rocket boosters and a liquid-fuel stage), the situation is more complicated.

27.7 Common misconceptions

When viewed as a variable-mass system, a rocket cannot be directly analyzed with Newton's second law of motion because the law is valid for constant-mass systems only.[5][6][7] It can cause confusion that the Tsiolkovsky rocket equation looks similar to the relativistic force equation $F = dp/dt = m \; dv/dt + v \; dm/dt$. Using this formula with $m(t)$ as the varying mass of the rocket seems to derive Tsiolkovsky rocket equation, but this derivation is not correct. Notice that the effective exhaust velocity v_e doesn't even appear in this formula.

27.8 See also

- Delta-v budget
- Oberth effect applying delta-v in a gravity well increases the final velocity
- Spacecraft propulsion
- Mass ratio
- Working mass
- Relativistic rocket
- Reversibility of orbits
- Variable-mass systems

27.9 References

[1] К. Э. Циолковский, Исследование мировых пространств реактивными приборами, 1903. It is available online here in a RARed PDF

[2] Moore, William; of the Military Academy at Woolwich (1813). *A Treatise on the Motion of Rockets. To which is added, An Essay on Naval Gunnery*. London: G. and S. Robinson.

[3] Johnson, W. (1995). "Contents and commentary on William Moore's a treatise on the motion of rockets and an essay on naval gunnery". *International Journal of Impact Engineering* **16** (3): 499–521. doi:10.1016/0734-743X(94)00052-X. ISSN 0734-743X.

[4] Forward, Robert L. "A Transparent Derivation of the Relativistic Rocket Equation" (see the right side of equation 15 on the last page, with R as the ratio of initial to final mass and w as the exhaust velocity, corresponding to v_e in the notation of this article)

[5] Plastino, Angel R.; Muzzio, Juan C. (1992). "On the use and abuse of Newton's second law for variable mass problems". *Celestial Mechanics and Dynamical Astronomy* (Netherlands: Kluwer Academic Publishers) **53** (3): 227–232. Bibcode:1992CeMDA..53..227P. doi:10.1007/BF00052611. ISSN 0923-2958. "We may conclude emphasizing that Newton's second law is valid for constant mass only. When the mass varies due to accretion or ablation, [an alternate equation explicitly accounting for the changing mass] should be used."

[6] Halliday; Resnick. *Physics* **1**. p. 199. ISBN 0-471-03710-9. It is important to note that we *cannot* derive a general expression for Newton's second law for variable mass systems by treating the mass in $\mathbf{F} = d\mathbf{P}/dt = d(M\mathbf{v})$ as a *variable*. [...] We *can* use $\mathbf{F} = d\mathbf{P}/dt$ to analyze variable mass systems *only* if we apply it to an *entire system of constant mass* having parts among which there is an interchange of mass. [Emphasis as in the original]

[7] Kleppner, Daniel; Robert Kolenkow (1973). *An Introduction to Mechanics*. McGraw-Hill. pp. 133–134. ISBN 0-07-035048-5. Recall that $\mathbf{F} = d\mathbf{P}/dt$ was established for a system composed of a certain set of particles[. ... I]t is essential to deal with the same set of particles throughout the time interval[. ...] Consequently, the mass of the system can not change during the time of interest.

27.10 External links

- How to derive the rocket equation
- Relativity Calculator - Learn Tsiolkovsky's rocket equations
- Tsiolkovsky's rocket equations plot and calculator

Chapter 28

Vis-viva equation

In astrodynamics, the *vis-viva equation*, also referred to as **orbital-energy-invariance law**, is one of the equations that model the motion of orbiting bodies. It is the direct result of the law of conservation of energy, which requires that the sum of kinetic and potential energy is constant at all points along the orbit.

Vis viva (Latin for "live force") is a term from the history of mechanics, and it survives in this sole context. It represents the principle that the difference between the aggregate work of the accelerating forces of a system and that of the retarding forces is equal to one half the *vis viva* accumulated or lost in the system while the work is being done.

28.1 Equation

For any Kepler orbit (elliptic, parabolic, hyperbolic, or radial), the *vis-viva* equation[1] is as follows:

$$v^2 = GM\left(\frac{2}{r} - \frac{1}{a}\right)$$

where:

- v is the relative speed of the two bodies
- r is the distance between the two bodies
- a is the semi-major axis ($a > 0$ for ellipses, $a = \infty$ or $1/a = 0$ for parabolas, and $a < 0$ for hyperbolas)
- G is the gravitational constant
- M is the mass of the central body

The product of GM can also be expressed using the Greek letter μ.

28.2 Derivation for Elliptic Orbits ($0 \leq$ eccentricity < 1)

In the vis-viva equation the mass m of the orbiting body (e.g., a spacecraft) is taken to be negligible in comparison to the mass M of the central body (e.g., the Earth). In the specific cases of an elliptical or circular orbit, the vis-viva equation may be readily derived from conservation of energy and momentum.

Specific total energy is constant throughout the orbit. Thus, using the subscripts a and p to denote apoapsis (apogee) and periapsis (perigee), respectively,

28.2. DERIVATION FOR ELLIPTIC ORBITS ($0 \leq$ ECCENTRICITY < 1)

$$\varepsilon = \frac{v_a^2}{2} - \frac{GM}{r_a} = \frac{v_p^2}{2} - \frac{GM}{r_p}$$

Rearranging,

$$\frac{v_a^2}{2} - \frac{v_p^2}{2} = \frac{GM}{r_a} - \frac{GM}{r_p}$$

Recalling that for an elliptical orbit (and hence also a circular orbit) the velocity and radius vectors are perpendicular at apoapsis and periapsis, conservation of angular momentum requires $h = r_p v_p = r_a v_a =$ constant, thus $v_p = \frac{r_a}{r_p} v_a$:

$$\frac{1}{2}\left(1 - \frac{r_a^2}{r_p^2}\right) v_a^2 = \frac{GM}{r_a} - \frac{GM}{r_p}$$

$$\frac{1}{2}\left(\frac{r_p^2 - r_a^2}{r_p^2}\right) v_a^2 = \frac{GM}{r_a} - \frac{GM}{r_p}$$

Isolating the kinetic energy at apoapsis and simplifying,

$$\frac{1}{2} v_a^2 = \left(\frac{GM}{r_a} - \frac{GM}{r_p}\right) \cdot \frac{r_p^2}{r_p^2 - r_a^2}$$

$$\frac{1}{2} v_a^2 = GM \left(\frac{r_p - r_a}{r_a r_p}\right) \frac{r_p^2}{r_p^2 - r_a^2}$$

$$\frac{1}{2} v_a^2 = GM \frac{r_p}{r_a(r_p + r_a)}$$

From the geometry of an ellipse, $2a = r_p + r_a$ where a is the length of the semimajor axis. Thus,

$$\frac{1}{2} v_a^2 = GM \frac{2a - r_a}{r_a(2a)}$$

Substituting this into our original expression for specific orbital energy,

$$\varepsilon = \frac{v_a^2}{2} - \frac{GM}{r_a} = GM \frac{2a - r_a}{2a r_a} - \frac{GM}{r_a}$$

$$\varepsilon = GM \left(\frac{2a - r_a}{2a r_a} - \frac{1}{r_a}\right) = -\frac{GM}{2a}$$

Thus, $\varepsilon = -\frac{GM}{2a}$ and the vis-viva equation may be written

$$\frac{v^2}{2} - \frac{GM}{r} = -\frac{GM}{2a}$$

or

$$v^2 = GM \left(\frac{2}{r} - \frac{1}{a}\right).$$

Therefore, the conserved angular momentum $L = mh$ is derived below using

$$r_a = a(1+e)$$
$$r_p = a(1-e)$$
$$r_a r_p = b^2$$

where a, b, e and m are semi-major axis, semi-minor axis, eccentricity and mass

$$v_a^2 = GM\left(\frac{2}{r_a} - \frac{1}{a}\right) = \frac{GM}{a}\left(\frac{2}{1+e} - \frac{1}{1}\right) = \frac{GM}{a} \cdot \frac{1-e}{1+e} = \frac{GM}{a} \cdot \frac{r_p}{r_a} = \frac{GM}{a} \cdot \frac{b^2}{r_a^2}$$

$$L = mh = v_a r_a m = mb\sqrt{\frac{GM}{a}}$$

28.3 Practical applications

Given the total mass and the scalars r and v at a single point of the orbit, one can compute r and v at any other point in the orbit.[2]

Given the total mass and the scalars r and v at a single point of the orbit, one can compute the specific orbital energy ε, allowing an object orbiting a larger object to be classified as having not enough energy to remain in orbit, hence being "suborbital" (a ballistic missile, for example), having enough energy to be "orbital", but without the possibility to complete a full orbit anyway because it eventually collides with the other body, or having enough energy to come from and/or go to infinity (as a meteor, for example).

28.4 References

[1] Tom Logsdon, Orbital Mechanics: theory and applications, John Wiley & Sons, 1998

[2] For the three-body problem there is hardly a comparable vis-viva equation: conservation of energy reduces the larger number of degrees of freedom by only one.

Chapter 29

Payload fraction

In aerospace engineering, **payload fraction** is a common term used to characterize the efficiency of a particular design. Payload fraction is calculated by dividing the weight of the payload by the takeoff weight of aircraft. Fuel represents a considerable amount of the overall takeoff weight, and for shorter trips it is quite common to load less fuel in order to carry a lighter load. For this reason the **useful load fraction** calculates a similar number, but based on the combined weight of the payload and fuel together.

Propeller-driven airliners had useful load fractions on the order of 25-35%. Modern jet airliners have considerably higher useful load fractions, on the order of 45-55%.

For spacecraft the payload fraction is often less than 1%, while the useful load fraction is perhaps 90%. In this case the useful load fraction is not a useful term, because spacecraft typically can't reach orbit without a full fuel load. For this reason the related term mass fraction, is used instead. However, if the latter is large, the payload can only be small.

29.1 Examples

Note: the above table may incorrectly include the mass of the empty upper stage or stages.

29.2 References

[1] Astronautix- Ariane 5g

[2] Astronautix - Saturn V

[3] Astronautix- Saturn IB

[4] Astronautix-V-2

[5] AIAA2001-4619 RLVs

Chapter 30

Propellant mass fraction

See also: Mass fraction (chemistry)

In aerospace engineering, the **propellant mass fraction** is the portion of a vehicle's mass which does not reach the destination, usually used as a measure of the vehicle's performance. In other words, the **propellant mass fraction** is the ratio between the propellant mass and the initial mass of the vehicle. In a spacecraft, the destination is usually an orbit, while for aircraft it is their landing location. A higher mass fraction represents less weight in a design. Another related measure is the payload fraction, which is the fraction of initial weight that is payload.

30.1 Formulation

The propellant mass fraction is given by:

$$\zeta = \frac{m_p}{m_0}$$

And because,

$$m_0 = m_f + m_p$$

it follows that:

$$\zeta = \frac{m_0 - m_f}{m_0} = \frac{m_p}{m_p + m_f} = 1 - \frac{m_f}{m_0}$$

Where:

ζ is the propellant mass fraction
m_p is the propellant mass
m_0 is the initial mass of the vehicle
m_f is the final mass of the vehicle

30.2 Significance

In rockets for a given target orbit, a rocket's mass fraction is the portion of the rocket's pre-launch mass (fully fueled) that does not reach orbit. The propellant mass fraction is the ratio of just the propellant to the entire mass of the vehicle at takeoff (propellant plus dry mass). In the cases of a single stage to orbit (SSTO) vehicle or suborbital vehicle, the mass fraction equals the propellant mass fraction; simply the fuel mass divided by the mass of the full spaceship. A rocket employing staging, which are the only designs to have reached orbit, has a mass fraction higher than the propellant mass fraction because parts of the rocket itself are dropped off en route. Propellant mass fractions are typically around 0.8 to 0.9.

In aircraft, mass fraction is related to range, an aircraft with a higher mass fraction can go farther. Aircraft mass fractions are typically around 0.5.

When applied to a rocket as a whole, a low mass fraction is desirable, since it indicates a greater capability for the rocket to deliver payload to orbit for a given amount of fuel. Conversely, when applied to a single stage, where the propellant mass fraction calculation doesn't include the payload, a higher propellant mass fraction corresponds to a more efficient design, since there is less non-propellant mass. Without the benefit of staging, SSTO designs are typically designed for mass fractions around 0.9. Staging increases the payload fraction, which is one of the reasons SSTO's appear difficult to build.

For example, the complete Space Shuttle system has:[1]

- fueled weight at liftoff: 1,708,500 kg
- dry weight at liftoff: 342,100 kg

Given these numbers, the propellant mass fraction is $1 - (342,100/1,708,500) = 0.7998$.

The mass fraction plays an important role in the rocket equation:

$$\Delta v = -v_e \ln \frac{m_f}{m_0}$$

Where m_f/m_0 is the ratio of final mass to initial mass (i.e., one minus the mass fraction), Δv is the change in the vehicle's velocity as a result of the fuel burn and v_e is the effective exhaust velocity (see below).

The term effective exhaust velocity is defined as:

$$v_e = g_n I_{sp}$$

where I_{sp} is the fuel's specific impulse in seconds and *gn* is the *standard acceleration of gravity* (note that this is not the local acceleration of gravity).

To make a powered landing from orbit on a celestial body without an atmosphere requires the same mass reduction as reaching orbit from its surface, if the speed at which the surface is reached is zero.

30.3 See also

- Fuel fraction
- Mass ratio

30.4 References

[1] Typical propellant mass fractions

Chapter 31

Mass ratio

In aerospace engineering, **mass ratio** is a measure of the efficiency of a rocket. It describes how much more massive the vehicle is with propellant than without; that is, the ratio of the rocket's *wet mass* (vehicle plus contents plus propellant) to its *dry mass* (vehicle plus contents). A more efficient rocket design requires less propellant to achieve a given goal, and would therefore have a lower mass ratio; however, for any given efficiency a higher mass ratio typically permits the vehicle to achieve higher delta-v.

The mass ratio is a useful quantity for back-of-the-envelope rocketry calculations: it is an easy number to derive from either Δv or from rocket and propellant mass, and therefore serves as a handy bridge between the two. It is also a useful for getting an impression of the size of a rocket: while two rockets with mass fractions of, say, 92% and 95% may appear similar, the corresponding mass ratios of 12.5 and 20 clearly indicate that the latter system requires much more propellant.

Typical multistage rockets have mass ratios in the range from 8 to 20. The Space Shuttle, for example, has a mass ratio around 16.

31.1 Derivation

The definition arises naturally from Tsiolkovsky's rocket equation:

$$\Delta v = v_e \ln \frac{m_0}{m_1}$$

where

- Δv is the desired change in the rocket's velocity
- ve is the effective exhaust velocity (see specific impulse)
- m_0 is the initial mass (rocket plus contents plus propellant)
- m_1 is the final mass (rocket plus contents)

This equation can be rewritten in the following equivalent form:

$$\frac{m_0}{m_1} = e^{\Delta v / v_e}$$

The fraction on the left-hand side of this equation is the rocket's mass ratio by definition.

This equation indicates that a Δv of n times the exhaust velocity requires a mass ratio of e^n. For instance, for a vehicle to achieve a Δv of 2.5 times its exhaust velocity would require a mass ratio of $e^{2.5}$ (approximately 12.2). One could say that a "velocity ratio" of n requires a mass ratio of e^n.

Sutton defines the mass ratio inversely as:[1]

$$M_R = \frac{m_1}{m_0}$$

In this case, the values for mass fraction are always less than 1.

31.2 See also

- Rocket fuel
- Propellant mass fraction
- Payload fraction

31.3 References

Zubrin, Robert (1999). *Entering Space: Creating a Spacefaring Civilization*. Tarcher/Putnam. ISBN 0-87477-975-8.

[1] Rocket Propulsion Elements, 7th Edition by George P. Sutton, Oscar Biblarz

Chapter 32

Apsis

"Apogee" redirects here. For the literary journal, see Perigee: Publication for the Arts. For other uses, see Apogee (disambiguation). For the architectural term, see Apse.
Not to be confused with Aspis.

The **apsis** (Greek ἀψίς), plural **apsides** (/ˈæpsɪdiːz/; Greek: ἀψίδες) is an extreme point in an object's orbit. For elliptic

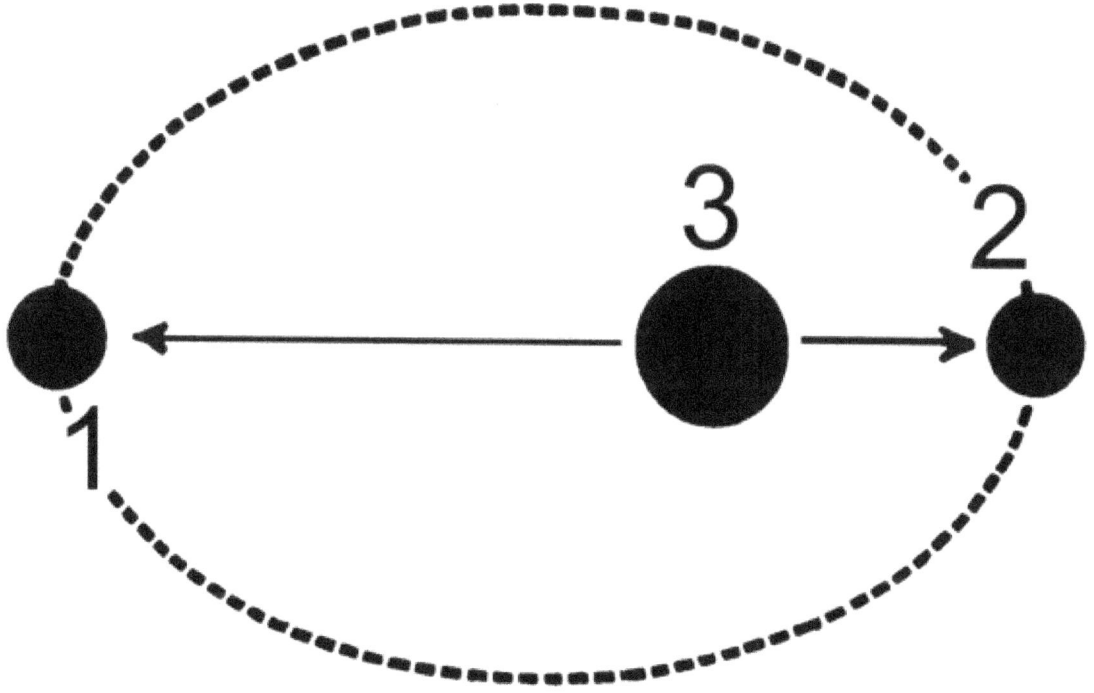

Apsides:
Apocenter (1), Pericenter (2) and Focus (3).

orbits about a larger body, there are two apsides, named with the prefixes *peri-* (from περί *(peri)*, meaning "near") and *ap-*, or *apo-*, (from ἀπό *(apó)*, meaning "away from") added to a description of the thing being orbited.

- For a body orbiting the Sun the point of least distance is the **perihelion** (/ˌpɛriˈhiːliən/) and the point of greatest distance is the **aphelion** (/əˈfiːliən/);[1]

- The terms become **periastron** and **apastron** when discussing orbits around other stars.

- For any satellite of Earth including the Moon the point of least distance is the **perigee** (/'pɛrɪdʒiː/) and greatest distance the **apogee**.

- For any orbits around a "center" of mass, there are the terms **pericenter** and **apocenter**. **Periapsis** and **apoapsis** (or **apapsis**) are equivalent alternatives.

A straight line connecting the pericenter and apocenter is the *line of apsides*. This is the major axis of the ellipse, its greatest diameter. For a two-body system the center of mass of the system lies on this line at one of the two foci of the ellipse. When one body is sufficiently larger than the other it may be taken to be at this focus. However whether or not this is the case, both bodies are in similar elliptical orbits each having one focus at the system's center of mass, with their respective lines of apsides being of length inversely proportional to their masses. Historically, in geocentric systems, apsides were measured from the center of the Earth. However, in the case of the Moon, the center of mass of the Earth–Moon system or Earth–Moon barycenter, as the common focus of both the Moon's and Earth's orbits about each other, is about 74% of the way from Earth's center to its surface.

In orbital mechanics, the apsis technically refers to the distance measured between the centers of mass of the central and orbiting body. However, in the case of spacecraft, the family of terms are commonly used to describe the orbital altitude of the spacecraft from the surface of the central body (assuming a constant, standard reference radius).

32.1 Mathematical formulae

These formulae characterize the pericenter and apocenter of an orbit:

- Pericenter: maximum speed $v_{\text{per}} = \sqrt{\frac{(1+e)\mu}{(1-e)a}}$ at minimum (pericenter) distance $r_{\text{per}} = (1-e)a$

- Apocenter: minimum speed $v_{\text{ap}} = \sqrt{\frac{(1-e)\mu}{(1+e)a}}$ at maximum (apocenter) distance $r_{\text{ap}} = (1+e)a$

while, in accordance with Kepler's laws of planetary motion (based on the conservation of angular momentum) and the conservation of energy, these two quantities are constant for a given orbit:

- specific relative angular momentum $h = \sqrt{(1-e^2)\mu a}$
- specific orbital energy $\epsilon = -\frac{\mu}{2a}$

where:

- a is the semi-major axis, equal to $\frac{r_{\text{per}} + r_{\text{ap}}}{2}$

- μ is the standard gravitational parameter

- e is the eccentricity, defined as $e = \frac{r_{\text{ap}} - r_{\text{per}}}{r_{\text{ap}} + r_{\text{per}}} = 1 - \frac{2}{\frac{r_{\text{ap}}}{r_{\text{per}}} + 1}$

Note that for conversion from heights above the surface to distances between an orbit and its primary, the radius of the central body has to be added, and conversely.

The arithmetic mean of the two limiting distances is the length of the semi-major axis a. The geometric mean of the two distances is the length of the semi-minor axis b.

The geometric mean of the two limiting speeds is $\sqrt{-2\epsilon} = \sqrt{\mu/a}$ which is the speed of a body in a circular orbit whose radius is a.

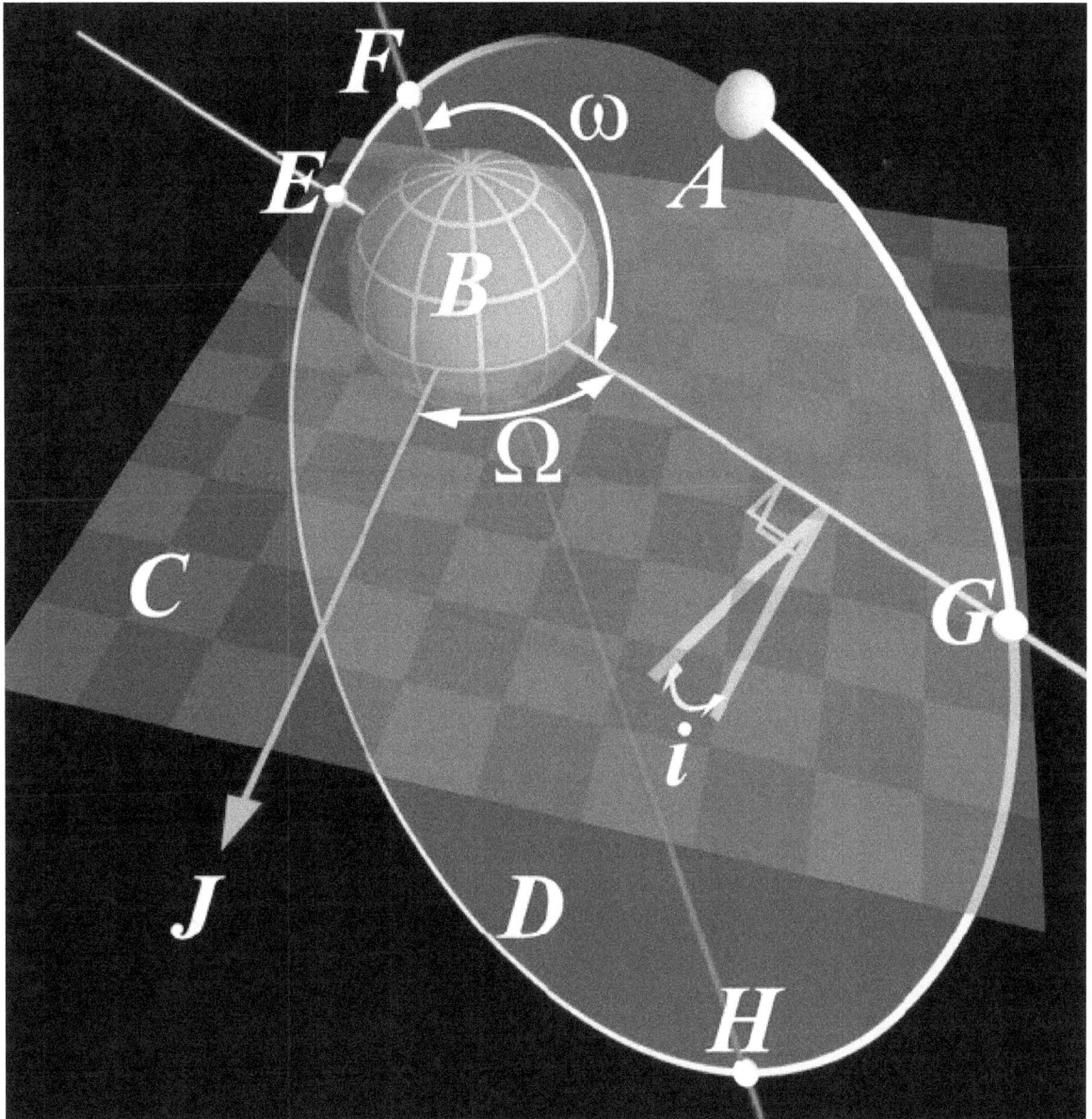

Keplerian orbital elements: point F is at the pericenter, point H is at the apocenter, and the red line between them is the line of apsides

32.2 Terminology

The words "pericenter" and "apocenter" are often seen, although periapsis/apoapsis are preferred in technical usage.

Various related terms are used for other celestial objects. The '-gee', '-helion' and '-astron' and '-galacticon' forms are frequently used in the astronomical literature when referring to the Earth, Sun, stars and the Galactic centre respectively. -jove is occasionally used, while '-saturnium' has very rarely been used in the last 50 years. The '-gee' form is commonly used as a generic 'closest approach to planet' term instead of specifically applying to the Earth. During the Apollo program, the terms *pericynthion* and *apocynthion* (referencing Cynthia, an alternative name for the Greek Moon goddess Artemis) were used when referring to the Moon.[2] Regarding black holes, the term *peri/apomelasma* (from a Greek root) was used by physicist Geoffrey A. Landis in 1998 before *peri/aponigricon* (from a Latin) appeared in the scientific literature in 2002.[3]

32.2.1 Terminology graph

The following suffixes are added to peri- and apo- to form the terms for the nearest and farthest orbital distances from these objects.

32.3 Perihelion and aphelion of the Earth

For the orbit of the Earth around the Sun, the time of apsis is often expressed in terms of a time relative to seasons, since this determines the contribution of the elliptical orbit to seasonal variations. The variation of the seasons is primarily controlled by the annual cycle of the elevation angle of the Sun, which is a result of the tilt of the axis of the Earth measured from the plane of the ecliptic. The Earth's eccentricity and other orbital elements are not constant, but vary slowly due to the perturbing effects of the planets and other objects in the solar system. See Milankovitch cycles.

Currently, the Earth reaches perihelion in early January, approximately 14 days after the December Solstice. At perihelion, the Earth's center is about 0.98329 astronomical units (AU) or 147,098,070 kilometers (about 91,402,500 miles) from the Sun's center.

The Earth reaches aphelion currently in early July, approximately 14 days after the June Solstice. The aphelion distance between the Earth's and Sun's centers is currently about 1.01671 AU or 152,097,700 kilometers (94,509,100 mi).

On a very long time scale, the dates of the perihelion and of the aphelion progress through the seasons, and they make one complete cycle in 22,000 to 26,000 years. There is a corresponding movement of the position of the stars as seen from Earth that is called the apsidal precession. (This is closely related to the precession of the axis.)

Astronomers commonly express the timing of perihelion relative to the vernal equinox not in terms of days and hours, but rather as an angle of orbital displacement, the so-called longitude of the pericenter. For the orbit of the Earth, this is called the *longitude of perihelion*, and in 2000 was about 282.895 degrees. By the year 2010, this had advanced by a small fraction of a degree to about 283.067 degrees.[4]

The dates and times of the perihelions and aphelions for several past and future years are listed in the following table:[5]

The dates and times of the perihelions and aphelions vary much more than those of the equinoxes and solstices due to the presence of the Moon. Because perihelion and aphelion are defined by the distance between the center of the Sun and the center of the Earth, the Earth's position in its monthly motion around the Earth–Moon barycenter greatly affects the time when the Earth is at its shortest or longest distance from the Sun.

32.4 Planetary perihelion and aphelion

The following table shows the distances of the planets and dwarf planets from the Sun at their perihelion and aphelion.[6]

The following chart shows the range of distances of the planets, dwarf planets and Halley's Comet from the Sun.

The images below show the perihelion (green dot) and aphelion (red dot) points of the inner and outer planets.

- Perihelion and aphelion points
- The perihelion and aphelion points of the inner planets of the Solar System
- The perihelion and aphelion points of the outer planets of the Solar System

32.5 See also

- Apsidal precession
- Eccentric anomaly

- Elliptic orbit
- Perifocal coordinate system
- Solstice

32.6 References

[1] This is the pronunciation given in the Oxford English Dictionary and other major English dictionaries, and is based on the usual pronunciation of scientific words derived from Greek. However, the pronunciation /æpˈhiːliən/ is often met, and is based on the misunderstanding that the elements ἀπό + ἥλιος remain separate.

[2] "Apollo 15 Mission Report". *Glossary*. Retrieved October 16, 2009.

[3] R. Schodel, T. Ott, R. Genzel, R. Hofmann, M. Lehnert, A. Eckart, N. Mouawad, T. Alexander, M.J. Reid, R. Lenzen, M. Hartung, F. Lacombe, D. Rouan, E. Gendron, G. Rousset, A.-M. Lagrange, W. Brandner, N. Ageorges, C. Lidman, A.F.M. Moorwood, J. Spyromilio, N. Hubin, and K.M. Menten, "Closest Star Seen Orbiting the Supermassive Black Hole at the Centre of the Milky Way," *Nature* 419, 694–696 (17 October 2002), doi:10.1038/nature01121.

[4] NASA.gov

[5] "Solex by Aldo Vitagliano". Retrieved 2012-07-09. (calculated by Solex 11)

[6] NASA planetary comparison chart http://solarsystem.nasa.gov/planets/compchart.cfm

32.7 External links

- Apogee – Perigee Photographic Size Comparison, perseus.gr
- Aphelion – Perihelion Photographic Size Comparison, perseus.gr
- Earth's Seasons: Equinoxes, Solstices, Perihelion, and Aphelion, 2000–2020, usno.navy.mil

Chapter 33

Eccentricity vector

In celestial mechanics, the **eccentricity vector** of a Kepler orbit is the vector that points towards the periapsis and has a magnitude equal to the orbit's scalar eccentricity. The magnitude is unitless. For Kepler orbits the **eccentricity vector** is a constant of motion. Its main use is in the analysis of almost circular orbits, as perturbing (non-Keplerian) forces on an actual orbit will cause the osculating eccentricity vector to change continuously. For the eccentricity and argument of periapsis parameters, eccentricity zero (circular orbit) corresponds to a singularity.

33.1 Calculation

The **eccentricity vector e** is: [1]

$$\mathbf{e} = \frac{\mathbf{v} \times \mathbf{h}}{\mu} - \frac{\mathbf{r}}{|\mathbf{r}|} = \left(\frac{|\mathbf{v}|^2}{\mu} - \frac{1}{|\mathbf{r}|} \right) \mathbf{r} - \frac{\mathbf{r} \cdot \mathbf{v}}{\mu} \mathbf{v}$$

which follows immediately from the vector identity:

$$\mathbf{v} \times (\mathbf{r} \times \mathbf{v}) = (\mathbf{v} \cdot \mathbf{v})\mathbf{r} - (\mathbf{r} \cdot \mathbf{v})\mathbf{v}$$

where:

- **v** is velocity vector
- **h** is specific angular momentum vector (equal to $\mathbf{r} \times \mathbf{v}$)
- **r** is position vector
- μ is standard gravitational parameter

33.2 See also

- Kepler orbit
- Orbit
- Eccentricity
- Laplace–Runge–Lenz vector

33.3 References

[1] Cordani, Bruno (2003). *The Kepler Problem*. Birkhaeuser. p. 22. ISBN 3-7643-6902-7.

Chapter 34

Non-inclined orbit

A **non-inclined orbit** is an orbit which is contained in the plane of reference. The inclination is 0 for prograde orbits, and π (180°) for retrograde orbits. If the plane of reference is the equator, these orbits are called **equatorial**; if the plane of reference is ecliptic, they are called **ecliptic**. As these orbits lack nodes, the ascending node is usually taken to lie in the reference direction (usually the vernal equinox), and thus the longitude of the ascending node is taken to be zero. Also, the argument of periapsis is undefined.

34.1 See also

- List of orbits

Chapter 35

Euler angles

This article is about the Euler angles used in mathematics. For the use of the term in physics and aerospace engineering, see Rigid body dynamics. For chained rotations, see chained rotations.

The **Euler angles** are three angles introduced by Leonhard Euler to describe the orientation of a rigid body.[1] To describe such an orientation in 3-dimensional Euclidean space three parameters are required. They can be given in several ways, Euler angles being one of them; see charts on SO(3) for others. Euler angles are also used to describe the orientation of a frame of reference (typically, a coordinate system or basis) relative to another. They are typically denoted as α, β, γ, or φ, θ, ψ.

Euler angles represent a sequence of three *elemental rotations*, i.e. rotations about the axes of a coordinate system. For instance, a first rotation about z by an angle α, a second rotation about x by an angle β, and a last rotation again about z, by an angle γ. These rotations start from a known standard orientation. In physics, this standard initial orientation is typically represented by a motionless (*fixed*, *global*, or *world*) coordinate system; in linear algebra, by a standard basis.

Any orientation can be achieved by composing three elemental rotations. The elemental rotations can either occur about the axes of the fixed coordinate system (extrinsic rotations) or about the axes of a rotating coordinate system, which is initially aligned with the fixed one, and modifies its orientation after each elemental rotation (intrinsic rotations). The rotating coordinate system may be imagined to be rigidly attached to a rigid body. In this case, it is sometimes called a *local* coordinate system. Without considering the possibility of using two different conventions for the definition of the rotation axes (intrinsic or extrinsic), there exist twelve possible sequences of rotation axes, divided in two groups:

- **Proper Euler angles** (*z-x-z, x-y-x, y-z-y, z-y-z, x-z-x, y-x-y*)
- **Tait–Bryan angles** (*x-y-z, y-z-x, z-x-y, x-z-y, z-y-x, y-x-z*).

Tait–Bryan angles are also called **Cardan angles**; **nautical angles**; **heading, elevation, and bank**; or **yaw, pitch, and roll**. Sometimes, both kinds of sequences are called "Euler angles". In that case, the sequences of the first group are called *proper* or *classic* Euler angles.

35.1 Proper Euler angles

35.1.1 Classic definition

Euler angles are a means of representing the spatial orientation of any reference frame (coordinate system or basis) as a composition of three elemental rotations starting from a known standard orientation, represented by another frame (sometimes referred to as the *original* or *fixed* reference frame, or standard basis). The reference orientation can be imagined to be an initial orientation from which the frame virtually rotates to reach its actual orientation. In the following,

35.1. PROPER EULER ANGLES

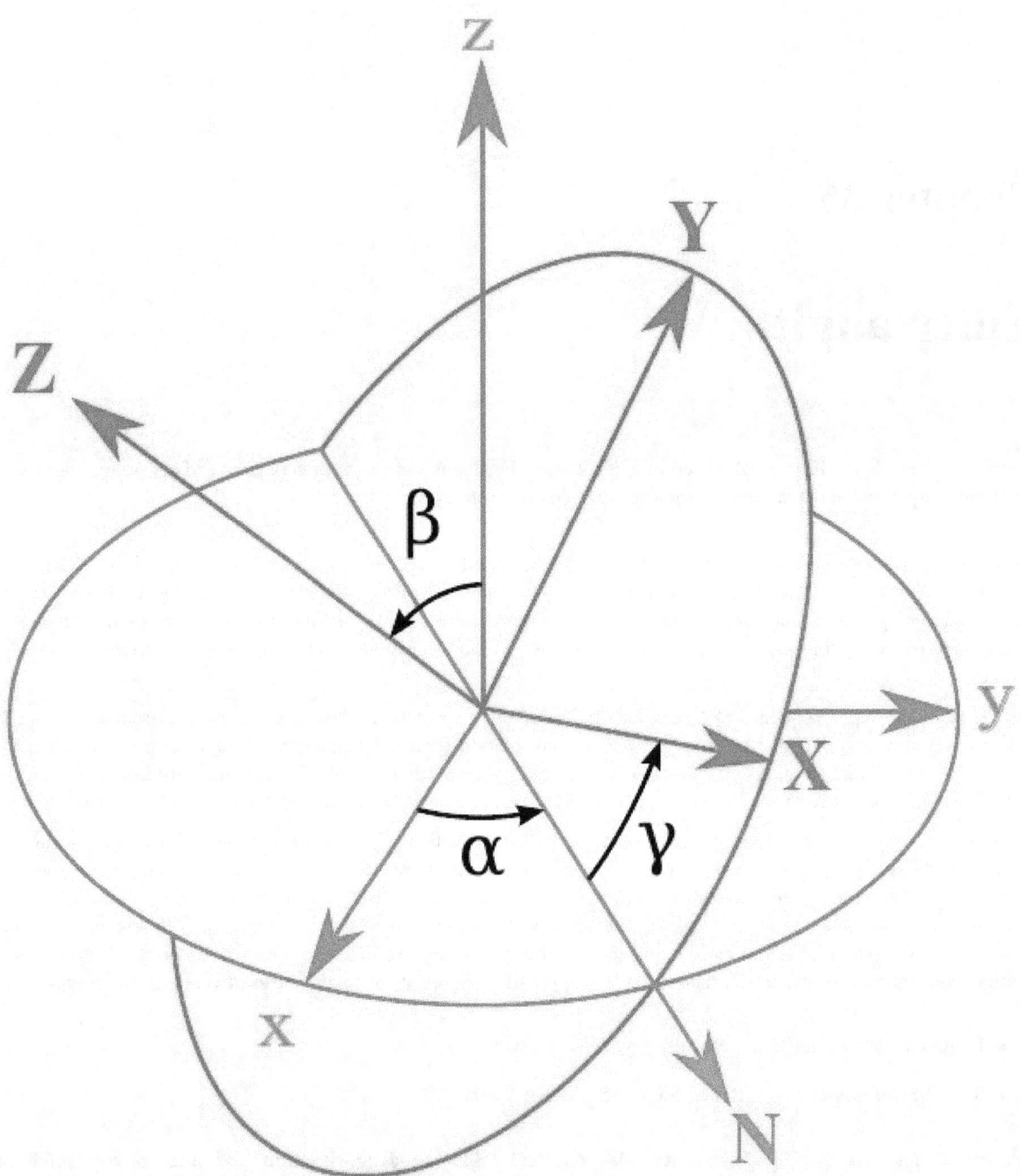

Proper Euler angles representing rotations about z, N, and Z axes. The xyz (original) system is shown in blue, the XYZ (rotated) system is shown in red. The line of nodes (N) is shown in green.

the axes of the original frame are denoted as x,y,z and the axes of the rotated frame are denoted as X,Y,Z. In geometry and physics, the rotated coordinate system is often imagined to be rigidly attached to a rigid body. In this case, it is called a "local" coordinate system, and it is meant to represent both the position and the orientation of the body.

The geometrical definition (referred sometimes as static) of the Euler angles is based on the axes of the above-mentioned (original and rotated) reference frames and an additional axis called the line of nodes. The line of nodes (N) is defined as the intersection of the xy and the XY coordinate planes. In other words, it is a line passing through the common origin of both frames, and perpendicular to the zZ plane, on which both z and Z lie. The three Euler angles are defined as follows:

- α (or φ) is the angle between the x axis and the N axis.
- β (or θ) is the angle between the z axis and the Z axis.
- γ (or ψ) is the angle between the N axis and the X axis.

This definition implies that:

- α represents a rotation around the z axis,
- β represents a rotation around the N axis,
- γ represents a rotation around the Z axis.

If β is zero, there is no rotation about N. As a consequence, Z coincides with z, α and γ represent rotations about the same axis (z), and the final orientation can be obtained with a single rotation about z, by an angle equal to $\alpha+\gamma$.

35.1.2 Alternative definition

The rotated frame XYZ may be imagined to be initially aligned with xyz, before undergoing the three elemental rotations represented by Euler angles. Its successive orientations may be denoted as follows:

- x-y-z, or x_0-y_0-z_0 (initial)
- x'-y'-z', or x_1-y_1-z_1 (after first rotation)
- x''-y''-z'', or x_2-y_2-z_2 (after second rotation)
- X-Y-Z, or x_3-y_3-z_3 (final)

For the above-listed sequence of rotations, the line of nodes N can be simply defined as the orientation of X after the first elemental rotation. Hence, N can be simply denoted x'. Moreover, since the third elemental rotation occurs about Z, it does not change the orientation of Z. Hence Z coincides with z''. This allows us to simplify the definition of the Euler angles as follows:

- α (or φ) represents a rotation around the z axis,
- β (or θ) represents a rotation around the x' axis,
- γ (or ψ) represents a rotation around the z'' axis.

35.1.3 Conventions

Different authors may use different sets of rotation axes to define Euler angles, or different names for the same angles. Therefore any discussion employing Euler angles should always be preceded by their definition.[2] Unless otherwise stated, this article will use the convention described above.

The three elemental rotations may occur either about the axes xyz of the original coordinate system, which is assumed to remain motionless (extrinsic rotations), or about the axes of the rotating coordinate system XYZ, which changes its orientation after each elemental rotation (intrinsic rotations). The definition above uses intrinsic rotations.

There are six possibilities of choosing the rotation axes for proper Euler angles. In all of them, the first and third rotation axes are the same. The six possible sequences are:

1. z-x'-z'' (intrinsic rotations) or z-x-z (extrinsic rotations)

35.1. PROPER EULER ANGLES

2. $x\text{-}y'\text{-}x''$ (intrinsic rotations) or $x\text{-}y\text{-}x$ (extrinsic rotations)
3. $y\text{-}z'\text{-}y''$ (intrinsic rotations) or $y\text{-}z\text{-}y$ (extrinsic rotations)
4. $z\text{-}y'\text{-}z''$ (intrinsic rotations) or $z\text{-}y\text{-}z$ (extrinsic rotations)
5. $x\text{-}z'\text{-}x''$ (intrinsic rotations) or $x\text{-}z\text{-}x$ (extrinsic rotations)
6. $y\text{-}x'\text{-}y''$ (intrinsic rotations) or $y\text{-}x\text{-}y$ (extrinsic rotations)

Euler angles between two reference frames are defined only if both frames have the same handedness.

35.1.4 Signs and ranges

Angles are commonly defined according to the right hand rule. Namely, they have positive values when they represent a rotation that appears clockwise when looking in the positive direction of the axis, and negative values when the rotation appears counter-clockwise. The opposite convention (left hand rule) is less frequently adopted.

About the ranges:

- for α and γ, the range is defined modulo 2π radians. A valid range could be $[-\pi, \pi]$.

- for β, the range covers π radians (but can't be said to be modulo π). For example could be $[0, \pi]$ or $[-\pi/2, \pi/2]$.

The angles α, β and γ are uniquely determined except for the singular case that the xy and the XY planes are identical, the z axis and the Z axis having the same or opposite directions. Indeed, if the z axis and the Z axis are the same, $\beta = 0$ and only $(\alpha + \gamma)$ is uniquely defined (not the individual values), and, similarly, if the z axis and the Z axis are opposite, $\beta = \pi$ and only $(\alpha - \gamma)$ is uniquely defined (not the individual values). These ambiguities are known as gimbal lock in applications.

35.1.5 Geometric derivation

The fastest way to get the Euler Angles of a given frame is to write the three given vectors as columns of a matrix and compare it with the expression of the theoretical matrix (see later table of matrices). Hence the three Euler Angles can be calculated. Nevertheless, the same result can be reached avoiding matrix algebra, which is more geometrical. Assuming a frame with unit vectors (X, Y, Z) as in the main diagram, it can be seen that:

$$\cos(\beta) = Z_3.$$

And, since

$$\sin^2 x = 1 - \cos^2 x,$$

we have

$$\sin(\beta) = \sqrt{1 - Z_3^2}.$$

As Z_2 is the double projection of a unitary vector,

$$\cos(\alpha) \cdot \sin(\beta) = Z_2,$$

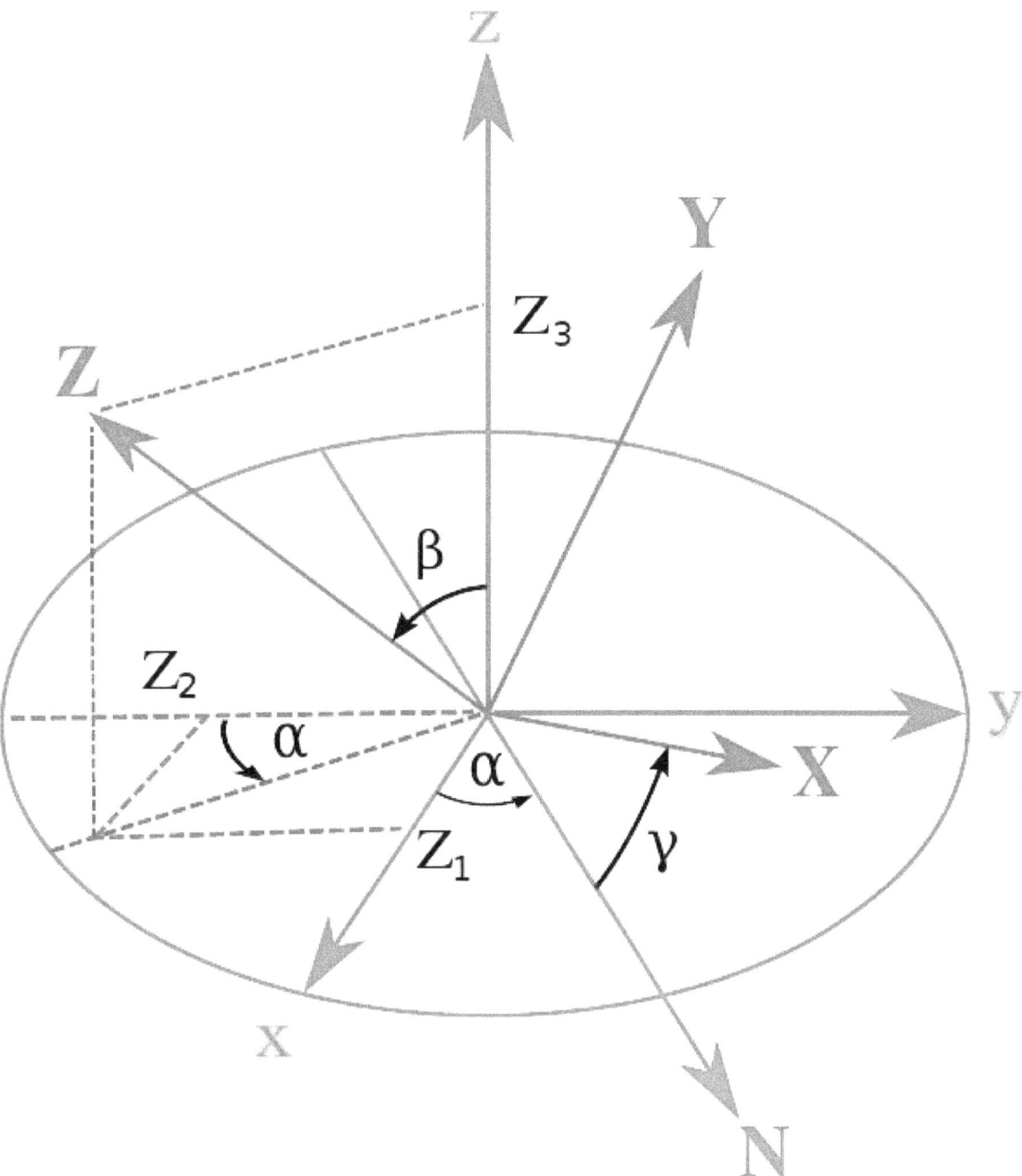

Projections of Z vector.

$$\cos(\alpha) = Z_2/\sqrt{1-Z_3^2}.$$

There is a similar construction for Y_3, projecting it first over the plane defined by the axis z and the line of nodes. As the angle between the planes is $\pi/2 - \beta$ and $\cos(\pi/2 - \beta) = \sin(\beta)$, this leads to:

$$\sin(\beta) \cdot \cos(\gamma) = Y_3,$$
$$\cos(\gamma) = Y_3/\sqrt{1-Z_3^2},$$

35.1. PROPER EULER ANGLES

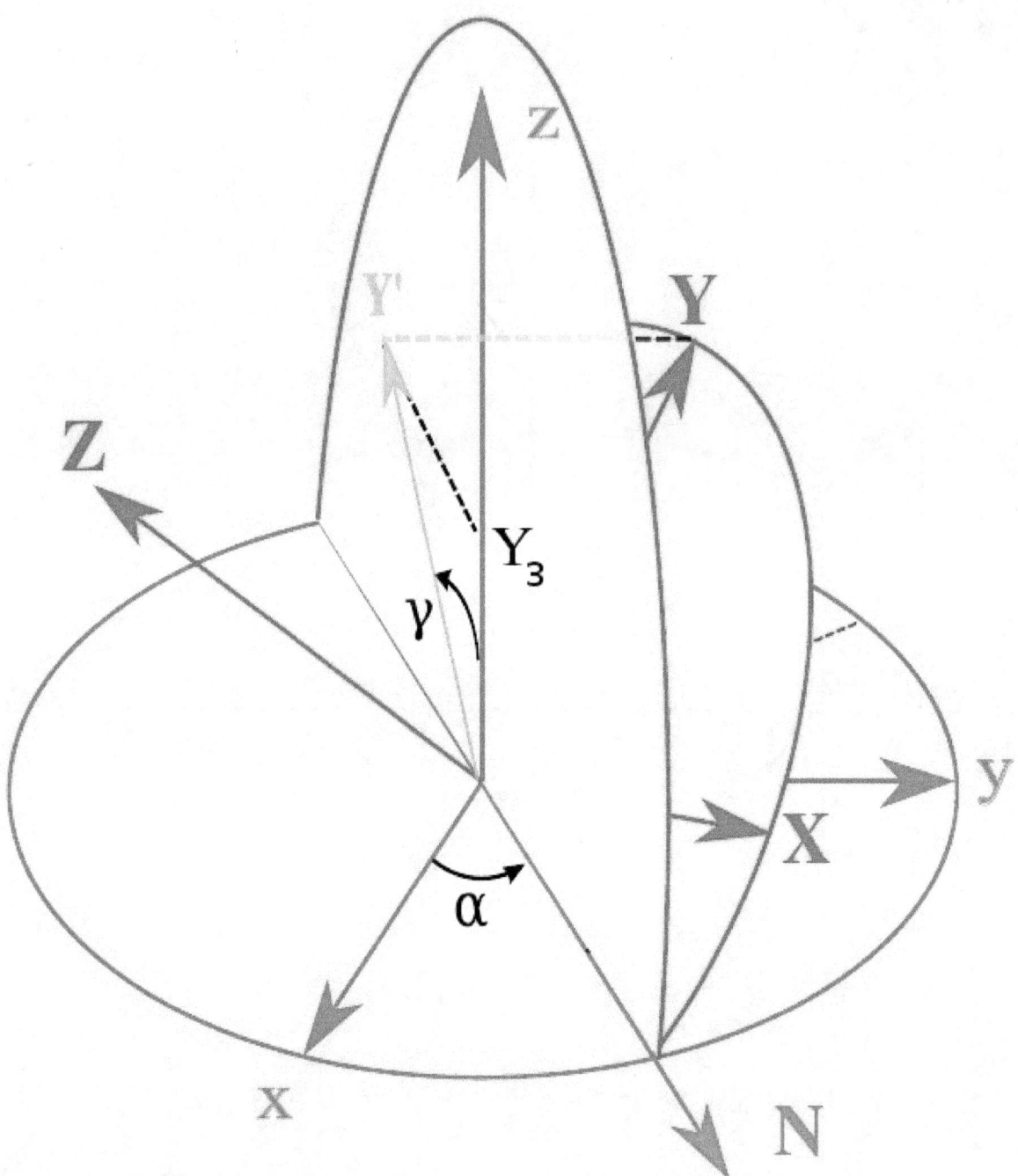

Projections of **Y** *vector.*

and finally, using the inverse cosine function,

$\alpha = \arccos(Z_2/\sqrt{1-Z_3^2})$,

$\beta = \arccos(Z_3)$,

$\gamma = \arccos(Y_3/\sqrt{1-Z_3^2})$.

It is interesting to note that the inverse cosine function yields two possible values for the argument. In this geometrical description only one of the solutions is valid. When Euler Angles are defined as a sequence of rotations, all the solutions can be valid, but there will be only one inside the angle ranges. This is because the sequence of rotations to reach the target frame is not unique if the ranges are not previously defined.[3]

For computational purposes, it may be useful to represent the angles using atan2(y,x):

$\alpha = \text{atan2}(Z_1, Z_2)$,

$\gamma = \text{atan2}(X_3, Y_3)$.

35.2 Tait–Bryan angles

The second type of formalism is called **Tait–Bryan angles**, after Peter Guthrie Tait and George H. Bryan.

The definitions and notations used for Tait-Bryan angles are similar to those described above for proper Euler angles (Classic definition, Alternative definition). The only difference is that Tait–Bryan angles represent rotations about three distinct axes (e.g. x-y-z, or x-y'-z''), while proper Euler angles use the same axis for both the first and third elemental rotations (e.g., z-x-z, or z-x'-z'').

This implies a different definition for the line of nodes. In the first case it was defined as the intersection between two homologous Cartesian planes (parallel when Euler angles are zero; e.g. xy and XY). In the second one, it is defined as the intersection of two non-homologous planes (perpendicular when Euler angles are zero; e.g. xy and YZ).

35.2.1 Conventions

The three elemental rotations may occur either about the axes of the original coordinate system, which remains motionless (extrinsic rotations), or about the axes of the rotating coordinate system, which changes its orientation after each elemental rotation (intrinsic rotations).

There are six possibilities of choosing the rotation axes for Tait–Bryan angles. The six possible sequences are:

1. x-y'-z'' (intrinsic rotations) or x-y-z (extrinsic rotations)

2. y-z'-x'' (intrinsic rotations) or y-z-x (extrinsic rotations)

3. z-x'-y'' (intrinsic rotations) or z-x-y (extrinsic rotations)

4. x-z'-y'' (intrinsic rotations) or x-z-y (extrinsic rotations)

5. z-y'-x'' (intrinsic rotations) or z-y-x (extrinsic rotations): the intrinsic rotations are known as: yaw, pitch and roll

6. y-x'-z'' (intrinsic rotations) or y-x-z (extrinsic rotations)

35.2.2 Alternative names

Tait-Bryan angles, following z-y'-x'' (intrinsic rotations) convention, are also known as **nautical angles**, because they can be used to describe the orientation of a ship or aircraft, or **Cardan angles**, after the Italian mathematician and physicist Gerolamo Cardano, who first described in detail the Cardan suspension and the Cardan joint. They are also called **heading, elevation and bank**, or **yaw, pitch and roll**. Notice that the second set of terms is also used for the three aircraft principal axes.

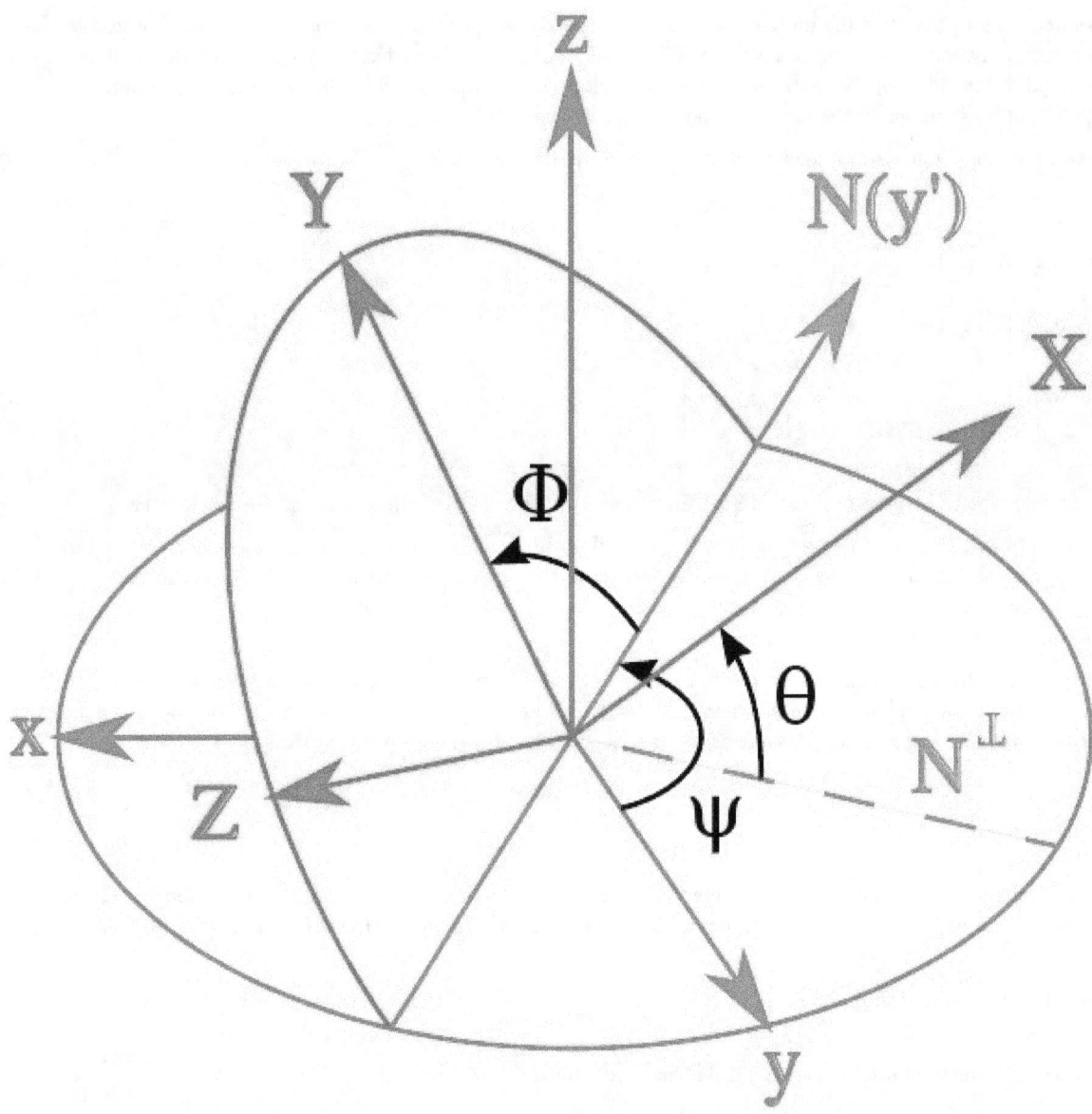

Tait–Bryan angles. z-y'-x" sequence (intrinsic rotations; N coincides with y'). The angle rotation sequence is ψ, θ, Φ. Note that in this case ψ > 90° and θ is a negative angle.

35.3 Relationship with physical motions

See also: Givens rotations and Davenport rotations

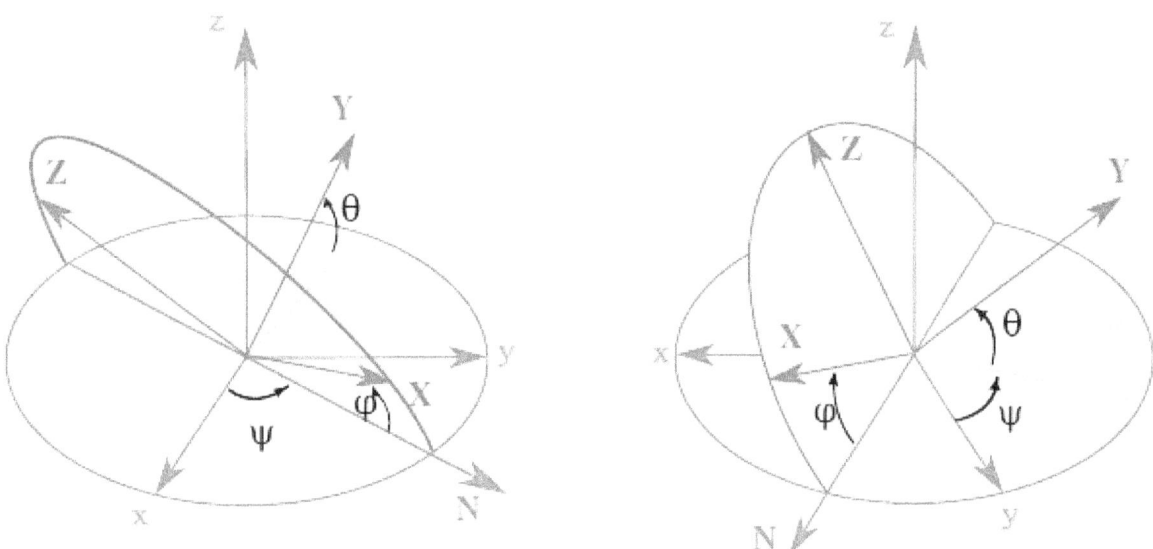

Tait–Bryan angles. z-x'-y" sequence (intrinsic rotations; N coincides with x')

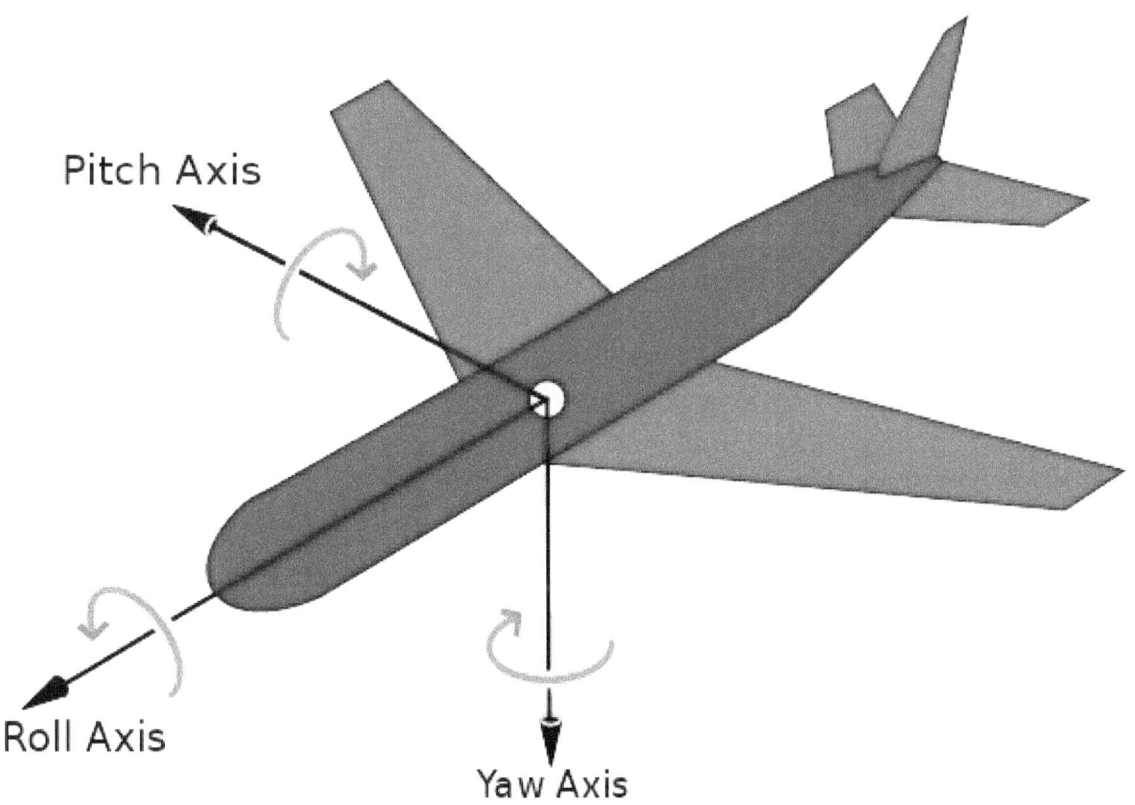

The principal axes of an aircraft

35.3.1 Intrinsic rotations

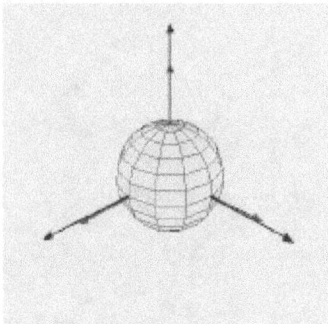

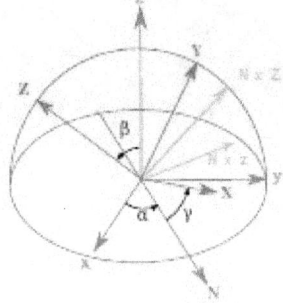

Any target orientation can be reached, starting from a known reference orientation, using a specific sequence of intrinsic rotations, whose magnitudes are the Euler angles of the target orientation. This example uses the z-x'-z'' sequence.

Intrinsic rotations are elemental rotations that occur about the axes of the rotating coordinate system XYZ, which changes its orientation after each elemental rotation. The XYZ system rotates, while xyz is fixed. Starting with XYZ overlapping xyz, a composition of three intrinsic rotations can be used to reach any target orientation for XYZ. The Euler or Tait-Bryan angles (α, β, γ) are the amplitudes of these elemental rotations. For instance, the target orientation can be reached as follows:

- The XYZ system rotates by α about the Z axis (which coincides with the z axis). The X axis now lies on the line of nodes.

- The XYZ system rotates about the now rotated X axis by β. The Z axis is now in its final orientation, and the X axis remains on the line of nodes.

- The XYZ system rotates a third time about the new Z axis by γ.

The above-mentioned notation allows us to summarize this as follows: the three elemental rotations of the XYZ-system occur about z, x' and z''. Indeed, this sequence is often denoted z-x'-z''. Sets of rotation axes associated with both proper Euler angles and Tait-Bryan angles are commonly named using this notation (see above for details). Sometimes, the same sequence is simply called z-x-z, Z-X-Z, or 3-1-3, but this notation may be ambiguous as it may be identical to that used for extrinsic rotations. In this case, it becomes necessary to separately specify whether the rotations are intrinsic or extrinsic.

Rotation matrices can be used to represent a sequence of intrinsic rotations. For instance,

$$R = X(\alpha)Y(\beta)Z(\gamma)$$

represents a composition of intrinsic rotations about axes x-y'-z'', if used to pre-multiply column vectors, while

$$R = Z(\gamma)Y(\beta)X(\alpha)$$

represents exactly the same composition when used to post-multiply row vectors. See Ambiguities in the definition of rotation matrices for more details.

35.3.2 Extrinsic rotations

Extrinsic rotations are elemental rotations that occur about the axes of the fixed coordinate system xyz. The XYZ system rotates, while xyz is fixed. Starting with XYZ overlapping xyz, a composition of three extrinsic rotations can be used to reach any target orientation for XYZ. The Euler or Tait-Bryan angles (α, β, γ) are the amplitudes of these elemental rotations. For instance, the target orientation can be reached as follows:

- The XYZ system rotates about the z axis by α. The X axis is now at angle α with respect to the x axis.
- The XYZ system rotates again about the x axis by β. The Z axis is now at angle β with respect to the z axis.
- The XYZ system rotates a third time about the z axis by γ.

In sum, the three elemental rotations occur about z, x and z. Indeed, this sequence is often denoted z-x-z (or 3-1-3). Sets of rotation axes associated with both proper Euler angles and Tait–Bryan angles are commonly named using this notation (see above for details).

Rotation matrices can be used to represent a sequence of extrinsic rotations. For instance,

$$R = Z(\gamma)Y(\beta)X(\alpha)$$

represents a composition of extrinsic rotations about axes x-y-z, if used to pre-multiply column vectors, while

$$R = X(\alpha)Y(\beta)Z(\gamma)$$

represents exactly the same composition when used to post-multiply row vectors. See Ambiguities in the definition of rotation matrices for more details.

35.3.3 Conversion between intrinsic and extrinsic rotations

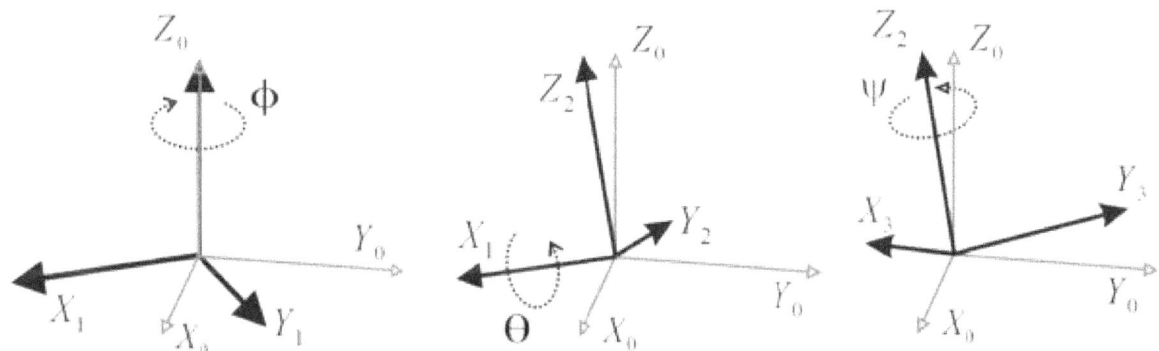

A rotation represented by Euler angles $(\alpha, \beta, \gamma) = (-60°, 30°, 45°)$, using z-x'-z'' intrinsic rotations

Any extrinsic rotation is equivalent to an intrinsic rotation by the same angles but with inverted order of elemental rotations, and vice versa. For instance, the intrinsic rotations x-y'-z'' by angles α, β, γ are equivalent to the extrinsic rotations z-y-x by angles γ, β, α. Both are represented by a matrix

$$R = X(\alpha)Y(\beta)Z(\gamma)$$

if R is used to pre-multiply column vectors, and by a matrix

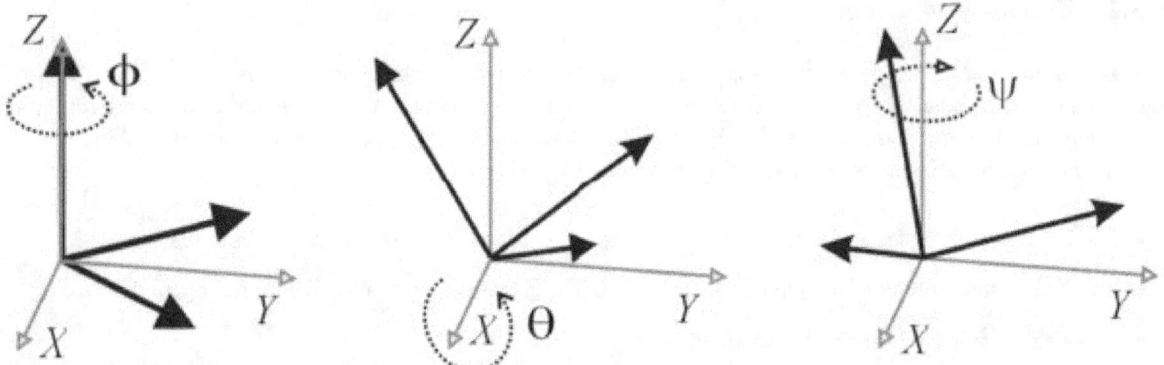

The same rotation represented by $(\gamma, \beta, \alpha) = (45°, 30°, -60°)$, using z-x-z extrinsic rotations

$R = Z(\gamma)Y(\beta)X(\alpha)$

if R is used to post-multiply row vectors. See Ambiguities in the definition of rotation matrices for more details.

35.4 Gimbal motion relationship

Euler basic motions are defined as the movements obtained by changing one of the Euler angles while leaving the other two constant. Euler rotations are never expressed in terms of the external frame, or in terms of the co-moving rotated body frame, but in a mixture. They constitute a **mixed axes of rotation** system, where the first angle moves the line of nodes around the external axis z, the second rotates around the line of nodes and the third one is an intrinsic rotation around an axis fixed in the body that moves.

These rotations are called precession, nutation, and intrinsic rotation (spin). As an example, consider a top. The top spins around its own axis of symmetry; this corresponds to its intrinsic rotation. It also rotates around its pivotal axis, with its center of mass orbiting the pivotal axis; this rotation is a precession. Finally, the top can wobble up and down; the inclination angle is the nutation angle. While all three are rotations when applied over individual frames, only precession is valid as a rotation operator, and only precession can be expressed in general as a matrix in the basis of the space.

35.4.1 Gimbal analogy

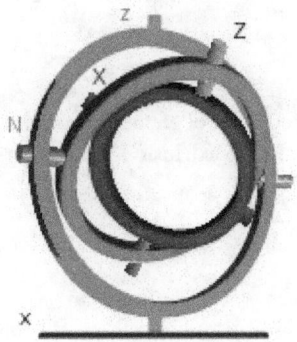

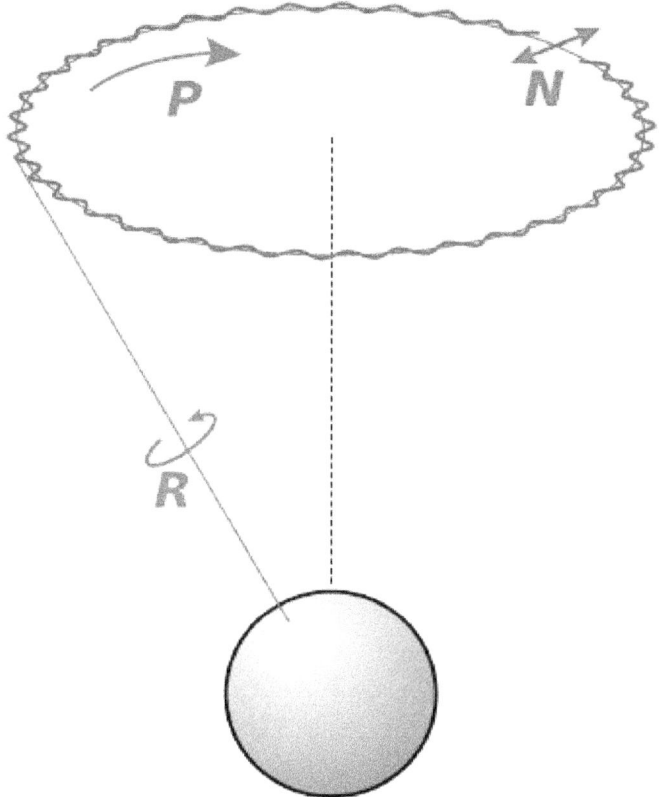

Euler basic motions of the Earth. Intrinsic (R), Precession (P) and Nutation (N)

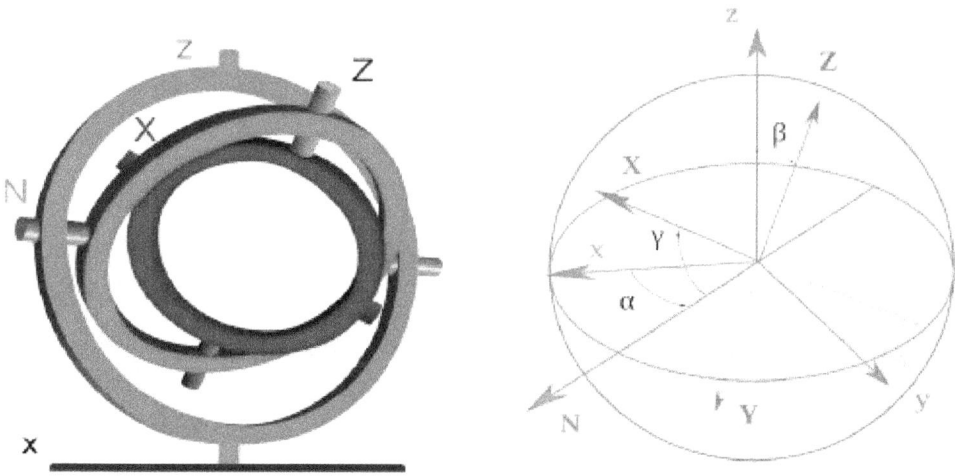

Left: A three axes z-x-z gimbal where the external frame and external axis x are not shown and axes Y are perpendicular to each gimbal ring. **Right:** A simple diagram showing the Euler angles and where the axes Y of intermediate frames are located.

If we suppose a set of frames, able to move each with respect to the former according to just one angle, like a gimbal, there will exist an external fixed frame, one final frame and two frames in the middle, which are called "intermediate frames". The two in the middle work as two gimbal rings that allow the last frame to reach any orientation in space.

In these conditions, each Euler rotation works on one of the rings, independently from the rest.

35.4.2 Intermediate frames

The gimbal rings indicate some intermediate frames. They can be defined statically too. Taking some vectors **i**, **j** and **k** over the axes x, y and z, and vectors **I**, **J**, **K** over X, Y and Z, and a vector **N** over the line of nodes, some intermediate frames can be defined using the vector cross product, as following:

- origin: [**i**,**j**,**k**] (where **k** = **i** × **j**)
- first: [**N**,**k** × **N**,**k**]
- second: [**N**,**K** × **N**,**K**]
- final: [**I**,**J**,**K**]

These intermediate frames are equivalent to those of the gimbal. They are such that they differ from the previous one in just a single elemental rotation. This proves that:

- Any target frame can be reached from the reference frame just composing three rotations.
- The values of these three rotations are exactly the Euler angles of the target frame.

35.5 Relationship to other representations

Main article: Rotation formalisms in three dimensions § Conversion formulae between formalisms

Euler angles are one way to represent orientations. There are others, and it is possible to change to and from other conventions.

35.5.1 Rotation matrix

Any orientation can be achieved by composing three elemental rotations, starting from a known standard orientation. Equivalently, any rotation matrix R can be decomposed as a product of three elemental rotation matrices. For instance:

$$R = X(\alpha)Y(\beta)Z(\gamma)$$

is a rotation matrix that may be used to represent a composition of intrinsic rotations about axes x-y'-z''. However, both the definition of the elemental rotation matrices X, Y, Z, and their multiplication order depend on the choices taken by the user about the definition of both rotation matrices and Euler angles (see, for instance, Ambiguities in the definition of rotation matrices). Unfortunately, different sets of conventions are adopted by users in different contexts. The following table was built according to this set of conventions:

1. Each matrix is meant to operate by pre-multiplying column vectors (see Ambiguities in the definition of rotation matrices)

2. Each matrix is meant to represent an active rotation (the composing and composed matrices are supposed to act on the coordinates of vectors defined in the initial fixed reference frame and give as a result the coordinates of a rotated vector defined in the same reference frame).

3. Each matrix is meant to represent the composition of intrinsic rotations (around the axes of the rotating reference frame).

4. Right handed reference frames are adopted, and the right hand rule is used to determine the sign of the angles α, β, γ.

For the sake of simplicity, the following table uses the following nomenclature:

1. 1, 2, 3 represent the angles α, β, γ.

2. X, Y, Z are the matrices representing the elemental rotations about the axes x, y, z of the fixed frame (e.g., X_1 represents a rotation about x by an angle α).

3. s and c represent sine and cosine (e.g., s_1 represents the sine of α).

4. Each matrix is denoted by the formula used to calculate it. If $R = Z_1 X_2 Z_3$, we name it $Z_1 X_2 Z_3$.

To change the formulas for the opposite direction of rotation, change the signs of the sine functions. To change the formulas for passive rotations, transpose the matrices (then each matrix transforms the initial coordinates of a vector remaining fixed to the coordinates of the same vector measured in the rotated reference system; same rotation axis, same angles, but now the coordinate system rotates, rather than the vector).

35.5.2 Quaternions

Unit quaternions, also known as Euler–Rodrigues parameters, provide another mechanism for representing 3D rotations. This is equivalent to the special unitary group description.

Expressing rotations in 3D as unit quaternions instead of matrices has some advantages:

- Concatenating rotations is computationally faster and numerically more stable.

- Extracting the angle and axis of rotation is simpler.

- Interpolation is more straightforward. See for example slerp.

- Quaternions don't suffer from gimbal lock unlike Euler angles.

35.5.3 Geometric algebra

Other representation comes from the Geometric algebra(GA). GA is a higher level abstraction, in which the quaternions are an even subalgebra. The principal tool in GA is the rotor $\mathbb{R} = [\cos(\theta/2) - Iu\sin(\theta/2)]$ where θ = angle of rotation, (u) = rotation axis (unitary vector) and (I) = pseudoscalar (trivector in $\mathbb{R}^3$)

35.6 Properties

See also: Charts on SO(3) and Quaternions and spatial rotation

The Euler angles form a chart on all of SO(3), the special orthogonal group of rotations in 3D space. The chart is smooth except for a polar coordinate style singularity along β=0. See charts on SO(3) for a more complete treatment.

The space of rotations is called in general "The Hypersphere of rotations", though this is a misnomer: the group Spin(3) is isometric to the hypersphere S^3, but the rotation space SO(3) is instead isometric to the real projective space $\mathbf{RP}^3$ which is a 2-fold quotient space of the hypersphere. This 2-to-1 ambiguity is the mathematical origin of spin in physics.

A similar three angle decomposition applies to SU(2), the special unitary group of rotations in complex 2D space, with the difference that β ranges from 0 to 2π. These are also called Euler angles.

The Haar measure for Euler angles has the simple form $\sin(\beta).d\alpha.d\beta.d\gamma$, usually normalized by a factor of $1/8\pi^2$.

For example, to generate uniformly randomized orientations, let α and γ be uniform from 0 to 2π, let z be uniform from −1 to 1, and let β = arccos(z).

35.7 Higher dimensions

It is possible to define parameters analogous to the Euler angles in dimensions higher than three.[4]

The number of degrees of freedom of a rotation matrix is always less than the dimension of the matrix squared. That is, the elements of a rotation matrix are not all completely independent. For example, the rotation matrix in dimension 2 has only one degree of freedom, since all four of its elements depend on a single angle of rotation. A rotation matrix in dimension 3 (which has nine elements) has three degrees of freedom, corresponding to each independent rotation, for example by its three Euler angles or a magnitude one (unit) quaternion.

In SO(4) the rotation matrix is defined by two quaternions, and is therefore 6-parametric (three degrees of freedom for every quaternion). The 4×4 rotation matrices have therefore 6 out of 16 independent components.

Any set of 6 parameters that define the rotation matrix could be considered an extension of Euler angles to dimension 4.

In general, the number of euler angles in dimension D is quadratic in D; since any one rotation consists of choosing two dimensions to rotate between, the total number of rotations available in dimension D is $N_{rot} = \binom{D}{2} = D(D-1)/2$, which for $D = 2, 3, 4$ yields $N_{rot} = 1, 3, 6$.

35.8 Applications

35.8.1 Vehicles and moving frames

Main article: rigid body
See also: axes conventions

Their main advantage over other orientation descriptions is that they are directly measurable from a gimbal mounted in a vehicle. As gyroscopes keep their rotation axis constant, angles measured in a gyro frame are equivalent to angles measured in the lab frame. Therefore gyros are used to know the actual orientation of moving spacecraft, and Euler angles are directly measurable. Intrinsic rotation angle cannot be read from a single gimbal, so there has to be more than one gimbal in a spacecraft. Normally there are at least three for redundancy. There is also a relation to the well-known gimbal lock problem of mechanical engineering [5].

The most popular application is to describe aircraft attitudes, normally using a Tait–Bryan convention so that zero degrees elevation represents the horizontal attitude. Tait–Bryan angles represent the orientation of the aircraft respect a reference axis system (*world frame*) with three angles which in the context of an aircraft are normally called Heading, Elevation and Bank. When dealing with vehicles, different axes conventions are possible.

A gyroscope keeps its rotation axis constant. Therefore, angles measured in this frame are equivalent to angles measured in the lab frame

When studying rigid bodies in general, one calls the *xyz* system *space coordinates*, and the *XYZ* system *body coordinates*. The space coordinates are treated as unmoving, while the body coordinates are considered embedded in the moving body. Calculations involving acceleration, angular acceleration, angular velocity, angular momentum, and kinetic energy are often easiest in body coordinates, because then the moment of inertia tensor does not change in time. If one also diagonalizes the rigid body's moment of inertia tensor (with nine components, six of which are independent), then one has a set of coordinates (called the principal axes) in which the moment of inertia tensor has only three components.

The angular velocity of a rigid body takes a simple form using Euler angles in the moving frame. Also the Euler's rigid body equations are simpler because the inertia tensor is constant in that frame.

35.8. APPLICATIONS

Industrial robot operating in a foundry.

35.8.2 Crystallographic texture

In materials science, crystallographic texture (or preferred orientation) can be described using Euler angles. In texture analysis, the Euler angles provide a mathematical depiction of the orientation of individual crystallites within a polycrystalline material, allowing for the quantitative description of the macroscopic material.[7] The most common definition of the angles is due to Bunge and corresponds to the ZXZ convention. It is important to note, however, that the application generally involves axis transformations of tensor quantities, i.e. passive rotations. Thus the matrix that corresponds to the Bunge Euler angles is the transpose of that shown in the table above.[8]

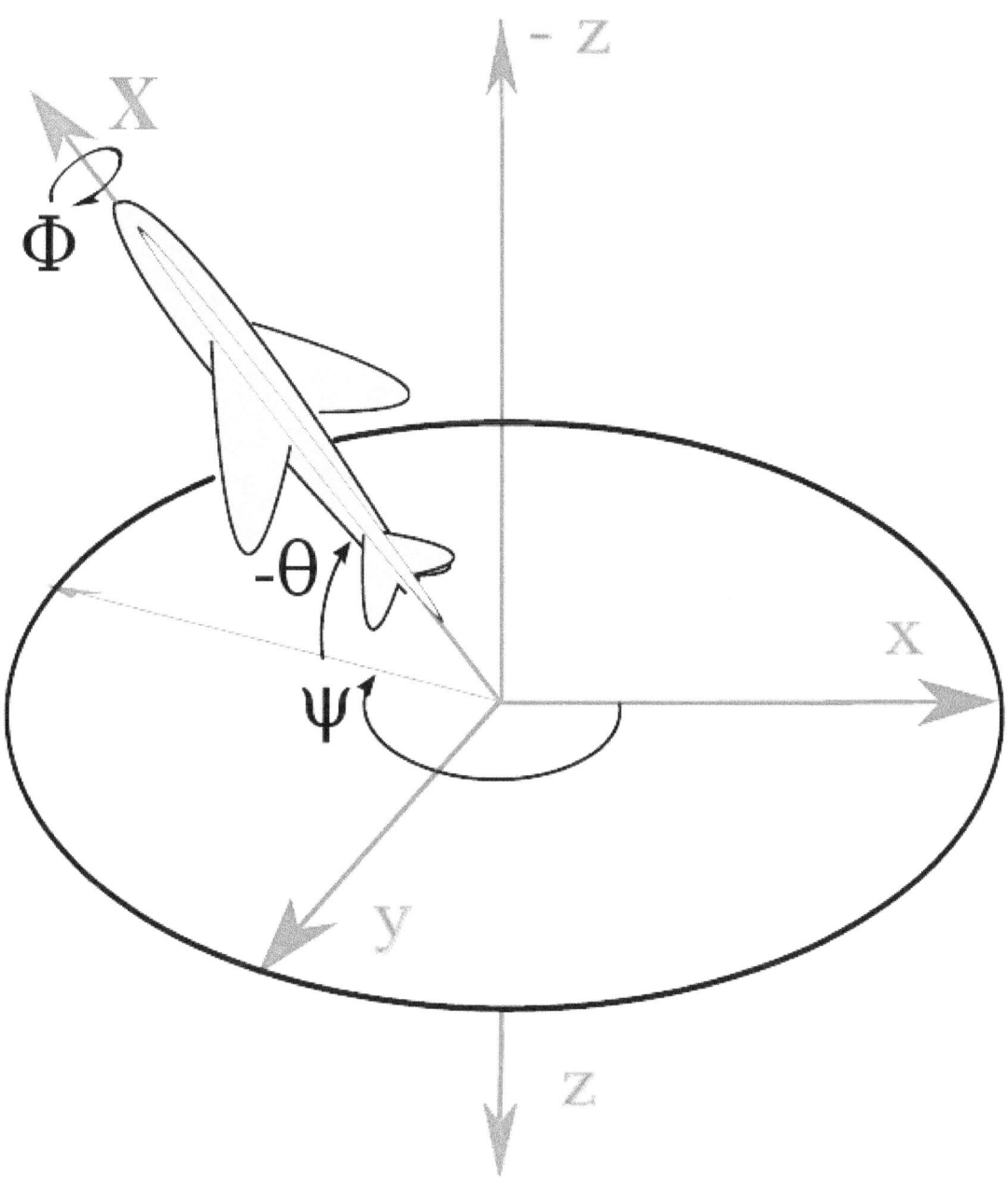

Heading, elevation and bank for an aircraft with axes DIN 9300

35.8.3 Others

Euler angles, normally in the Tait–Bryan convention, are also used in robotics for speaking about the degrees of freedom of a wrist. They are also used in Electronic stability control in a similar way.

Gun fire control systems require corrections to gun-order angles (bearing and elevation) to compensate for deck tilt (pitch and roll). In traditional systems, a stabilizing gyroscope with a vertical spin axis corrects for deck tilt, and stabilizes the optical sights and radar antenna. However, gun barrels point in a direction different from the line of sight to the target, to anticipate target movement and fall of the projectile due to gravity, among other factors. Gun mounts roll and pitch with

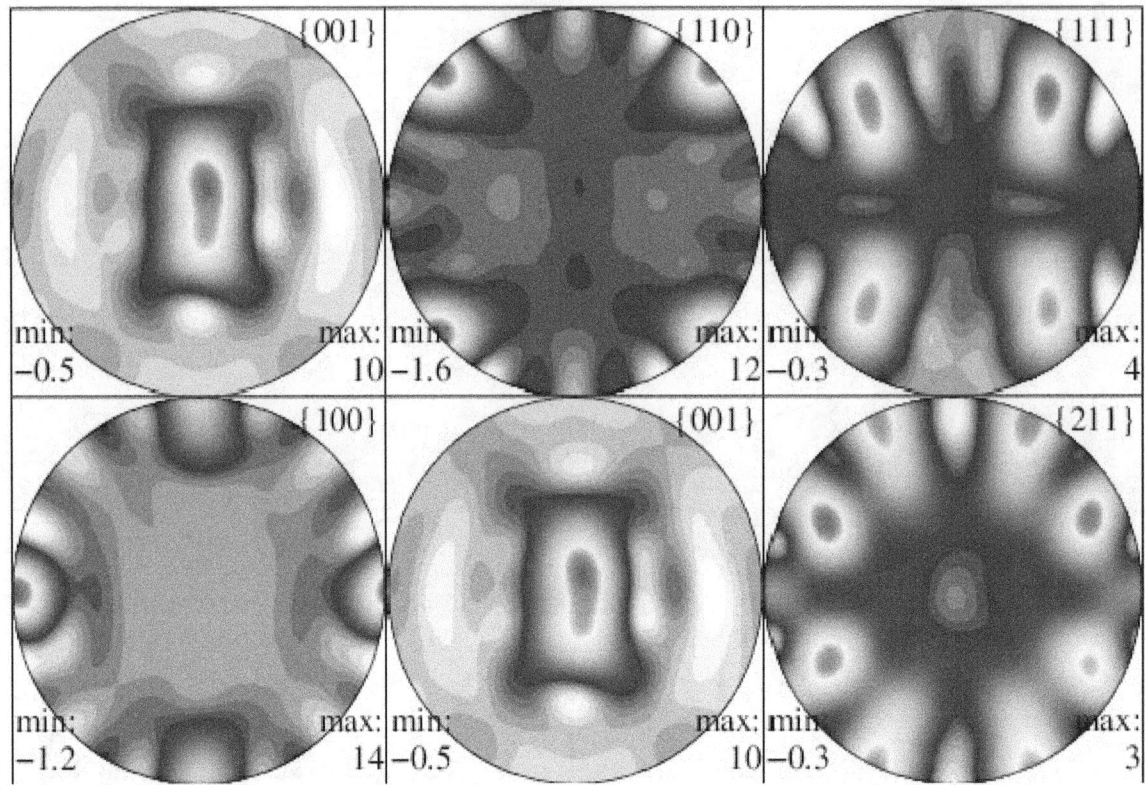

Pole figures displaying crystallographic texture of gamma-TiAl in an alpha2-gamma alloy, as measured by high energy X-rays.[6]

the deck plane, but also require stabilization. Gun orders include angles computed from the vertical gyro data, and those computations involve Euler angles.

Euler angles are also used extensively in the quantum mechanics of angular momentum. In quantum mechanics, explicit descriptions of the representations of *SO(3)* are very important for calculations, and almost all the work has been done using Euler angles. In the early history of quantum mechanics, when physicists and chemists had a sharply negative reaction towards abstract group theoretic methods (called the *Gruppenpest*), reliance on Euler angles was also essential for basic theoretical work.

Many mobile computing devices contain accelerometers which can determine these devices' Euler angles with respect to the earth's gravitational attraction. These are used in applications such as games, bubble level simulations, and kaleidoscopes.

35.9 See also

- 3D projection
- Axis-angle representation
- Conversion between quaternions and Euler angles
- Euler's rotation theorem
- Quaternion
- Quaternions and spatial rotation
- Rotation formalisms in three dimensions

- Spherical coordinate system

35.10 References

[1] Novi Commentarii academiae scientiarum Petropolitanae 20, 1776, pp. 189–207 (E478) pdf

[2] Mathworld does a good job describing this issue

[3] Gregory G. Slabaugh, Computing Euler angles from a rotation matrix

[4] (Italian) A generalization of Euler Angles to n-dimensional real spaces

[5] The relation between the Euler angles and the Cardan suspension is explained in chap. 11.7 of the following textbook: U. Krey, A. Owen, *Basic Theoretical Physics – A Concise Overview*, New York, London, Berlin, Heidelberg, Springer (2007).

[6] Liss KD, Bartels A, Schreyer A, Clemens H (2003). "High energy X-rays: A tool for advanced bulk investigations in materials science and physics". *Textures Microstruct.* 35 (3/4): 219–52. doi:10.1080/07303300310001634952.

[7] Kocks, U.F.; Tomé, C.N.; Wenk, H.-R. (2000), *Texture and Anisotropy: Preferred Orientations in Polycrystals and their effect on Materials Properties*, Cambridge, ISBN 978-0-521-79420-6

[8] Bunge, H. (1993), *Texture Analysis in Materials Science: Mathematical Methods*, CUVILLIER VERLAG, ASIN B0014XV9HU

35.11 Bibliography

- Biedenharn, L. C.; Louck, J. D. (1981), *Angular Momentum in Quantum Physics*, Reading, MA: Addison–Wesley, ISBN 978-0-201-13507-7

- Goldstein, Herbert (1980), *Classical Mechanics* (2nd ed.), Reading, MA: Addison–Wesley, ISBN 978-0-201-02918-5

- Gray, Andrew (1918), *A Treatise on Gyrostatics and Rotational Motion*, London: Macmillan (published 2007), ISBN 978-1-4212-5592-7

- Rose, M. E. (1957), *Elementary Theory of Angular Momentum*, New York, NY: John Wiley & Sons (published 1995), ISBN 978-0-486-68480-2

- Symon, Keith (1971), *Mechanics*, Reading, MA: Addison-Wesley, ISBN 978-0-201-07392-8

- Landau, L.D.; Lifshitz, E. M. (1996), *Mechanics* (3rd ed.), Oxford: Butterworth-Heinemann, ISBN 978-0-7506-2896-9

35.12 External links

- Weisstein, Eric W., "Euler Angles", *MathWorld*.

- Java applet for the simulation of Euler angles available at http://www.parallemic.org/Java/EulerAngles.html.

- EulerAngles - An iOS app for visualizing in 3D the three rotations associated with Euler angles.

- http://sourceforge.net/projects/orilib – A collection of routines for rotation / orientation manipulation, including special tools for crystal orientations.

- Online tool to compose rotation matrices available at http://www.vectoralgebra.info/eulermatrix.html

Chapter 36

Drag (physics)

In fluid dynamics, **drag** (sometimes called **air resistance**, a type of friction, or **fluid resistance**, another type of friction or fluid friction) refers to forces acting opposite to the relative motion of any object moving with respect to a surrounding fluid.[1] This can exist between two fluid layers (or surfaces) or a fluid and a solid surface. Unlike other resistive forces, such as dry friction, which are nearly independent of velocity, drag forces depend on velocity.[2][3] Drag force is proportional to the velocity for a laminar flow and the squared velocity for a turbulent flow. Even though the ultimate cause of a drag is viscous friction, the turbulent drag is independent of viscosity.[4]

Drag forces always decrease fluid velocity relative to the solid object in the fluid's path.

36.1 Examples of drag

Examples of drag include the component of the net aerodynamic or hydrodynamic force acting opposite to the direction of movement of the solid object relative to the Earth as for cars, aircraft[3] and boat hulls; or acting in the same geographical direction of motion as the solid, as for sails attached to a down wind sail boat, or in intermediate directions on a sail depending on points of sail.[5][6][7][8] In the case of viscous drag of fluid in a pipe, drag force on the immobile pipe decreases fluid velocity relative to the pipe.[9][10]

36.2 Types of drag

Types of drag are generally divided into the following categories:

- parasitic drag, consisting of
    - form drag,
    - skin friction,
    - interference drag,
- lift-induced drag, and
- wave drag (aerodynamics) or wave resistance (ship hydrodynamics).

The phrase *parasitic drag* is mainly used in aerodynamics, since for lifting wings drag is in general small compared to lift. For flow around bluff bodies, drag is most often dominating, and then the qualifier "parasitic" is meaningless. Form drag, skin friction and interference drag on bluff bodies are not coined as being elements of "parasitic drag", but directly as elements of drag.

Further, lift-induced drag is only relevant when wings or a lifting body are present, and is therefore usually discussed either in the aviation perspective of drag, or in the design of either semi-planing or planing hulls. Wave drag occurs when a solid object is moving through a fluid at or near the speed of sound in that fluid—or in case there is a freely-moving fluid surface with surface waves radiating from the object, e.g. from a ship.

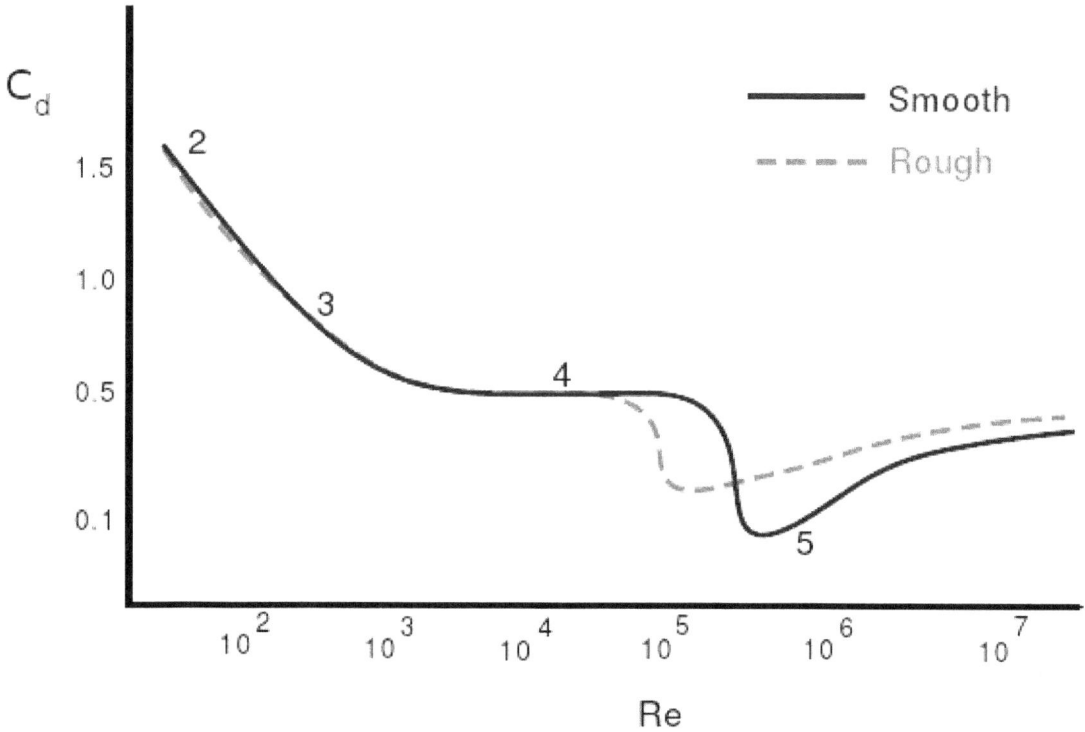

Drag coefficient C_d for a sphere as a function of Reynolds number Re, as obtained from laboratory experiments. The solid line is for a sphere with a smooth surface, while the dashed line is for the case of a rough surface.

Drag depends on the properties of the fluid and on the size, shape, and speed of the object. One way to express this is by means of the drag equation:

$$F_D = \tfrac{1}{2} \rho v^2 C_D A$$

where

    F_D is the **drag force**,

    ρ is the density of the fluid,[11]

    v is the speed of the object relative to the fluid,

    A is the cross sectional area, and

    C_D is the drag coefficient – a dimensionless number.

The drag coefficient depends on the shape of the object and on the Reynolds number:

$$R_e = \frac{vD}{\nu}$$

36.3. DRAG AT HIGH VELOCITY

where D is some characteristic diameter or linear dimension and ν is the kinematic viscosity of the fluid (equal to the viscosity μ divided by the density). At low Reynolds number, the drag coefficient is asymptotically proportional to the inverse of the Reynolds number, which means that the drag is proportional to the speed. At high Reynolds number, the drag coefficient is more or less constant. The graph to the right shows how the drag coefficient varies with Reynolds number for the case of a sphere.

For high velocities (or more precisely, at high Reynolds number) drag will vary as the square of velocity. Thus, the resultant power needed to overcome this drag will vary as the cube of velocity. The standard equation for drag is one half the coefficient of drag multiplied by the fluid mass density, the cross sectional area of the specified item, and the square of the velocity.

Wind resistance is a layman's term for drag. Its use is often vague, and is usually used in a relative sense (e.g. a badminton shuttlecock has more *wind resistance* than a squash ball).

36.3 Drag at high velocity

Main article: Drag equation

As mentioned, the drag equation with a constant drag coefficient gives the force experienced by an object moving through a

Explanation of drag by NASA.

fluid at relatively large velocity (i.e. high Reynolds number, Re > ~1000). This is also called *quadratic drag*. The equation is attributed to Lord Rayleigh, who originally used L^2 in place of A (L being some length).

$$F_D = \tfrac{1}{2}\rho v^2 C_d A,$$

see derivation

The reference area A is often orthographic projection of the object—on a plane perpendicular to the direction of motion—e.g. for objects with a simple shape, such as a sphere, this is the cross sectional area. Sometimes different reference areas are given for the same object in which case a drag coefficient corresponding to each of these different areas must be given.

In case of a wing, comparison of the drag to the lift force is easiest when the reference areas are the same, since then the ratio of drag to lift force is just the ratio of drag to lift coefficient.[12] Therefore, the reference for a wing is often the lifting area ("wing area") rather than the frontal area.[13]

For an object with a smooth surface, and non-fixed separation points—like a sphere or circular cylinder—the drag coefficient may vary with Reynolds number Re, even up to very high values (Re of the order 10^7). [14] [15] For an object with well-defined fixed separation points, like a circular disk with its plane normal to the flow direction, the drag coefficient is constant for $Re > 3{,}500$.[15] Further the drag coefficient Cd is, in general, a function of the orientation of the flow with respect to the object (apart from symmetrical objects like a sphere).

36.3.1 Power

The power required to overcome the aerodynamic drag is given by:

$$P_d = \mathbf{F}_d \cdot \mathbf{v} = \tfrac{1}{2}\rho v^3 A C_d$$

Note that the power needed to push an object through a fluid increases as the cube of the velocity. A car cruising on a highway at 50 mph (80 km/h) may require only 10 horsepower (7.5 kW) to overcome air drag, but that same car at 100 mph (160 km/h) requires 80 hp (60 kW). With a doubling of speed the drag (force) quadruples per the formula. Exerting four times the force over a fixed distance produces four times as much work. At twice the speed the work (resulting in displacement over a fixed distance) is done twice as fast. Since power is the rate of doing work, four times the work done in half the time requires eight times the power.

36.3.2 Velocity of a falling object

Main article: Terminal velocity

The velocity as a function of time for an object falling through a non-dense medium, and released at zero relative-velocity $v = 0$ at time $t = 0$, is roughly given by a function involving a hyperbolic tangent (tanh):

$$v(t) = \sqrt{\frac{2mg}{\rho A C_d}} \tanh\left(t\sqrt{\frac{g\rho C_d A}{2m}}\right).$$

The hyperbolic tangent has a limit value of one, for large time t. In other words, velocity asymptotically approaches a maximum value called the terminal velocity vt:

$$v_t = \sqrt{\frac{2mg}{\rho A C_d}}.$$

For a potato-shaped object of average diameter d and of density ϱobj, terminal velocity is about

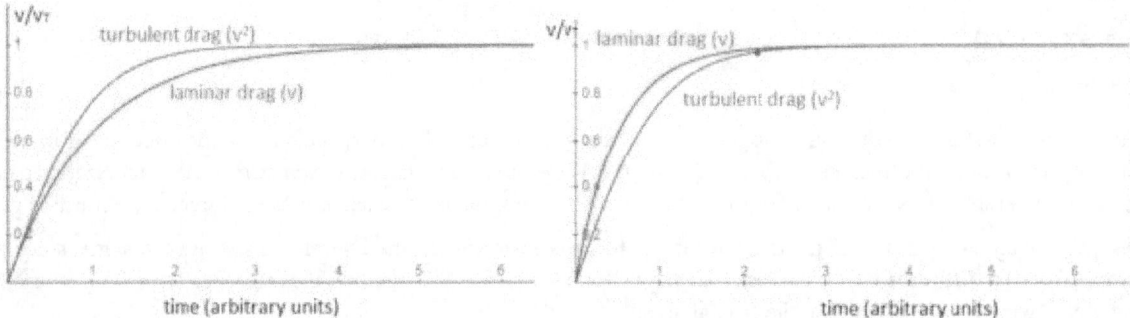

An object falling through viscous medium accelerates quickly towards its terminal speed, approaching gradually as the speed gets nearer to the terminal speed. Whether the object experiences turbulent or laminar drag changes the characteristic shape of the graph with turbulent flow resulting in a constant acceleration for a larger fraction of its accelerating time.

$$v_t = \sqrt{gd\frac{\rho_{obj}}{\rho}}.$$

For objects of water-like density (raindrops, hail, live objects—mammals, birds, insects, etc.) falling in air near the surface of the Earth at sea level, terminal velocity is roughly equal to

$$v_t = 90\sqrt{d},$$

with d in metre and vt in m/s. For example, for a human body (d ~ 0.6 m) v_t ~ 70 m/s, for a small animal like a cat (d ~ 0.2 m) v_t ~ 40 m/s, for a small bird (d ~ 0.05 m) v_t ~ 20 m/s, for an insect (d ~ 0.01 m) v_t ~ 9 m/s, and so on. Terminal velocity for very small objects (pollen, etc.) at low Reynolds numbers is determined by Stokes law.

Terminal velocity is higher for larger creatures, and thus potentially more deadly. A creature such as a mouse falling at its terminal velocity is much more likely to survive impact with the ground than a human falling at its terminal velocity. A small animal such as a cricket impacting at its terminal velocity will probably be unharmed. This, combined with the relative ratio of limb cross-sectional area vs. body mass (commonly referred to as the Square-cube law), explains why very small animals can fall from a large height and not be harmed.[16]

36.4 Very low Reynolds numbers: Stokes' drag

Main article: Stokes' law

The equation for **viscous resistance** or **linear drag** is appropriate for objects or particles moving through a fluid at relatively slow speeds where there is no turbulence (i.e. low Reynolds number, $R_e < 1$).[17] Note that purely laminar flow only exists up to Re = 0.1 under this definition. In this case, the force of drag is approximately proportional to velocity, but opposite in direction. The equation for viscous resistance is:[18]

$$\mathbf{F}_d = -b\mathbf{v}$$

where:

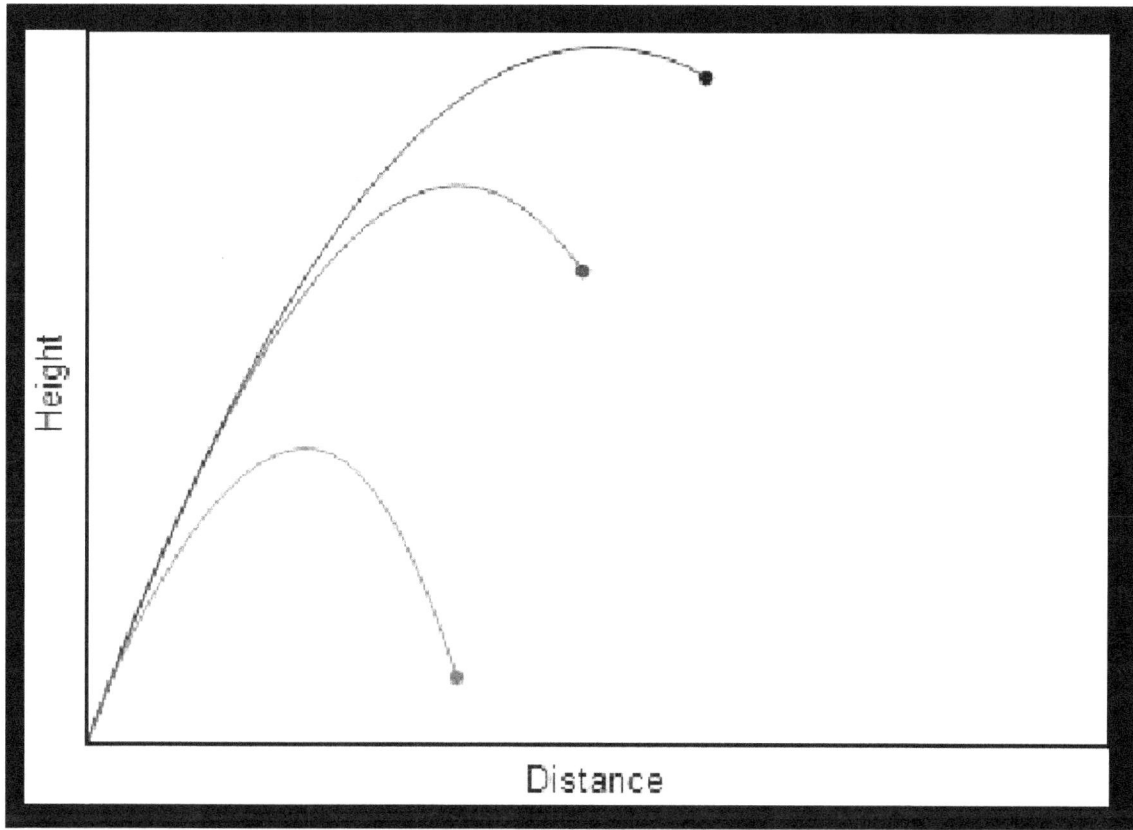

Trajectories of three objects thrown at the same angle (70°). The black object does not experience any form of drag and moves along a parabola. The blue object experiences Stokes' drag, and the green object Newton drag.

b is a constant that depends on the properties of the fluid and the dimensions of the object, and
v is the velocity of the object

When an object falls from rest, its velocity will be

$$v(t) = \frac{(\rho - \rho_0)Vg}{b}\left(1 - e^{-bt/m}\right)$$

which asymptotically approaches the terminal velocity $v_t = \frac{(\rho - \rho_0)Vg}{b}$. For a given b, heavier objects fall more quickly. For the special case of small spherical objects moving slowly through a viscous fluid (and thus at small Reynolds number), George Gabriel Stokes derived an expression for the drag constant:

$$b = 6\pi\eta r$$

where:

r is the Stokes radius of the particle, and η is the fluid viscosity.

The resulting expression for the drag is known as Stokes' drag:[19]

$$\mathbf{F}_d = -6\pi \eta r \, \mathbf{v}.$$

For example, consider a small sphere with radius r = 0.5 micrometre (diameter = 1.0 µm) moving through water at a velocity v of 10 µm/s. Using 10^{-3} Pa·s as the dynamic viscosity of water in SI units, we find a drag force of 0.09 pN. This is about the drag force that a bacterium experiences as it swims through water.

36.5 Drag in aerodynamics

Main article: Aerodynamic drag

36.5.1 Lift-induced drag

Main article: Lift-induced drag

Lift-induced drag (also called **induced drag**) is drag which occurs as the result of the creation of lift on a three-dimensional lifting body, such as the wing or fuselage of an airplane. Induced drag consists of two primary components, including drag due to the creation of vortices (**vortex drag**) and the presence of additional viscous drag (**lift-induced viscous drag**). The vortices in the flow-field, present in the wake of a lifting body, derive from the turbulent mixing of air of varying pressure on the upper and lower surfaces of the body, which is a necessary condition for the creation of lift.

With other parameters remaining the same, as the lift generated by a body increases, so does the lift-induced drag. For an aircraft in flight, this means that as the angle of attack, and therefore the lift coefficient, increases to the point of stall, so does the lift-induced drag. At the onset of stall, lift is abruptly decreased, as is lift-induced drag, but viscous pressure drag, a component of parasite drag, increases due to the formation of turbulent unattached flow on the surface of the body.

36.5.2 Parasitic drag

Main article: parasitic drag

Parasitic drag (also called **parasite drag**) is drag caused by moving a solid object through a fluid. Parasitic drag is made up of multiple components including viscous pressure drag (**form drag**), and drag due to surface roughness (**skin friction drag**). Additionally, the presence of multiple bodies in relative proximity may incur so called **interference drag**, which is sometimes described as a component of parasitic drag.

In aviation, induced drag tends to be greater at lower speeds because a high angle of attack is required to maintain lift, creating more drag. However, as speed increases the induced drag becomes much less, but parasitic drag increases because the fluid is flowing more quickly around protruding objects increasing friction or drag. At even higher speeds in the transonic, wave drag enters the picture. Each of these forms of drag changes in proportion to the others based on speed. The combined overall drag curve therefore shows a minimum at some airspeed - an aircraft flying at this speed will be at or close to its optimal efficiency. Pilots will use this speed to maximize endurance (minimum fuel consumption), or maximize gliding range in the event of an engine failure.

36.5.3 Power curve in aviation

The interaction of parasitic and induced drag *vs.* airspeed can be plotted as a characteristic curve, illustrated here. In aviation, this is often referred to as the *power curve*, and is important to pilots because it shows that, below a certain airspeed, maintaining airspeed counterintuitively requires *more* thrust as speed decreases, rather than less. The consequences of

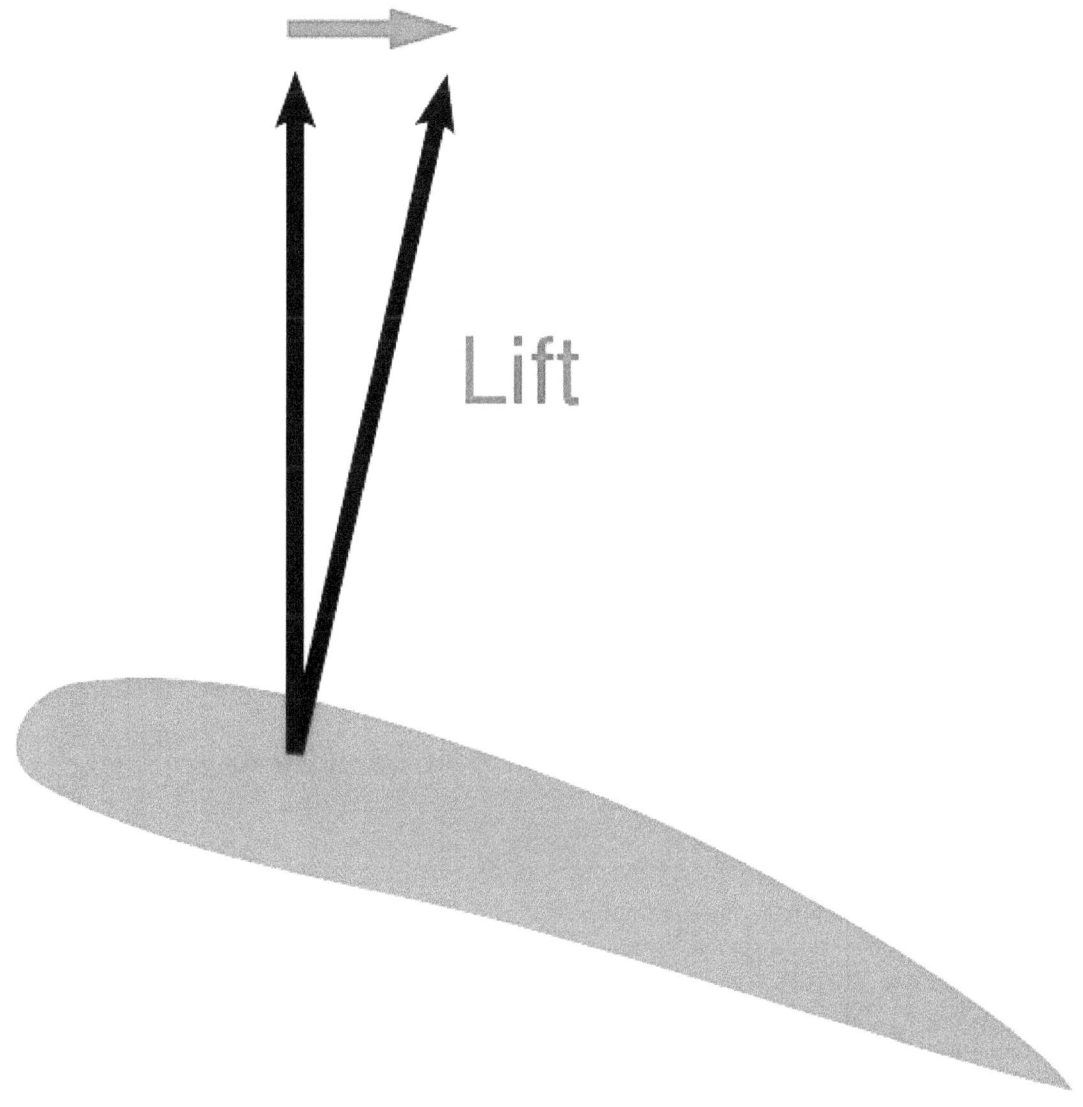

Induced drag vs. lift[20][21]

being "behind the curve" in flight are important and are taught as part of pilot training. At the subsonic airspeeds where the "U" shape of this curve is significant, wave drag has not yet become a factor, and so it is not shown in the curve.

36.5.4 Wave drag in transonic and supersonic flow

36.5. DRAG IN AERODYNAMICS

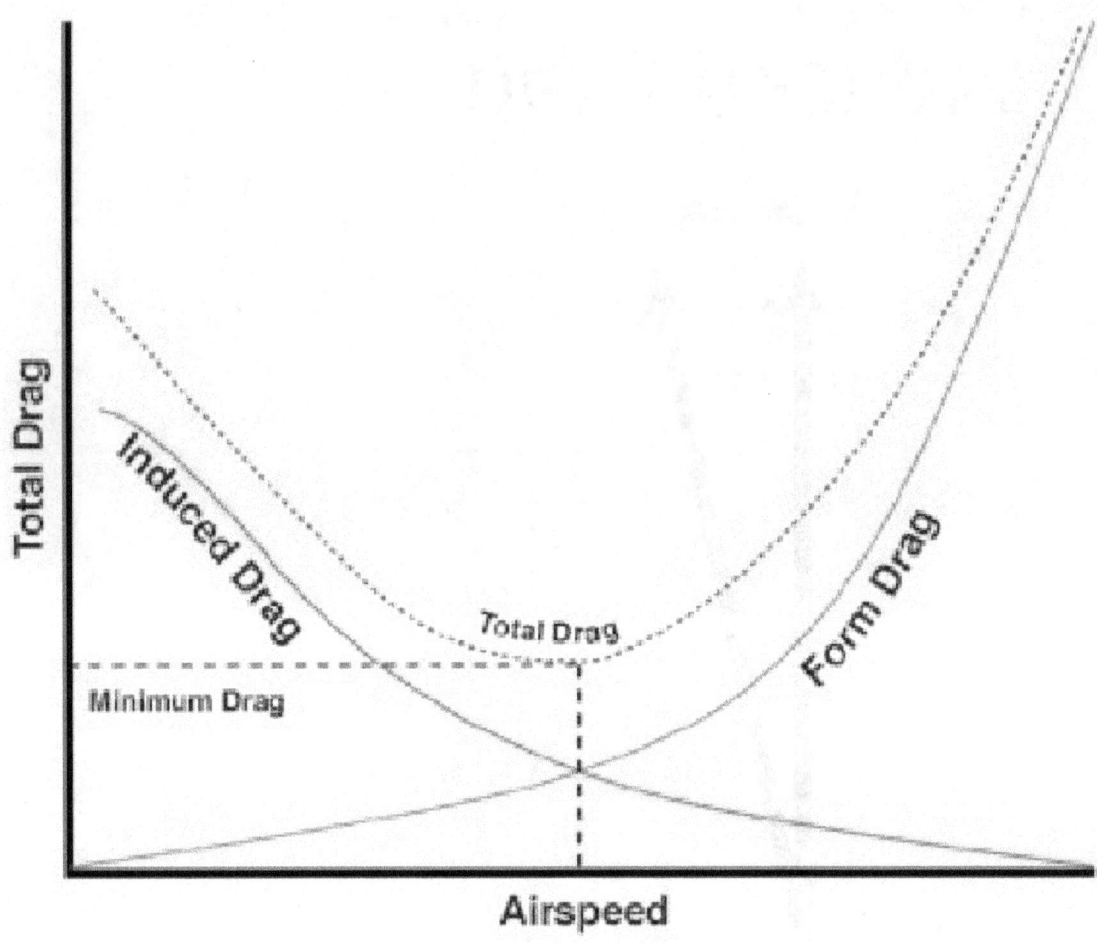

The power curve: *form and induced drag vs. airspeed*

Wave drag (also called **compressibility drag**) is drag which is created by the presence of a body moving at high speed through a compressible fluid. In aerodynamics, Wave drag consists of multiple components depending on the speed regime of the flight.

In transonic flight (Mach numbers greater than about 0.8 and less than about 1.4), wave drag is the result of the formation of shockwaves on the body, formed when areas of local supersonic (Mach number greater than 1.0) flow are created. In practice, supersonic flow occurs on bodies traveling well below the speed of sound, as the local speed of air on a body increases when it accelerates over the body, in this case above Mach 1.0. However, full supersonic flow over the vehicle will not develop until well past Mach 1.0. Aircraft flying at transonic speed often incur wave drag through the normal course of operation. In transonic flight, wave drag is commonly referred to as **transonic compressibility drag**. Transonic compressibility drag increases significantly as the speed of flight increases towards Mach 1.0, dominating other forms of drag at these speeds.

In supersonic flight (Mach numbers greater than 1.0), **wave drag** is the result of shockwaves present on the body, typically **oblique shockwaves** formed at the leading and trailing edges of the body. In highly supersonic flows, or in bodies with turning angles sufficiently large, **unattached shockwaves**, or **bow waves** will instead form. Additionally, local areas of transonic flow behind the initial shockwave may occur at lower supersonic speeds, and can lead to the development of additional, smaller shockwaves present on the surfaces of other lifting bodies, similar to those found in transonic flows. In supersonic flow regimes, **wave drag** is commonly separated into two components, **supersonic lift-dependent wave drag** and **supersonic volume-dependent wave drag**.

The closed form solution for the minimum wave drag of a body of revolution with a fixed length was found by Sears and

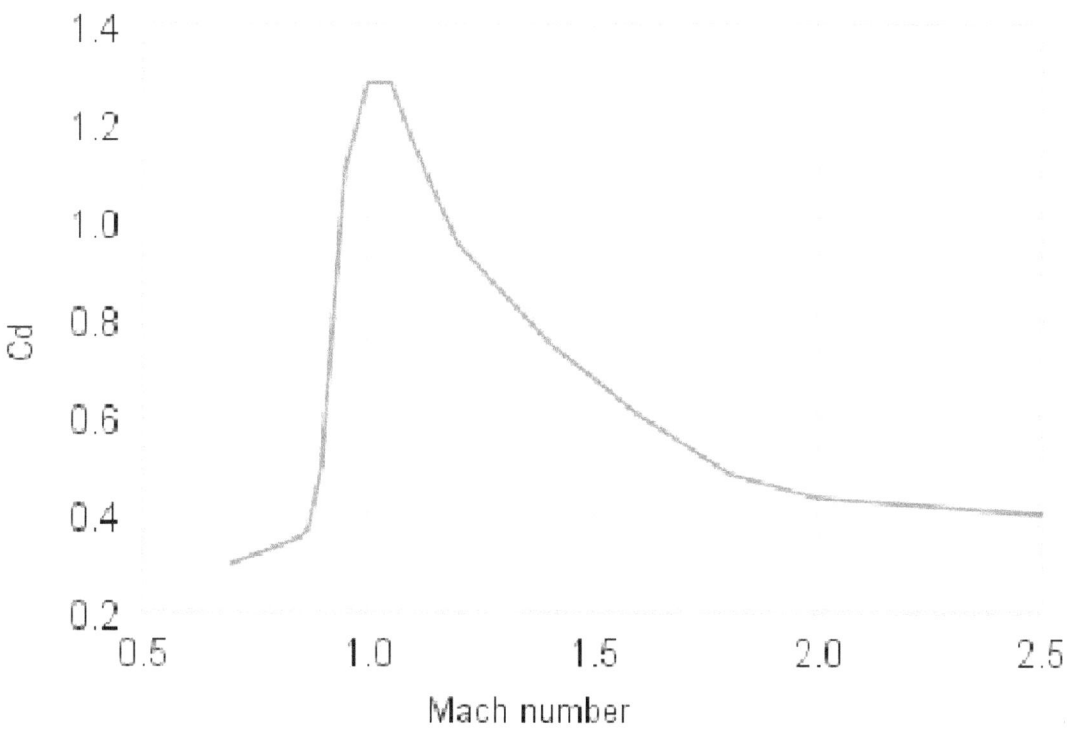

Qualitative variation in Cd factor with Mach number for aircraft

Haack, and is known as the **Sears-Haack Distribution**. Similarly, for a fixed volume, the shape for minimum wave drag is the **Von Karman Ogive**.

Busemann's Biplane is not, in principle, subject to wave drag at all when operated at its design speed, but is incapable of generating lift.

36.6 d'Alembert's paradox

In 1752 d'Alembert proved that potential flow, the 18th century state-of-the-art inviscid flow theory amenable to mathematical solutions, resulted in the prediction of zero drag. This was in contradiction with experimental evidence, and became known as d'Alembert's paradox. In the 19th century the Navier–Stokes equations for the description of viscous flow were developed by Saint-Venant, Navier and Stokes. Stokes derived the drag around a sphere at very low Reynolds numbers, the result of which is called Stokes law.[22]

In the limit of high-Reynolds numbers the Navier–Stokes equations approach the inviscid Euler equations; of which the potential-flow solutions considered by d'Alembert are solutions. However, at high Reynolds numbers all experiments showed there is drag. Attempts to construct inviscid steady flow solutions to the Euler equations, other than the potential flow solutions, did not result in realistic results.[22]

The notion of boundary layers—introduced by Prandtl in 1904, founded on both theory and experiments—explained the causes of drag at high Reynolds numbers. The boundary layer is the thin layer of fluid close to the object's boundary, where viscous effects remain important when the viscosity becomes very small (or equivalently the Reynolds number becomes very large).[22]

36.7 See also

- Added mass
- Aerodynamic force
- Angle of attack
- Boundary layer
- Coandă effect
- Drag crisis
- Drag coefficient
- Drag equation
- Gravity drag
- Keulegan–Carpenter number
- Morison equation
- Nose cone design
- Parasitic drag
- Ram pressure
- Reynolds number
- Stall (fluid mechanics)
- Stokes' law
- Terminal velocity
- Windage

36.8 Notes

[1] http://www.merriam-webster.com/dictionary/drag

[2] French (1970), p. 211, Eq. 7-20

[3] "What is Drag?".

[4] G. Falkovich (2011). *Fluid Mechanics (A short course for physicists)*. Cambridge University Press. ISBN 978-1-107-00575-4.

[5] Eiffel, Gustave (1913). *The Resistance of The Air and Aviation*. London: Constable &Co Ltd.

[6] Marchaj, C. A. (2003). *Sail performance : techniques to maximise sail power* (Rev. ed.). London: Adlard Coles Nautical. pp. 147 figure 127 lift vs drag polar curves. ISBN 978-0-7136-6407-2.

[7] Forces on sails#Relationship of lift coefficient to angle of incidence: polar diagram

[8] Drayton, Fabio Fossati ; translated by Martyn (2009). *Aero-hydrodynamics and the performance of sailing yachts : the science behind sailing yachts and their design*. Camden, Maine: International Marine /McGraw-Hill. pp. 98 Fig 5.17 Chapter five Sailing Boat Aerodynamics. ISBN 978-0-07-162910-2.

[9] "Calculating Viscous Flow: Velocity Profiles in Rivers and Pipes" (PDF). Retrieved 16 October 2011.

[10] "Viscous Drag Forces". Retrieved 16 October 2011.

[11] Note that for the Earth's atmosphere, the air density can be found using the barometric formula. It is 1.293 kg/m^3 at 0 °C and 1 atmosphere.

[12] *Size effects on drag*, from NASA Glenn Research Center.

[13] *Wing geometry definitions*, from NASA Glenn Research Center.

[14] Roshko, Anatol (1961). "Experiments on the flow past a circular cylinder at very high Reynolds number". *Journal of Fluid Mechanics* **10** (3): 345–356. Bibcode:1961JFM....10..345R. doi:10.1017/S0022112061000950.

[15] Batchelor (1967), p. 341.

[16] Haldane, J.B.S., "On Being the Right Size"

[17] Drag Force

[18] Air friction, from Department of Physics and Astronomy, Georgia State University

[19] Collinson, Chris; Roper, Tom (1995). *Particle Mechanics*. Butterworth-Heinemann. p. 30. ISBN 9780080928593.

[20] Clancy, L.J. (1975) *Aerodynamics* Fig 5.24. Pitman Publishing Limited, London. ISBN 0-273-01120-0

[21] Hurt, H. H. (1965) *Aerodynamics for Naval Aviators*, Figure 1.30, NAVWEPS 00-80T-80

[22] Batchelor (2000), pp. 337–343.

36.9 References

- French, A. P. (1970). *Newtonian Mechanics (The M.I.T. Introductory Physics Series)* (1st ed.). W. W. Norton & Company Inc., New York. ISBN 978-0-393-09958-4.

- G. Falkovich (2011). *Fluid Mechanics (A short course for physicists)*. Cambridge University Press. ISBN 978-1-107-00575-4.

- Serway, Raymond A.; Jewett, John W. (2004). *Physics for Scientists and Engineers* (6th ed.). Brooks/Cole. ISBN 978-0-534-40842-8.

- Tipler, Paul (2004). *Physics for Scientists and Engineers: Mechanics, Oscillations and Waves, Thermodynamics* (5th ed.). W. H. Freeman. ISBN 978-0-7167-0809-4.

- Huntley, H. E. (1967). *Dimensional Analysis*. Dover. LOC 67-17978.

- Batchelor, George (2000). *An introduction to fluid dynamics*. Cambridge Mathematical Library (2nd ed.). Cambridge University Press. ISBN 978-0-521-66396-0. MR 1744638.

- Clancy, L.J. (1975), *Aerodynamics*, Pitman Publishing Limited, London. ISBN 978-0-273-01120-0

36.10 External links

- Educational materials on air resistance
- Aerodynamic Drag and its effect on the acceleration and top speed of a vehicle.
- Vehicle Aerodynamic Drag calculator based on drag coefficient, frontal area and speed.
- Smithsonian National Air and Space Museum's How Things Fly website
- Effect of dimples on a golf ball and a car

Chapter 37

Theory of relativity

This article is about the scientific concept. For philosophical or ontological theories about relativity, see Relativism. For the silent film, see The Einstein Theory of Relativity.

The **theory of relativity**, or simply **relativity** in physics, usually encompasses two theories by Albert Einstein: special

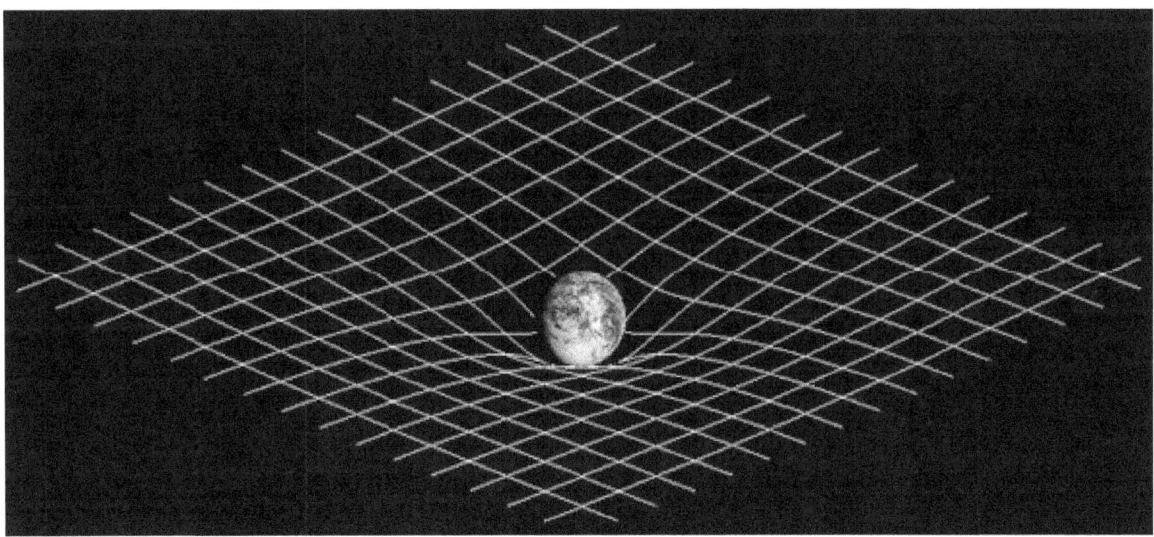

Two-dimensional projection of a three-dimensional analogy of spacetime curvature described in general relativity

relativity and general relativity.[1]

Concepts introduced by the theories of relativity include:

- Measurements of various quantities are *relative* to the velocities of observers. In particular, space contracts and time dilates.

- Spacetime: space and time should be considered together and in relation to each other.

- Space is a physical entity that can be changed, space is not just nothing, space can affect mass (gravity)[2]

- The speed of light is nonetheless invariant, the same for all observers.

The term "theory of relativity" was based on the expression "relative theory" (German: *Relativtheorie*) used in 1906 by Max Planck, who emphasized how the theory uses the principle of relativity. In the discussion section of the same paper, Alfred Bucherer used for the first time the expression "theory of relativity" (German: *Relativitätstheorie*).[3][4]

37.1 Scope

The theory of relativity transformed theoretical physics and astronomy during the 20th century. When first published, relativity superseded a 200-year-old theory of mechanics created primarily by Isaac Newton.[5][6][7]

In the field of physics, relativity improved the science of elementary particles and their fundamental interactions, along with ushering in the nuclear age. With relativity, cosmology and astrophysics predicted extraordinary astronomical phenomena such as neutron stars, black holes and gravitational waves.[5][6][7]

37.1.1 Two-theory view

The theory of relativity was representative of more than a single new physical theory. There are some explanations for this. First, special relativity was published in 1905, and the final form of general relativity was published in 1916.[5]

Second, special relativity applies to elementary particles and their interactions, whereas general relativity applies to the cosmological and astrophysical realm, including astronomy.[5]

Third, special relativity was accepted in the physics community by 1920. This theory rapidly became a significant and necessary tool for theorists and experimentalists in the new fields of atomic physics, nuclear physics, and quantum mechanics. Conversely, general relativity did not appear to be as useful. There appeared to be little applicability for experimentalists as most applications were for astronomical scales. It seemed limited to only making minor corrections to predictions of Newtonian gravitation theory.[5]

Finally, the mathematics of general relativity appeared to be very difficult. Consequently, it was thought that a small number of people in the world, at that time, could fully understand the theory in detail, but this has been discredited by Richard Feynman. Then, at around 1960 a critical resurgence in interest occurred which has resulted in making general relativity central to physics and astronomy. New mathematical techniques applicable to the study of general relativity substantially streamlined calculations. From this, physically discernible concepts were isolated from the mathematical complexity. Also, the discovery of exotic astronomical phenomena, in which general relativity was relevant, helped to catalyze this resurgence. The astronomical phenomena included quasars (1963), the 3-kelvin microwave background radiation (1965), pulsars (1967), and the discovery of the first black hole candidates (1981).[5]

37.2 On the theory of relativity

Einstein stated that the theory of relativity belongs to a class of "principle-theories". As such it employs an analytic method. This means that the elements which comprise this theory are not based on hypothesis but on empirical discovery. The empirical discovery leads to understanding the general characteristics of natural processes. Mathematical models are then developed to describe accurately the observed natural processes. Therefore, by analytical means the necessary conditions that have to be satisfied are deduced. Separate events must satisfy these conditions. Experience should then match the conclusions.[8]

The special theory of relativity and the general theory of relativity are connected. As stated below, special theory of relativity applies to all physical phenomena except gravity. The general theory provides the law of gravitation, and its relation to other forces of nature.[8]

37.3 Special relativity

Main articles: Special relativity and Introduction to special relativity
 Special relativity is a theory of the structure of spacetime. It was introduced in Einstein's 1905 paper "On the Electrodynamics of Moving Bodies" (for the contributions of many other physicists see History of special relativity). Special relativity is based on two postulates which are contradictory in classical mechanics:

  1. The laws of physics are the same for all observers in uniform motion relative to one another (principle of relativity).

USSR stamp dedicated to Albert Einstein

2. The speed of light in a vacuum is the same for all observers, regardless of their relative motion or of the motion of the light source.

The resultant theory copes with experiment better than classical mechanics. For instance, postulate 2 explains the results of the Michelson–Morley experiment. Moreover, the theory has many surprising and counterintuitive consequences. Some of these are:

- Relativity of simultaneity: Two events, simultaneous for one observer, may not be simultaneous for another observer if the observers are in relative motion.

- Time dilation: Moving clocks are measured to tick more slowly than an observer's "stationary" clock.

- Relativistic mass

- Length contraction: Objects are measured to be shortened in the direction that they are moving with respect to the observer.

- Mass–energy equivalence: $E = mc^2$, energy and mass are equivalent and transmutable.

- Maximum speed is finite: No physical object, message or field line can travel faster than the speed of light in a vacuum.

- The effect of Gravity can only travel through space at the speed of light, not faster or instantaneously.[2]

The defining feature of special relativity is the replacement of the Galilean transformations of classical mechanics by the Lorentz transformations. (See Maxwell's equations of electromagnetism).

37.4 General relativity

General relativity is a theory of gravitation developed by Einstein in the years 1907–1915. The development of general relativity began with the equivalence principle, under which the states of accelerated motion and being at rest in a gravitational field (for example when standing on the surface of the Earth) are physically identical. The upshot of this is that free fall is inertial motion: an object in free fall is falling because that is how objects move when there is no force being exerted on them, instead of this being due to the force of gravity as is the case in classical mechanics. This is incompatible with classical mechanics and special relativity because in those theories inertially moving objects cannot accelerate with respect to each other, but objects in free fall do so. To resolve this difficulty Einstein first proposed that spacetime is curved. In 1915, he devised the Einstein field equations which relate the curvature of spacetime with the mass, energy, and momentum within it.

Some of the consequences of general relativity are:

- Clocks run slower in deeper gravitational wells.[9] This is called gravitational time dilation.
- Orbits precess in a way unexpected in Newton's theory of gravity. (This has been observed in the orbit of Mercury and in binary pulsars).
- Rays of light bend in the presence of a gravitational field.
- Rotating masses "drag along" the spacetime around them; a phenomenon termed "frame-dragging".
- The universe is expanding, and the far parts of it are moving away from us faster than the speed of light.

Technically, general relativity is a theory of gravitation whose defining feature is its use of the Einstein field equations. The solutions of the field equations are metric tensors which define the topology of the spacetime and how objects move inertially.

37.5 Experimental evidence

37.5.1 Tests of special relativity

Main article: Tests of special relativity

Like all falsifiable scientific theories, relativity makes predictions that can be tested by experiment. In the case of special relativity, these include the principle of relativity, the constancy of the speed of light, and time dilation.[10] The predictions of special relativity have been confirmed in numerous tests since Einstein published his paper in 1905, but three experiments conducted between 1881 and 1938 were critical to its validation. These are the Michelson–Morley experiment, the Kennedy–Thorndike experiment, and the Ives–Stilwell experiment. Einstein derived the Lorentz transformations from first principles in 1905, but these three experiments allow the transformations to be induced from experimental evidence.

Maxwell's equations – the foundation of classical electromagnetism – describe light as a wave which moves with a characteristic velocity. The modern view is that light needs no medium of transmission, but Maxwell and his contemporaries were convinced that light waves were propagated in a medium, analogous to sound propagating in air, and ripples propagating on the surface of a pond. This hypothetical medium was called the luminiferous aether, at rest relative to the "fixed stars" and through which the Earth moves. Fresnel's partial ether dragging hypothesis ruled out the measurement of first-order (v/c) effects, and although observations of second-order effects (v^2/c^2) were possible in principle, Maxwell thought they were too small to be detected with then-current technology.[11][12]

The Michelson–Morley experiment was designed to detect second order effects of the "aether wind" – the motion of the aether relative to the earth. Michelson designed an instrument called the Michelson interferometer to accomplish this. The apparatus was more than accurate enough to detect the expected effects, but he obtained a null result when

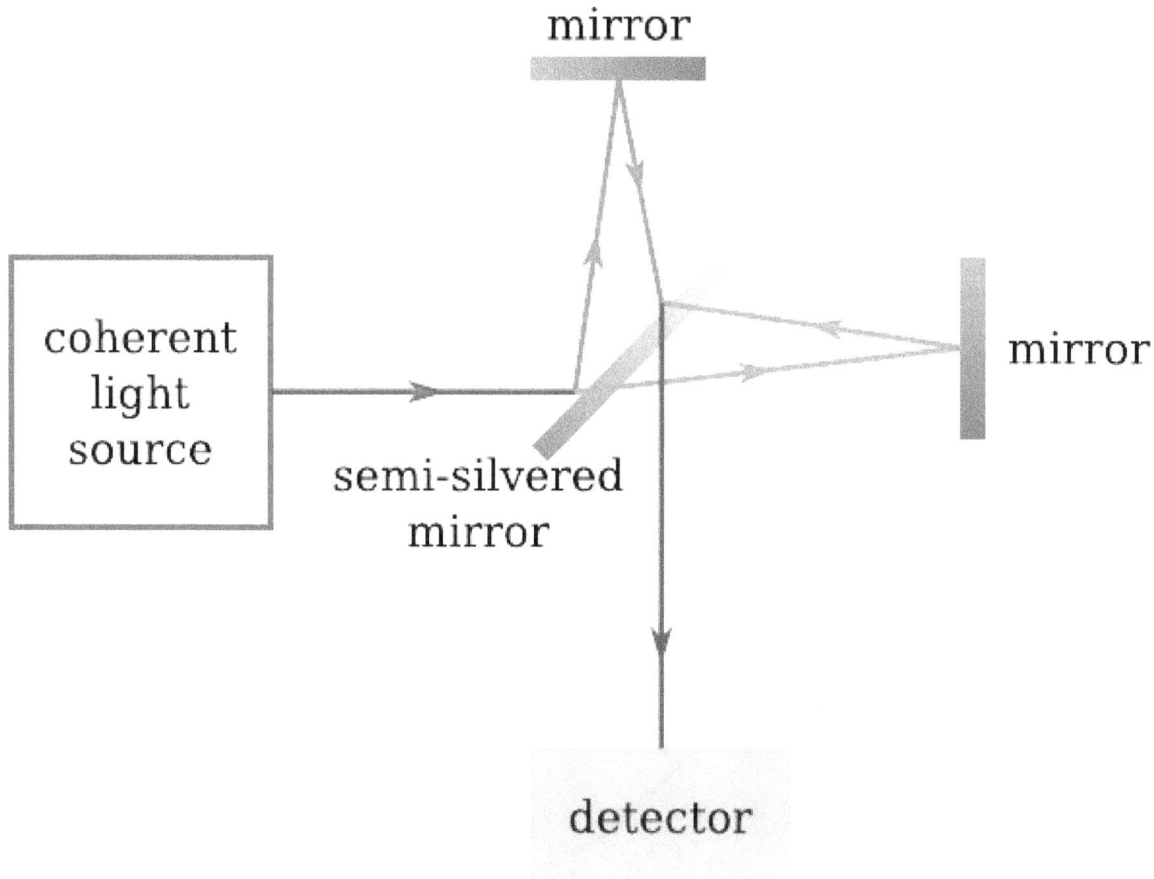

A diagram of the Michelson–Morley experiment

the first experiment was conducted in 1881,[13] and again in 1887.[14] Although the failure to detect an aether wind was a disappointment, the results were accepted by the scientific community.[12] In an attempt to salvage the aether paradigm, Fitzgerald and Lorentz independently created an *ad hoc* hypothesis in which the length of material bodies changes according to their motion through the aether.[15] This was the origin of FitzGerald–Lorentz contraction, and their hypothesis had no theoretical basis. The interpretation of the null result of the Michelson–Morley experiment is that the round-trip travel time for light is isotropic (independent of direction), but the result alone is not enough to discount the theory of the aether or validate the predictions of special relativity.[16][17]

While the Michelson–Morley experiment showed that the velocity of light is isotropic, it said nothing about how the magnitude of the velocity changed (if at all) in different inertial frames. The Kennedy–Thorndike experiment was designed to do that, and was first performed in 1932 by Roy Kennedy and Edward Thorndike.[18] They obtained a null result, and concluded that "there is no effect ... unless the velocity of the solar system in space is no more than about half that of the earth in its orbit".[17][19] That possibility was thought to be too coincidental to provide an acceptable explanation, so from the null result of their experiment it was concluded that the round-trip time for light is the same in all inertial reference frames.[16][17]

The Ives–Stilwell experiment was carried out by Herbert Ives and G.R. Stilwell first in 1938[20] and with better accuracy in 1941.[21] It was designed to test the transverse Doppler effect – the redshift of light from a moving source in a direction perpendicular to its velocity – which had been predicted by Einstein in 1905. The strategy was to compare observed Doppler shifts with what was predicted by classical theory, and look for a Lorentz factor correction. Such a correction was observed, from which was concluded that the frequency of a moving atomic clock is altered according to special relativity.[16][17]

Those classic experiments have been repeated many times with increased precision. Other experiments include, for

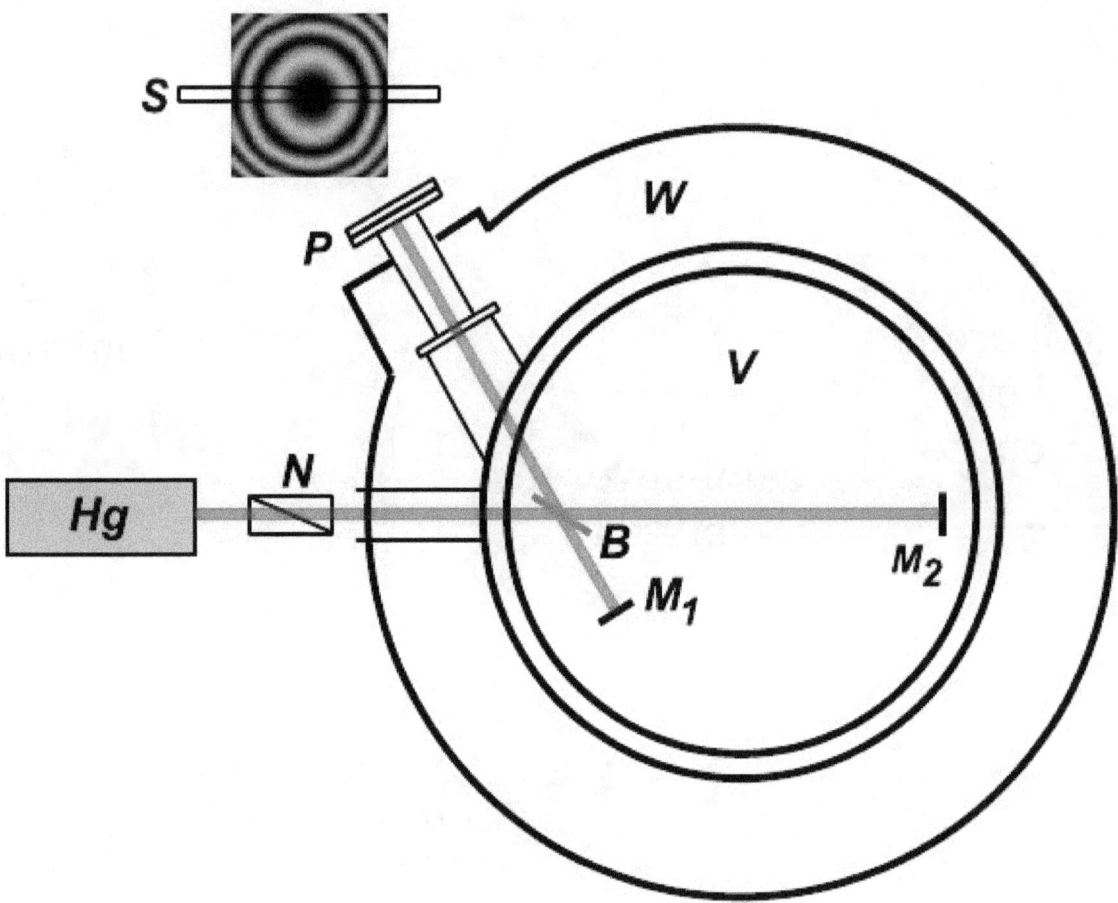

The Kennedy–Thorndike experiment shown with interference fringes.

instance, relativistic energy and momentum increase at high velocities, time dilation of moving particles, and modern searches for Lorentz violations.

37.5.2 Tests of general relativity

General relativity has also been confirmed many times, the classic experiments being the perihelion precession of Mercury's orbit, the deflection of light by the Sun, and the gravitational redshift of light. Other tests confirmed the equivalence principle and frame dragging.

37.6 History

The history of special relativity consists of many theoretical results and empirical findings obtained by Albert A. Michelson, Hendrik Lorentz, Henri Poincaré and others. It culminated in the theory of special relativity proposed by Albert Einstein, and subsequent work of Max Planck, Hermann Minkowski and others.

General relativity (GR) is a theory of gravitation that was developed by Albert Einstein between 1907 and 1915, with contributions by many others after 1915.

Currently, it can be said that far from being simply of theoretical scientific interest or requiring experimental verification, the analysis of relativistic effects on time measurement is an important practical engineering concern in the operation of the global positioning systems such as GPS, GLONASS, and the forthcoming Galileo, as well as in the high precision dissemination of time.[22] Instruments ranging from electron microscopes to particle accelerators simply will not work if relativistic considerations are omitted.

37.7 Everyday applications

The theory of relativity is used in many of our modern electronics such as the Global Positioning System (GPS). GPS systems are made up of three components, the control component, the space component, and the user component. The space component consists of satellites that are placed in specific orbits. The control component consists of a station to which all of the data from the space component is sent. Many relativistic effects occur in GPS systems. Since each of the components is in different reference frames, all of the relativistic effects need to be accounted for so that the GPS works with precision. The clocks used in the GPS systems need to be synchronized. In GPS systems, the gravitational field of the Earth has to be accounted for. There are relativistic effects within the satellite that is in space that need to be accounted for too. GPS systems work with such precision because of the Theory of Relativity.[23]

37.8 See also

- Doubly special relativity
- Galilean invariance
- General relativity references
- Scale relativity
- Special relativity references

37.9 References

[1] Einstein A. (1916), *Relativity: The Special and General Theory* (Translation 1920), New York: H. Holt and Company

[2] Greene, Brian. "The Theory of Relativity, Then and Now". Retrieved 2015-09-26.

[3] Planck, Max (1906), "Die Kaufmannschen Messungen der Ablenkbarkeit der β-Strahlen in ihrer Bedeutung für die Dynamik der Elektronen (The Measurements of Kaufmann on the Deflectability of β-Rays in their Importance for the Dynamics of the Electrons)", *Physikalische Zeitschrift* 7: 753–761

[4] Miller, Arthur I. (1981), *Albert Einstein's special theory of relativity. Emergence (1905) and early interpretation (1905–1911)*, Reading: Addison–Wesley, ISBN 0-201-04679-2

[5] Will, Clifford M (August 1, 2010). "Relativity". *Grolier Multimedia Encyclopedia*. Retrieved 2010-08-01.

[6] Will, Clifford M (August 1, 2010). "Space-Time Continuum". *Grolier Multimedia Encyclopedia*. Retrieved 2010-08-01.

[7] Will, Clifford M (August 1, 2010). "Fitzgerald–Lorentz contraction". *Grolier Multimedia Encyclopedia*. Retrieved 2010-08-01.

[8] Einstein, Albert (November 28, 1919). "Time, Space, and Gravitation". *The Times*.

[9] Feynman, Richard Phillips; Morínigo, Fernando B.; Wagner, William; Pines, David; Hatfield, Brian (2002). *Feynman Lectures on Gravitation*. Westview Press. p. 68. ISBN 0-8133-4038-1., Lecture 5

[10] Roberts, T; Schleif, S; Dlugosz, JM (ed.) (2007). "What is the experimental basis of Special Relativity?". *Usenet Physics FAQ*. University of California, Riverside. Retrieved 2010-10-31.

[11] Maxwell, James Clerk (1880), "On a Possible Mode of Detecting a Motion of the Solar System through the Luminiferous Ether", *Nature* **21**: 314–315, Bibcode:1880Natur..21S.314., doi:10.1038/021314c0

[12] Pais, Abraham (1982). *"Subtle is the Lord ...": The Science and the Life of Albert Einstein* (1st ed.). Oxford: Oxford Univ. Press. pp. 111–113. ISBN 0192806726.

[13] Michelson, Albert A. (1881). "The Relative Motion of the Earth and the Luminiferous Ether". *American Journal of Science* **22**: 120–129. doi:10.2475/ajs.s3-22.128.120.

[14] Michelson, Albert A. & Morley, Edward W. (1887). "On the Relative Motion of the Earth and the Luminiferous Ether". *American Journal of Science* **34**: 333–345. doi:10.2475/ajs.s3-34.203.333.

[15] Pais, Abraham (1982). *"Subtle is the Lord ...": The Science and the Life of Albert Einstein* (1st ed.). Oxford: Oxford Univ. Press. p. 122. ISBN 0192806726.

[16] Robertson, H.P. (July 1949). "Postulate versus Observation in the Special Theory of Relativity". *Reviews of Modern Physics* **21** (3): 378–382. Bibcode:1949RvMP...21..378R. doi:10.1103/RevModPhys.21.378.

[17] Taylor, Edwin F.; John Archibald Wheeler (1992). *Spacetime physics: Introduction to Special Relativity* (2nd ed.). New York: W.H. Freeman. pp. 84–88. ISBN 0716723271.

[18] Kennedy, R. J.; Thorndike, E. M. (1932). "Experimental Establishment of the Relativity of Time". *Physical Review* **42** (3): 400–418. Bibcode:1932PhRv...42..400K. doi:10.1103/PhysRev.42.400.

[19] Robertson, H.P. (July 1949). "Postulate versus Observation in the Special Theory of Relativity". *Reviews of Modern Physics* **21** (3): 381. Bibcode:1949RvMP...21..378R. doi:10.1103/revmodphys.21.378.

[20] Ives, H. E.; Stilwell, G. R. (1938). "An experimental study of the rate of a moving atomic clock". *Journal of the Optical Society of America* **28** (7): 215. Bibcode:1938JOSA...28..215I. doi:10.1364/JOSA.28.000215.

[21] Ives, H. E.; Stilwell, G. R. (1941). "An experimental study of the rate of a moving atomic clock. II". *Journal of the Optical Society of America* **31** (5): 369. Bibcode:1941JOSA...31..369I. doi:10.1364/JOSA.31.000369.

[22] Francis, S.; B. Ramsey; S. Stein; Leitner, J.; M. Moreau. J. M.; Burns, R.; Nelson, R. A.; Bartholomew, T. R.; Gifford, A. (2002). "Timekeeping and Time Dissemination in a Distributed Space-Based Clock Ensemble" (PDF). *Proceedings 34th Annual Precise Time and Time Interval (PTTI) Systems and Applications Meeting*: 201–214. Retrieved 14 April 2013.

[23] http://relativity.livingreviews.org/Articles/lrr-2003-1/download/lrr-2003-1Color.pdf

37.10 Further reading

- Einstein, Albert (2005). *Relativity: The Special and General Theory*. Translated by Robert W. Lawson (The masterpiece science ed.). New York: Pi Press. ISBN 0131862618.

- Einstein, Albert; trans. Schilpp; Paul Arthur (1979). *Albert Einstein, Autobiographical Notes* (A Centennial ed.). La Salle, Ill.: Open Court Publishing Co. ISBN 0875483526.

- Einstein, Albert (2009). *Einstein's Essays in Science*. Translated by Alan Harris (Dover ed.). Mineola, N.Y.: Dover Publications. ISBN 9780486470115.

- Einstein, Albert (1956) [1922]. *The Meaning of Relativity* (5 ed.). Princeton University Press.

37.11 External links

- Theory of relativity at DMOZ
- Relativity Milestones: Timeline of Notable Relativity Scientists and Contributions
- The dictionary definition of theory of relativity at Wiktionary
- Media related to Theory of relativity at Wikimedia Commons

Chapter 38

Radiation pressure

Radiation pressure is the pressure exerted upon any surface exposed to electromagnetic radiation. Radiation pressure implies an interaction between electromagnetic radiation and bodies of various types, including clouds of particles or gases. The interactions can be absorption, reflection, or some of both (the common case). Bodies also emit radiation and thereby experience a resulting pressure.

The forces generated by radiation pressure are generally too small to be detected under everyday circumstances; however, they do play a crucial role in some settings, such as astronomy and astrodynamics. For example, had the effects of the sun's radiation pressure on the spacecraft of the Viking program been ignored, the spacecraft would have missed Mars orbit by about 15,000 kilometers.[1]

This article addresses the macroscopic aspects of radiation pressure. Detailed quantum mechanical aspects of interactions are addressed in specialized articles on the subject. The details of how photons of various wavelengths interact with atoms can be explored through links in the *See also* section.

38.1 Discovery

Johannes Kepler put forward the concept of radiation pressure back in 1619 to explain the observation that a tail of a comet always points away from the Sun.[2]

The assertion that light, as electromagnetic radiation, has the property of momentum and thus exerts a pressure upon any surface exposed to it was published by James Clerk Maxwell in 1862, and proven experimentally by Russian physicist Pyotr Lebedev in 1900[3] and by Ernest Fox Nichols and Gordon Ferrie Hull in 1901.[4] The pressure is very feeble, but can be detected by allowing the radiation to fall upon a delicately poised vane of reflective metal in a Nichols radiometer (this should not be confused with the Crookes radiometer, whose characteristic motion is *not* caused by radiation pressure but by impacting gas molecules).

38.2 Theory

See also: Electromagnetic radiation and Speed of light

Radiation pressure can be analyzed as interactions by either electromagnetic waves or particles (photons). The waves and photons both have the property of momentum, which allows their interchangeability under classical conditions.

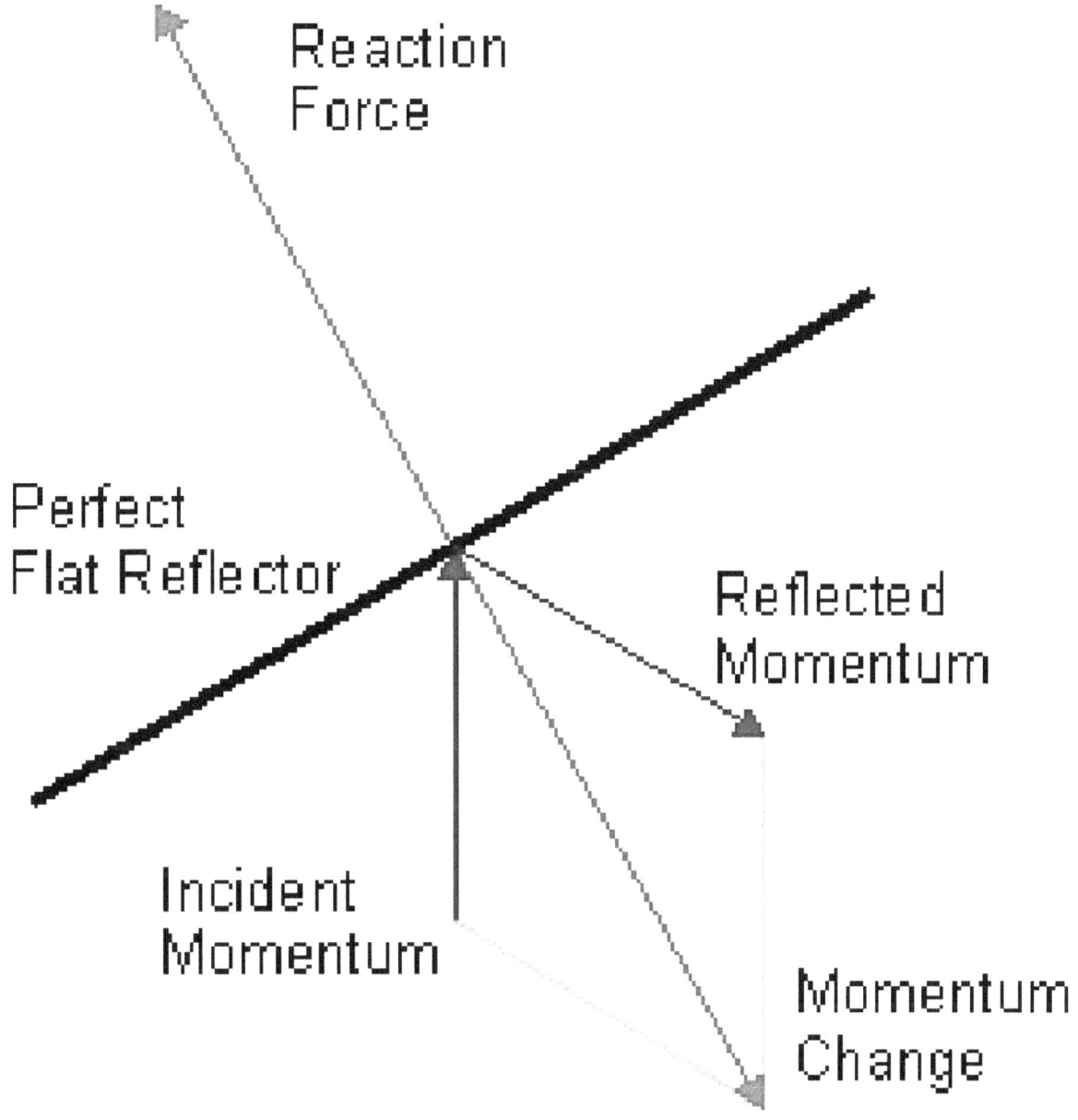

Force on a reflector results from reflecting the photon flux

38.2.1 Radiation pressure by absorption (using classical electromagnetism: waves)

Main article: Poynting vector

According to Maxwell's theory of electromagnetism, an electromagnetic wave carries momentum, which can be transferred to a reflecting or absorbing surface hit by the wave.

The energy flux (intensity) is expressed by the Poynting vector $\mathbf{S} = \mathbf{E} \times \mathbf{H}$, whose magnitude we denote by S. S divided by the square of the speed of light in free space is the density of the linear momentum of the electromagnetic field. The time-averaged intensity $\langle S \rangle$ divided by the speed of light in free space is the radiation pressure exerted by an electromagnetic wave on the surface of a target, if the wave is completely absorbed:

$$P_{absorb} = \frac{\langle S \rangle}{c} = \frac{E_f}{c} \; (\text{N·m}^{-2} \text{ or Pa})$$

38.2. THEORY

where P is pressure, Ef is energy flux (intensity) in W/m^2, c is speed of light in vacuum.

If the absorbing surface is planar at an angle α to the radiation source, the intensity across the surface will be reduced by the angle of the flux and a reduction in the frontal area:

$$P_{absorb} = \frac{E_f}{c} \cos^2 \alpha \; (\text{N·m}^{-2} \text{ or Pa})$$

For total absorption on an inclined surface, the case assumed here, the momentum of the flux is delivered entirely to the surface in the same direction that the flux had. The component of the momentum normal to the surface creates the pressure on the surface, as given above. The component tangent to the surface does not contribute to the pressure.[5]

38.2.2 Radiation pressure by reflection (using particle model: photons)

See also: Photons and Momentum

Electromagnetic radiation is quantized in particles called photons, the particle aspect of its wave–particle duality. Photons are best explained by quantum mechanics. Although photons are zero-rest mass particles, they have the properties of energy and momentum, thus exhibit the property of mass as they travel at light speed. The momentum of a photon is given by:

$$p = \frac{h}{\lambda} = mc$$

where p is momentum, h is Planck's constant, λ is wavelength, m is mass, and c is speed of light in vacuum. This expression shows the wave–particle duality.

$$E = mc^2 = pc$$

is the mass-energy relationship where E is the energy. Then

$$p = \frac{E}{c}$$

The generation of radiation pressure results from the momentum property of photons, specifically, changing the momentum when incident radiation strikes a surface. The surface exerts a force on the photons in changing their momentum by Newton's Second Law. A reactive force is applied to the body by Newton's Third Law.

The orientation of a reflector determines the component of momentum normal to its surface, and also affects the frontal area of the surface facing the energy source. Each factor contributes a cosine function, reducing the pressure on the surface.[6] The pressure experienced by a perfectly reflecting planar surface is then:

$$P_{reflect} = \frac{2E_f}{c} \cos^2 \alpha \; (\text{N·m}^{-2} \text{ or Pa})$$

where P is pressure, Ef is the energy flux (intensity) in W/m^2, c is speed of light in vacuum, α is the angle between the surface normal and the incident radiation

[5]

38.2.3 Radiation pressure by emission

See also: Emissivity

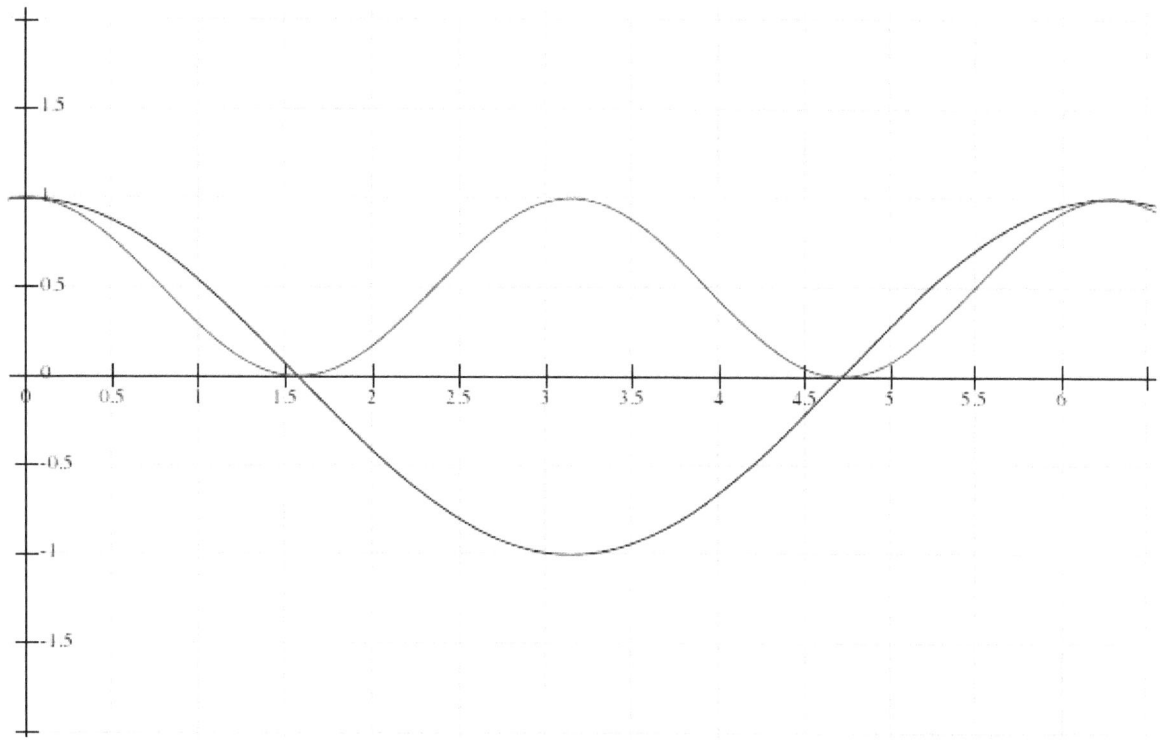

main peak when alpha is zero(front forward), also when alpha is at pi (back to front), 2pi (face forward again)

Bodies radiate thermal energy according to their temperature. The emissions are electromagnetic radiation, and therefore have the properties of energy and momentum. The energy leaving a body tends to reduce its temperature. The momentum of the radiation causes a reactive force, expressed as a pressure across the radiating surface.

The Stefan–Boltzmann law describes the power radiated from a black body. The law states that the total energy radiated per unit surface area of a black body across all wavelengths per unit time (also known as the black-body radiant exitance or emissive power) is directly proportional to the fourth power of the body's absolute temperature. The emissions from 'gray' bodies can be approximated by this law.

The emissions by other bodies are treated in an empirical manner, relying on in particular the coefficient of emission (emissivity), which is determined by measurements.

A body that does not absorb all incident radiation (sometimes known as a gray body) emits less total energy than a black body and is characterized by an emissivity, $\varepsilon < 1$, so the emitted energy flux (intensity) is:

$$E_f = \varepsilon \sigma T^4 \ (\text{J·s}^{-1} \text{·m}^{-2} \text{ or W·m}^{-2})$$

where σ is the Stefan–Boltzmann constant and T is absolute temperature. The emissivity depends on the wavelength, $\varepsilon = \varepsilon(\lambda)$.

The radiation pressure on an emitting surface by emitted radiation is then:

$$P_{emission} = \frac{E_f}{c} = \frac{\varepsilon \sigma}{c} T^4 \ (\text{N·m}^{-2} \text{ or Pa})$$

38.2.4 Moderating factors

Several factors affect the radiation pressure on a body or a cloud of particles or gases. The most prominent are the surface reflectivity, absorptivity, and emissivity. The values of these parameters vary across the spectrum, so a representative

38.3. SOLAR RADIATION PRESSURE

value is typically used in calculations. Calculations are also affected by surface curvature and roughness on a wide range of scales. Rotation of a body can also be an important factor.

38.2.5 Compression in a uniform radiation field

A body in a uniform radiation field (equal intensities from all directions) will experience a compressive pressure. It may be shown by electromagnetic theory, by quantum theory, or by thermodynamics, making no assumptions as to the nature of the radiation (other than isotropy), that the pressure against a surface exposed in a space traversed by radiation uniformly in all directions is equal to one third of the total radiant energy per unit volume within that space.[7][8][9][10]

Quantitatively, this can be expressed as [11]

$$P_{compress} = \frac{u}{3} = \frac{4\sigma}{3c}T^4 \ (\ \text{N·m}^{-2} \text{ or Pa }\ )$$

for a radiation energy density u (J·m^{-3}). The second equality holds if we are considering uniform thermal radiation at a temperature T. There σ is the Stefan–Boltzmann constant and c is the speed of light in vacuum.

38.3 Solar radiation pressure

Solar radiation pressure is exerted by solar radiation on objects within the solar system. While it acts on all bodies within the system, the smaller bodies are most affected. All spacecraft experience the pressure.

Solar radiation pressure is calculated on an irradiance (solar constant or radiant flux) value of **1361 W/m^2** at 1 AU, as revised in 2011.[12]

All stars have a spectral energy distribution that depends on their surface temperature. The distribution is approximately that of black-body radiation. This distribution is important in selecting reflector materials best suited for the application.

38.3.1 Pressures of absorption and reflection

Solar radiation pressure is calculated from the solar constant. It varies inversely by the square of the distance from the sun. The pressure experienced by a perfectly absorbing planar surface that may be at an angle to the source is:

$$P_{absorb} = \frac{W}{cR^2} \cos^2 \alpha \ (\ \text{N·m}^{-2} \text{ or Pa }\ )$$

$$P_{absorb} = \frac{4.54}{R^2} \cos^2 \alpha \ (\ \mu\text{N·m}^{-2} \text{ or } \mu\text{Pa }\ )$$

The pressure experienced by a perfectly reflecting planar surface is:

$$P_{reflect} = \frac{2W}{cR^2} \cos^2 \alpha \ (\ \text{N·m}^{-2} \text{ or Pa }\ )$$

$$P_{reflect} = \frac{9.08}{R^2} \cos^2 \alpha \ (\ \mu\text{N·m}^{-2} \text{ or } \mu\text{Pa }\ )$$

where P is pressure, W is the solar constant (W·m^{-2}), c is speed of light in vacuum, R is solar distance in AU treated as a dimensionless number, and α is the angle between the surface normal and the incident radiation. The expressions are for cases where the frontal area changes with angle, as with a solar sail.[5][13]

38.3.2 Radiation pressure perturbations

See also: Yarkovsky effect, YORP effect and Poynting–Robertson effect

Solar radiation pressure is a source of orbital perturbations. It affects the orbits and trajectories of small bodies and all spacecraft.

Solar radiation pressure affects bodies throughout much of the Solar System. Small bodies are more affected than large because of their lower mass and inertial properties. Spacecraft are affected along with natural bodies (comets, asteroids, dust grains, gas molecules).

The radiation pressure results in forces and torques on the bodies that can change their translational and rotational motions. Translational changes affect the orbits of the bodies. Rotational rates may increase or decrease. Loosely aggregated bodies may break apart under high rotation rates. Dust grains can either leave the Solar System or spiral into the Sun.

A whole body is typically composed of numerous surfaces that have different orientations on the body. The facets may be flat or curved. They will have different areas. They may have optical properties differing from other facets.

At any particular time, some facets will be exposed to the Sun and some will be in shadow. Each surface exposed to the Sun will be reflecting, absorbing, and emitting radiation. Facets in shadow will be emitting radiation. The summation of pressures across all of the facets will define the net force and torque on the body. These can be calculated using the equations in the preceding sections.[5][13]

The Yarkovsky effect affects the translation of a small body. It results from a face leaving solar exposure being at a higher temperature than a face approaching solar exposure. The radiation emitted from the warmer face will be more intense than that of the opposite face, resulting in a net force on the body that will affect its motion.

The YORP effect is a collection of effects expanding upon the earlier concept of the Yarkovsky effect, but of a similar nature. It affects the spin properties of bodies.

The Poynting–Robertson effect applies to grain-size particles. From the perspective of a grain of dust circling the Sun, the Sun's radiation appears to be coming from a slightly forward direction (aberration of light). Therefore the absorption of this radiation leads to a force with a component against the direction of movement. (The angle of aberration is extremely small since the radiation is moving at the speed of light while the dust grain is moving many orders of magnitude slower than that.) The result is a slow spiral of dust grains into the Sun. Over long periods of time this effect cleans out much of the dust in the Solar System.

While rather small in comparison to other forces, the radiation pressure force is inexorable. Over long periods of time, the net effect of the force is substantial. Such feeble pressures are able to produce marked effects upon minute particles like gas ions and electrons, and are important in the theory of electron emission from the Sun, of cometary material, and so on.

Because the ratio of surface area to volume (and thus mass) increases with decreasing particle size, dusty (micrometre-size) particles are susceptible to radiation pressure even in the outer solar system. For example, the evolution of the outer rings of Saturn is significantly influenced by radiation pressure.

38.3.3 Solar sails

Main article: Solar sail

Solar sailing, an experimental method of spacecraft propulsion, uses radiation pressure from the Sun as a motive force. The idea of interplanetary travel by light was mentioned by Jules Verne in *From the Earth to the Moon*.

A sail reflects about 90% of the incident radiation. The 10% that is absorbed is radiated away from both surfaces, with the proportion radiated from the unlit surface depending on the thermal conductivity of the sail. A sail has curvature, surface irregularities, and other minor factors that affect its performance.

The Japan Aerospace Exploration Agency (JAXA) has successfully unfurled a solar sail in space which has already succeeded in propelling its payload with the IKAROS project.

38.4 Cosmic effects of radiation pressure

Radiation pressure has had a major effect on the development of the cosmos, from the birth of the universe to ongoing formation of stars and shaping of clouds of dust and gasses on a wide range of scales.

38.4.1 The early universe

The photon epoch is a phase when the energy of the universe was dominated by photons, between 10 seconds and 380,000 years after the Big Bang.

38.4.2 Galaxy formation and evolution

The process of galaxy formation and evolution began early in the history of the cosmos. Observations of the early universe strongly suggest that objects grew from bottom-up (i.e., smaller objects merging to form larger ones).

Early in the universe, galaxies were composed mostly of gas and dark matter. As a galaxy gained mass by accretion of smaller galaxies, the dark matter stayed mostly in the outer parts of the galaxy. The gas, however, contracted, causing the galaxy to rotate faster, until the result was a thin, rotating disk.

Astronomers do not currently know what process stopped the contraction. Theories of galaxy formation are not successful at producing the rotation speed and size of disk galaxies. It has been suggested that the radiation from bright newly formed stars, or from an active galactic nuclei, could have slowed the contraction of a forming disk. It has also been suggested that the dark matter halo could pull on galactic matter, stopping disk contraction.

38.4.3 Clouds of dust and gases

The gravitational compression of clouds of dust and gases is strongly influenced by radiation pressure, especially when the condensations lead to star births. The larger young stars forming within the compressed clouds emit intense levels of radiation that shift the clouds, causing either dispersion or condensations in nearby regions, which influences birth rates in those nearby regions.

38.4.4 Clusters of stars

Stars predominantly form in regions of large clouds of dust and gases, giving rise to star clusters. Radiation pressure from the member stars eventually disperses the clouds, which can have a profound effect on the evolution of the cluster.

Many open clusters are inherently unstable, with a small enough mass that the escape velocity of the system is lower than the average velocity of the constituent stars. These clusters will rapidly disperse within a few million years. In many cases, the stripping away of the gas from which the cluster formed by the radiation pressure of the hot young stars reduces the cluster mass enough to allow rapid dispersal.

38.4.5 Star formation

Star formation is the process by which dense regions within molecular clouds in interstellar space collapse to form stars. As a branch of astronomy, star formation includes the study of the interstellar medium and giant molecular clouds (GMC) as precursors to the star formation process, and the study of protostars and young stellar objects as its immediate products. Star formation theory, as well as accounting for the formation of a single star, must also account for the statistics of binary stars and the initial mass function.

The Pillars of Creation *clouds within the Eagle Nebula shaped by radiation pressure and stellar winds.*

38.4.6 Stellar planetary systems

Planetary systems are generally believed to form as part of the same process that results in star formation. A protoplanetary disk forms by gravitational collapse of a molecular cloud, called a solar nebula, and then evolves into a planetary system by collisions and gravitational capture. Radiation pressure can clear a region in the immediate vicinity of the star. As the formation process continues, radiation pressure continues to play a role in affecting the distribution of matter. In particular, dust and grains can spiral into the star or escape the stellar system under the action of radiation pressure.

38.4.7 Stellar interiors

In stellar interiors the temperatures are very high. Stellar models predict a temperature of 15 MK in the center of the Sun, and at the cores of supergiant stars the temperature may exceed 1 GK. As the radiation pressure scales as the fourth power of the temperature, it becomes important at these high temperatures. In the Sun, radiation pressure is still quite small when compared to the gas pressure. In the heaviest non-degenerate stars, radiation pressure is the dominant pressure component.[14]

38.4.8 Comets

Solar radiation pressure strongly affects comet tails. Solar heating causes gases to be released from the comet nucleus, which also carry away dust grains. Radiation pressure and solar wind then drive the dust and gases away from the Sun's

38.5. LASER APPLICATIONS OF RADIATION PRESSURE

A protoplanetary disk with a cleared central region.

direction. The gases form a generally straight tail, while slower moving dust particles create a broader, curving tail.

38.5 Laser applications of radiation pressure

Laser cooling is applied to cooling materials very close to absolute zero. Atoms traveling towards a laser light source perceive a doppler effect tuned to the absorption frequency of the target element. The radiation pressure on the atom slows movement in a particular direction until the Doppler effect moves out of the frequency range of the element, causing an overall cooling effect.

Large lasers operating in space have been suggested as a means of propelling sail craft in beam-powered propulsion.

The reflection of a laser pulse from the surface of an elastic solid gives rise to various types of elastic waves that propagate inside the solid. The weakest waves are generally those that are generated by the radiation pressure acting during the reflection of the light. Recently, such light-pressure-induced elastic waves were observed inside an ultrahigh-reflectivity dielectric mirror.[15] These waves are the most basic fingerprint of a light-solid matter interaction on the macroscopic scale.

Comet Hale–Bopp (C/1995 O1). Radiation pressure and solar wind effects on the dust and gas tails are clearly seen.

38.6 See also

- Compton scattering
- De Broglie wavelength
- Electromagnetic radiation
- Irradiance
- Light absorption
- Photoelectric effect
- Photons

- Poynting–Robertson effect
- Poynting vector
- Quantum mechanics
- Solar constant
- Solar sail
- Speed of light
- Sunlight
- Wavelength
- Wave-particle duality
- Yarkovsky effect
- YORP effect

38.7 References

[1] Eugene Hecht, "Optics", 4th edition

[2] Johannes Kepler (1619). *De Cometis Libelli Tres.*

[3] P. Lebedev, 1901, "Untersuchungen über die Druckkräfte des Lichtes", Annalen der Physik, 1901

[4] Nichols, E.F & Hull, G.F. (1903) The Pressure due to Radiation, *The Astrophysical Journal*,Vol.17 No.5, p.315-351

[5] Wright, Jerome L. (1992), *Space Sailing*, Gordon and Breach Science Publishers

[6] T. Požar (2014), *Oblique reflection of a laser pulse from a perfect elastic mirror.* Optics Letter 39 (1), 48-51

[7] Shankar R., *Principles of Quantum Mechanics*, 2nd edition.

[8] Carroll, Bradley W. & Dale A. Ostlie, *An Introduction to Modern Astrophysics*, 2nd edition.

[9] Jackson, John David, (1999) *Classical Electrodynamics.*

[10] Kardar, Mehran. "Statistical Physics of Particles".

[11] Planck's law

[12] Kopp, G.; Lean, J. L. (2011). "A new, lower value of total solar irradiance: Evidence and climate significance". Geophysical Research Letters 38.

[13] Georgevic, R. M. (1973) "The Solar Radiation Pressure Forces and Torques Model", *The Journal of the Astronautical Sciences*, Vol. 27, No. 1, Jan–Feb. First known publication describing how solar radiation pressure creates forces and torques that affect spacecraft.

[14] Dale A. Ostlie and Bradley W. Carroll, *An Introduction to Modern Astrophysics* (2nd edition), page 341, Pearson, San Francisco, 2007

[15] T. Požar and J. Možina (2013), *Measurement of elastic waves induced by the reflection of light.* Physical Review Letters 111 (18), 185501

38.8 Further reading

- Demir, Dilek,"A table-top demonstration of radiation pressure",2011, Diplomathesis, E-Theses univie (http://othes.univie.ac.at/16381/)
- R. Shankar, "Principles of Quantum Mechanics", 2nd edition.

Chapter 39

Electromagnetism

Electromagnetism is a branch of physics which involves the study of the **electromagnetic force**, a type of physical interaction that occurs between electrically charged particles. The electromagnetic force usually shows electromagnetic fields, such as electric fields, magnetic fields, and light. The electromagnetic force is one of the four fundamental interactions in nature. The other three fundamental interactions are the strong interaction, the weak interaction, and gravitation.[1]

Lightning is an electrostatic discharge that travels between two charged regions.

The word *electromagnetism* is a compound form of two Greek terms, ἤλεκτρον, *ēlektron*, "amber", and μαγνῆτις λίθος *magnētis lithos*, which means "magnesian stone", a type of iron ore. The science of electromagnetic phenomena is defined in terms of the electromagnetic force, sometimes called the Lorentz force, which includes both electricity and magnetism as elements of one phenomenon.

The electromagnetic force plays a major role in determining the internal properties of most objects encountered in daily life. Ordinary matter takes its form as a result of intermolecular forces between individual molecules in matter. Electrons are bound by electromagnetic wave mechanics into orbitals around atomic nuclei to form atoms, which are the building blocks of molecules. This governs the processes involved in chemistry, which arise from interactions between the electrons of neighboring atoms, which are in turn determined by the interaction between electromagnetic force and the momentum of the electrons.

There are numerous mathematical descriptions of the electromagnetic field. In classical electrodynamics, electric fields are described as electric potential and electric current in Ohm's law, magnetic fields are associated with electromagnetic induction and magnetism, and Maxwell's equations describe how electric and magnetic fields are generated and altered by each other and by charges and currents.

The theoretical implications of electromagnetism, in particular the establishment of the speed of light based on properties of the "medium" of propagation (permeability and permittivity), led to the development of special relativity by Albert Einstein in 1905.

Although electromagnetism is considered one of the four fundamental forces, at high energy the weak force and electromagnetism are unified. In the history of the universe, during the quark epoch, the electroweak force split into the electromagnetic and weak forces.

39.1 History of the theory

See also: History of electromagnetic theory

Originally electricity and magnetism were thought of as two separate forces. This view changed, however, with the publication of James Clerk Maxwell's 1873 *A Treatise on Electricity and Magnetism* in which the interactions of positive and negative charges were shown to be regulated by one force. There are four main effects resulting from these interactions, all of which have been clearly demonstrated by experiments:

1. Electric charges attract or repel one another with a force inversely proportional to the square of the distance between them: unlike charges attract, like ones repel.

2. Magnetic poles (or states of polarization at individual points) attract or repel one another in a similar way and always come in pairs: every north pole is yoked to a south pole.

3. An electric current inside a wire creates a corresponding circular magnetic field outside the wire. Its direction (clockwise or counter-clockwise) depends on the direction of the current in the wire.

4. A current is induced in a loop of wire when it is moved towards or away from a magnetic field, or a magnet is moved towards or away from it, the direction of current depending on that of the movement.

While preparing for an evening lecture on 21 April 1820, Hans Christian Ørsted made a surprising observation. As he was setting up his materials, he noticed a compass needle deflected away from magnetic north when the electric current from the battery he was using was switched on and off. This deflection convinced him that magnetic fields radiate from all sides of a wire carrying an electric current, just as light and heat do, and that it confirmed a direct relationship between electricity and magnetism.

At the time of discovery, Ørsted did not suggest any satisfactory explanation of the phenomenon, nor did he try to represent the phenomenon in a mathematical framework. However, three months later he began more intensive investigations. Soon thereafter he published his findings, proving that an electric current produces a magnetic field as it flows through a wire. The CGS unit of magnetic induction (oersted) is named in honor of his contributions to the field of electromagnetism.

Hans Christian Ørsted.

His findings resulted in intensive research throughout the scientific community in electrodynamics. They influenced French physicist André-Marie Ampère's developments of a single mathematical form to represent the magnetic forces between current-carrying conductors. Ørsted's discovery also represented a major step toward a unified concept of energy.

This unification, which was observed by Michael Faraday, extended by James Clerk Maxwell, and partially reformulated by Oliver Heaviside and Heinrich Hertz, is one of the key accomplishments of 19th century mathematical physics. It had far-reaching consequences, one of which was the understanding of the nature of light. Unlike what was proposed in Electromagnetism, light and other electromagnetic waves are at the present seen as taking the form of quantized, self-propagating oscillatory electromagnetic field disturbances which have been called photons. Different frequencies of oscillation give rise to the different forms of electromagnetic radiation, from radio waves at the lowest frequencies, to visible light at intermediate frequencies, to gamma rays at the highest frequencies.

Ørsted was not the only person to examine the relationship between electricity and magnetism. In 1802, Gian Domenico Romagnosi, an Italian legal scholar, deflected a magnetic needle using electrostatic charges. Actually, no galvanic current existed in the setup and hence no electromagnetism was present. An account of the discovery was published in 1802 in an Italian newspaper, but it was largely overlooked by the contemporary scientific community.[2]

39.2 Fundamental forces

The electromagnetic force is one of the four known fundamental forces. The other fundamental forces are:

- the weak nuclear force, which binds to all known particles in the Standard Model, and causes certain forms of radioactive decay. (In particle physics though, the electroweak interaction is the unified description of two of the four known fundamental interactions of nature: electromagnetism and the weak interaction);
- the strong nuclear force, which binds quarks to form nucleons, and binds nucleons to form nuclei
- the gravitational force.

All other forces (e.g., friction) are derived from these four fundamental forces (including momentum which is carried by the movement of particles).

The electromagnetic force is the one responsible for practically all the phenomena one encounters in daily life above the nuclear scale, with the exception of gravity. Roughly speaking, all the forces involved in interactions between atoms can be explained by the electromagnetic force acting on the electrically charged atomic nuclei and electrons inside and around the atoms, together with how these particles carry momentum by their movement. This includes the forces we experience in "pushing" or "pulling" ordinary material objects, which come from the intermolecular forces between the individual molecules in our bodies and those in the objects. It also includes all forms of chemical phenomena.

A necessary part of understanding the intra-atomic to intermolecular forces is the effective force generated by the momentum of the electrons' movement, and that electrons move between interacting atoms, carrying momentum with them. As a collection of electrons becomes more confined, their minimum momentum necessarily increases due to the Pauli exclusion principle. The behaviour of matter at the molecular scale including its density is determined by the balance between the electromagnetic force and the force generated by the exchange of momentum carried by the electrons themselves.

39.3 Classical electrodynamics

Main article: Classical electrodynamics

The scientist William Gilbert proposed, in his *De Magnete* (1600), that electricity and magnetism, while both capable of causing attraction and repulsion of objects, were distinct effects. Mariners had noticed that lightning strikes had the ability to disturb a compass needle, but the link between lightning and electricity was not confirmed until Benjamin Franklin's proposed experiments in 1752. One of the first to discover and publish a link between man-made electric current and magnetism was Romagnosi, who in 1802 noticed that connecting a wire across a voltaic pile deflected a nearby compass needle. However, the effect did not become widely known until 1820, when Ørsted performed a similar experiment.[3] Ørsted's work influenced Ampère to produce a theory of electromagnetism that set the subject on a mathematical foundation.

A theory of electromagnetism, known as classical electromagnetism, was developed by various physicists over the course of the 19th century, culminating in the work of James Clerk Maxwell, who unified the preceding developments into a single theory and discovered the electromagnetic nature of light. In classical electromagnetism, the electromagnetic field obeys a set of equations known as Maxwell's equations, and the electromagnetic force is given by the Lorentz force law.

One of the peculiarities of classical electromagnetism is that it is difficult to reconcile with classical mechanics, but it is compatible with special relativity. According to Maxwell's equations, the speed of light in a vacuum is a universal constant, dependent only on the electrical permittivity and magnetic permeability of free space. This violates Galilean invariance, a long-standing cornerstone of classical mechanics. One way to reconcile the two theories (electromagnetism and classical mechanics) is to assume the existence of a luminiferous aether through which the light propagates. However, subsequent experimental efforts failed to detect the presence of the aether. After important contributions of Hendrik Lorentz and Henri Poincaré, in 1905, Albert Einstein solved the problem with the introduction of special relativity, which replaces classical kinematics with a new theory of kinematics that is compatible with classical electromagnetism. (For more information, see History of special relativity.)

In addition, relativity theory shows that in moving frames of reference a magnetic field transforms to a field with a nonzero electric component and vice versa; thus firmly showing that they are two sides of the same coin, and thus the term "electromagnetism". (For more information, see Classical electromagnetism and special relativity and Covariant formulation of classical electromagnetism.

39.4 Quantum mechanics

39.4.1 Photoelectric effect

Main article: Photoelectric effect

In another paper published in 1905, Albert Einstein undermined the very foundations of classical electromagnetism. In his theory of the photoelectric effect (for which he won the Nobel prize in physics) and inspired by the idea of Max Planck's "quanta", he posited that light could exist in discrete particle-like quantities as well, which later came to be known as photons. Einstein's theory of the photoelectric effect extended the insights that appeared in the solution of the ultraviolet catastrophe presented by Max Planck in 1900. In his work, Planck showed that hot objects emit electromagnetic radiation in discrete packets ("quanta"), which leads to a finite total energy emitted as black body radiation. Both of these results were in direct contradiction with the classical view of light as a continuous wave. Planck's and Einstein's theories were progenitors of quantum mechanics, which, when formulated in 1925, necessitated the invention of a quantum theory of electromagnetism. This theory, completed in the 1940s-1950s, is known as quantum electrodynamics (or "QED"), and, in situations where perturbation theory is applicable, is one of the most accurate theories known to physics.

39.4.2 Quantum electrodynamics

Main article: Quantum electrodynamics

All electromagnetic phenomena are underpinned by quantum mechanics, specifically by quantum electrodynamics (which includes classical electrodynamics as a limiting case) and this accounts for almost all physical phenomena observable to the unaided human senses, including light and other electromagnetic radiation, all of chemistry, most of mechanics (excepting gravitation), and, of course, magnetism and electricity.

39.4.3 Electroweak interaction

Main article: Electroweak interaction

The **electroweak interaction** is the unified description of two of the four known fundamental interactions of nature: electromagnetism and the weak interaction. Although these two forces appear very different at everyday low energies, the theory models them as two different aspects of the same force. Above the unification energy, on the order of 100 GeV, they would merge into a single **electroweak force**. Thus if the universe is hot enough (approximately 10^{15} K, a temperature exceeded until shortly after the Big Bang) then the electromagnetic force and weak force merge into a combined electroweak force. During the electroweak epoch, the electroweak force separated from the strong force. During the quark epoch, the electroweak force split into the electromagnetic and weak force.

39.5 Quantities and units

See also: List of physical quantities and List of electromagnetism equations

Electromagnetic units are part of a system of electrical units based primarily upon the magnetic properties of electric currents, the fundamental SI unit being the ampere. The units are:

- ampere (electric current)
- coulomb (electric charge)
- farad (capacitance)
- henry (inductance)
- ohm (resistance)
- tesla (magnetic flux density)
- volt (electric potential)
- watt (power)
- weber (magnetic flux)

In the electromagnetic cgs system, electric current is a fundamental quantity defined via Ampère's law and takes the permeability as a dimensionless quantity (relative permeability) whose value in a vacuum is unity. As a consequence, the square of the speed of light appears explicitly in some of the equations interrelating quantities in this system.

Formulas for physical laws of electromagnetism (such as Maxwell's equations) need to be adjusted depending on what system of units one uses. This is because there is no one-to-one correspondence between electromagnetic units in SI and those in CGS, as is the case for mechanical units. Furthermore, within CGS, there are several plausible choices of electromagnetic units, leading to different unit "sub-systems", including Gaussian, "ESU", "EMU", and Heaviside–Lorentz. Among these choices, Gaussian units are the most common today, and in fact the phrase "CGS units" is often used to refer specifically to CGS-Gaussian units.

39.6 See also

- Abraham–Lorentz force
- Aeromagnetic surveys
- Computational electromagnetics
- Double-slit experiment
- Electromagnet

- Electromagnetic induction
- Electromagnetic wave equation
- Electromechanics
- Magnetostatics
- Magnetoquasistatic field
- Optics
- Relativistic electromagnetism
- Wheeler–Feynman absorber theory

39.7 References

[1] Ravaioli, Fawwaz T. Ulaby, Eric Michielssen, Umberto (2010). *Fundamentals of applied electromagnetics* (6th ed.). Boston: Prentice Hall. p. 13. ISBN 978-0-13-213931-1.

[2] Martins, Roberto de Andrade. "Romagnosi and Volta's Pile: Early Difficulties in the Interpretation of Voltaic Electricity". In Fabio Bevilacqua and Lucio Fregonese (eds). *Nuova Voltiana: Studies on Volta and his Times* (PDF). vol. 3. Università degli Studi di Pavia. pp. 81–102. Retrieved 2010-12-02.

[3] Stern, Dr. David P.; Peredo, Mauricio (2001-11-25). "Magnetic Fields -- History". NASA Goddard Space Flight Center. Retrieved 2009-11-27.

[4] International Union of Pure and Applied Chemistry (1993). *Quantities, Units and Symbols in Physical Chemistry*, 2nd edition, Oxford: Blackwell Science. ISBN 0-632-03583-8. pp. 14–15. Electronic version.

39.8 Further reading

39.8.1 Web sources

- Nave, R. "Electricity and magnetism". *HyperPhysics*. Georgia State University. Retrieved 2013-11-12.

39.8.2 Lecture notes

- Littlejohn, Robert (Spring 2011). "Emission and absorption of radiation" (PDF). *Physics 221B: Quantum mechanics*. University of California Berkeley. Retrieved 2013-11-12.
- Littlejohn, Robert (Spring 2011). "The Classical Electromagnetic Field Hamiltonian" (PDF). *Physics 221B: Quantum mechanics*. University of California Berkeley. Retrieved 2013-11-12.

39.8.3 Textbooks

- G.A.G. Bennet (1974). *Electricity and Modern Physics* (2nd ed.). Edward Arnold (UK). ISBN 0-7131-2459-8.
- Dibner, Bern (2012). *Oersted and the discovery of electromagnetism*. Literary Licensing, LLC. ISBN 9781258335557.
- Durney, Carl H. and Johnson, Curtis C. (1969). *Introduction to modern electromagnetics*. McGraw-Hill. ISBN 0-07-018388-0.
- Feynman, Richard P. (1970). *The Feynman Lectures on Physics Vol II*. Addison Wesley Longman. ISBN 978-0-201-02115-8.

- Fleisch, Daniel (2008). *A Student's Guide to Maxwell's Equations*. Cambridge, UK: Cambridge University Press. ISBN 978-0-521-70147-1.

- I.S. Grant, W.R. Phillips, Manchester Physics (2008). *Electromagnetism* (2nd ed.). John Wiley & Sons. ISBN 978-0-471-92712-9.

- Griffiths, David J. (1998). *Introduction to Electrodynamics* (3rd ed.). Prentice Hall. ISBN 0-13-805326-X.

- Jackson, John D. (1998). *Classical Electrodynamics* (3rd ed.). Wiley. ISBN 0-471-30932-X.

- Moliton, André (2007). *Basic electromagnetism and materials. 430 pages* (New York City: Springer-Verlag New York, LLC). ISBN 978-0-387-30284-3.

- Purcell, Edward M. (1985). *Electricity and Magnetism Berkeley Physics Course Volume 2 (2nd ed.)*. McGraw-Hill. ISBN 0-07-004908-4.

- Rao, Nannapaneni N. (1994). *Elements of engineering electromagnetics (4th ed.)*. Prentice Hall. ISBN 0-13-948746-8.

- Rothwell, Edward J.; Cloud, Michael J. (2001). *Electromagnetics*. CRC Press. ISBN 0-8493-1397-X.

- Tipler, Paul (1998). *Physics for Scientists and Engineers: Vol. 2: Light, Electricity and Magnetism* (4th ed.). W. H. Freeman. ISBN 1-57259-492-6.

- Wangsness, Roald K.; Cloud, Michael J. (1986). *Electromagnetic Fields (2nd Edition)*. Wiley. ISBN 0-471-81186-6.

39.8.4 General references

- A. Beiser (1987). *Concepts of Modern Physics* (4th ed.). McGraw-Hill (International). ISBN 0-07-100144-1.

- L.H. Greenberg (1978). *Physics with Modern Applications*. Holt-Saunders International W.B. Saunders and Co. ISBN 0-7216-4247-0.

- R.G. Lerner, G.L. Trigg (2005). *Encyclopaedia of Physics* (2nd ed.). VHC Publishers, Hans Warlimont, Springer. pp. 12–13. ISBN 978-0-07-025734-4.

- J.B. Marion, W.F. Hornyak (1984). *Principles of Physics*. Holt-Saunders International Saunders College. ISBN 4-8337-0195-2.

- H.J. Pain (1983). *The Physics of Vibrations and Waves* (3rd ed.). John Wiley & Sons,. ISBN 0-471-90182-2.

- C.B. Parker (1994). *McGraw Hill Encyclopaedia of Physics* (2nd ed.). McGraw Hill. ISBN 0-07-051400-3.

- R. Penrose (2007). *The Road to Reality*. Vintage books. ISBN 0-679-77631-1.

- P.A. Tipler, G. Mosca (2008). *Physics for Scientists and Engineers: With Modern Physics* (6th ed.). W.H. Freeman and Co. ISBN 9-781429-202657.

- P.M. Whelan, M.J. Hodgeson (1978). *Essential Principles of Physics* (2nd ed.). John Murray. ISBN 0-7195-3382-1.

39.9 External links

- Oppelt, Arnulf (2006-11-02). "magnetic field strength". Retrieved 2007-06-04.
- "magnetic field strength converter". Retrieved 2007-06-04.
- Electromagnetic Force - from Eric Weisstein's World of Physics
- Goudarzi, Sara (2006-08-15). "Ties That Bind Atoms Weaker Than Thought". *LiveScience.com*. Retrieved 2013-11-12.
- Quarked Electromagnetic force - A good introduction for kids
- The Deflection of a Magnetic Compass Needle by a Current in a Wire (video) on YouTube
- Electromagnetism abridged

André-Marie Ampère

Michael Faraday

James Clerk Maxwell

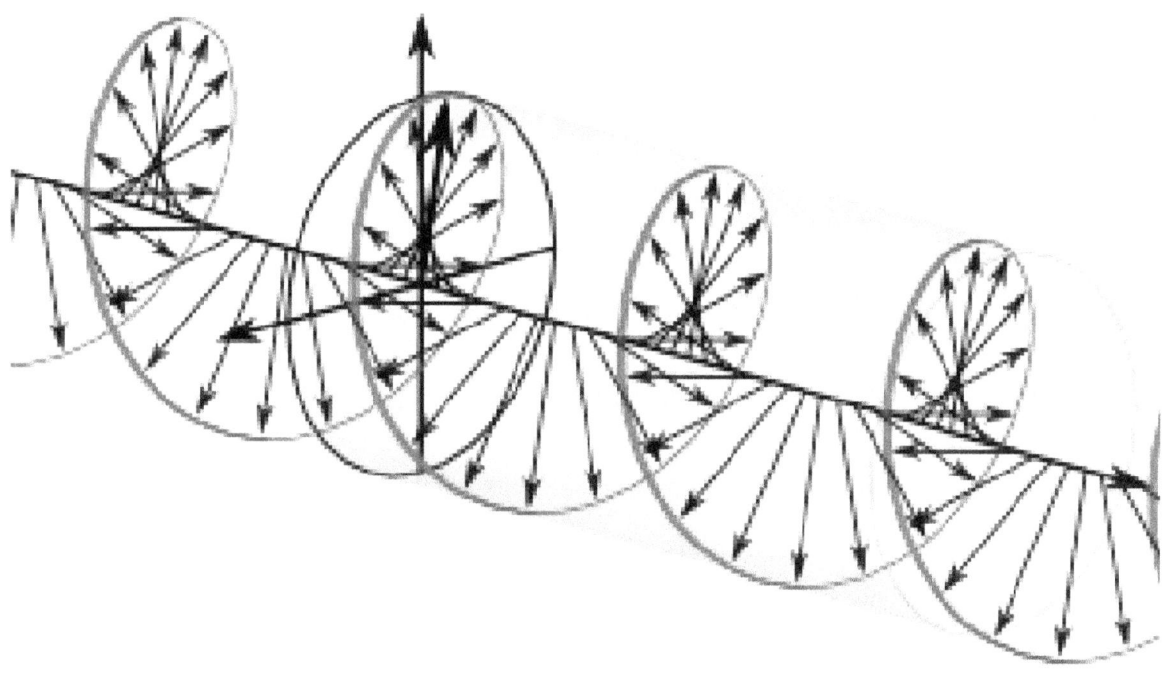

Representation of the electric field vector of a wave of circularly polarized electromagnetic radiation.

Chapter 40

Two-line element set

A **two-line element set** (**TLE**) is a data format encoding a list of orbital elements of an Earth-orbiting object for a given point in time, the *epoch*. Using suitable prediction formula, the state (position and velocity) at any point in the past or future can be estimated to some accuracy. The TLE data representation is specific to the simplified perturbations models (SGP, SGP4, SDP4, SGP8 and SDP8), so any algorithm using a TLE as a data source must implement one of the SGP models to correctly compute the state at a time of interest.

The format uses two lines of 80-column ASCII text to store the data, having originated as punch card format with one line per card. The United States Air Force tracks all detectable objects in Earth orbit, creating a corresponding TLE for each object, and makes available TLEs for non-classified objects on the website Space Track.[1][2] The TLE format is a *de facto* standard for distribution of an Earth-orbiting object's orbital elements. TLEs can describe the trajectories only of earth orbiting objects.

A TLE set may include a title line proceeding the element data, so each listing may take up three lines in the file. The title is not required, as each data line includes a unique object identifier code.

40.1 History

The two-line format traces its history to seminal work by Max Lane in the early 1960s to develop mathematical models for predicting the locations of satellites based on a minimal set of data elements. His first paper on the topic, published in 1965, introduced the Analytical Drag Theory, which concerned itself primarily with the effects of drag caused by a spherically-symmetric non-rotating atmosphere.[3] Joined by K. Cranford, the two published a greatly improved model in 1969 that added various harmonic effects due to Earth-Moon-Sun interactions and various other inputs.[4]

Lane's models were widely used by the military and NASA starting in the late 1960s. The improved version became the standard model for NORAD in the early 1970s, which ultimately led to the creation of the TLE format. At the time there were two formats designed for punch cards, an "internal format" that used three cards encoding complete details for the satellite (including name and other data), and the two card "transmission format" that listed only those elements that were subject to change.[5] The later saved on cards and produced smaller decks when updating the databases.

Cranford continued to work on the modelling, eventually leading Lane to publish *Spacetrack Report #2* detailing the Air Force General Perturbation theory, or AFGP4. The paper also described two simplified versions of the system, IGP4 which used a simplified drag model, and SGP4 (Simplified General Perturbations) which used IGP4's drag model along with a simplified gravity model.[6] The differences between the three models were slight for most objects. One year later, *Spacetrack Report #3* was released, included full FORTRAN source code for the SGP4 model.[7] This quickly became the *de facto* standard model, both in the industry as well as the astronomy field.

Shortly after the publication of *Report #3*, NASA began posting elements for a variety of visible and other well known objects in their periodic *NASA Prediction Bulletins*, which consisted of the transmission format data in printed form. After trying for some time to convince NASA to release these in electronic form, Ted Kelso took matters into his own hands and began manually copying the listings into text files which he distributed through his CelesTrak bulletin board system. This revealed a problem in NASA's checksum system, which traced back to the lack of the plus character (+) on the teletype machines used at NASA, which ultimately turned out to be a problem from the punch card era that occurred when NORAD updated from the BCD to EBCDIC character set on the computer sending out the updates. This problem went away when Kelso began to receive data directly from NORAD in 1989.[8]

The SGP4 model was later extended with corrections for deep space objects, creating SDP4, which used the same TLE input data. Over the years a number of more advanced prediction models have been created, but these have not seen widespread use. This is due to the TLE not containing the additional information needed by some of these formats, which makes it difficult to find the elements needed to take advantages of the improved model. More subtly, the TLE data is massaged in a fashion to improve the results when used with the SGP series models, which may cause the predictions of other models to be less accurate that SGP when used with common TLEs. The only new model to see widespread use is SGP8/SDP8, which were designed to use the same data inputs and are relatively minor corrections to the SGP4 model.

40.2 Format

Originally there were two data formats used with the SGP models, one containing complete details on the object known as the "internal format", and the "transmission format" used to provide updates to that data.

The internal format used three 80-column punch cards. Each card started with a card number, 1, 2 or 3, and ended with the letter "G". For this reason the system was often known as the "G-card format". In addition to the orbital elements, the G-card included various flags like the launching country and orbit type (geostationary, etc.), calculated values like the perigee altitude and visual magnitude, and a 38-character comments field.

The transmission format is essentially a cut-down version of the G-card format, removing any data that is not subject to common change, or can be calculated using other values. For instance, the perigee altitude from the G-card is not included as this can be calculated from the other elements. What remains is the set of data needed to update the original G-card data as additional measurements are made. The data is fit into 70 columns, and does not include a trailing character. TLEs are simply the transmission format data rendered as ASCII text.

An example TLE for the International Space Station:

ISS (ZARYA) 1 25544U 98067A 08264.51782528 −.00002182 00000-0 −11606-4 0 2927 2 25544 51.6416 247.4627 0006703 130.5360 325.0288 15.72125391563537

The meaning of this data is as follows:[9]

Title line

TLE title

LINE 1

TLE first row

LINE 2

TLE second row

Where decimal points are assumed, they are leading decimal points. The last two symbols in Fields 10 and 11 of the first line give powers of 10 to apply to the preceding decimal. Thus, for example, Field 11 (−11606-4) translates to −0.11606E-4 (−0.11606×10^{-4}).

The checksums for each line are calculated by adding all numerical digits on that line, including the line number. One is added to the checksum for each negative sign (−) on that line. All other non-digit characters are ignored.

For a body in a typical Low Earth orbit, the accuracy that can be obtained with the SGP4 orbit model is on the order of 1 km within a few days of the epoch of the element set.[11] The term "low orbit" may refer to either the altitude (minimal or global) or orbital period of the body. Historically, the SGP algorithms defines low orbit as an orbit of less-than 225 minutes.

40.3 Applications

TLEs are widely used as input for projecting the future orbital tracks of space debris for purposes of characterizing "future debris events to support risk analysis, close approach analysis, collision avoidance maneuvering" and forensic analysis.[12]

40.4 References

[1] "Introduction and sign in to Space-Track.Org". Space-track.org. Retrieved 28 November 2014.

[2] "Celestrak homepage". Celestrak.com. Retrieved 28 November 2014.

[3] Vallado, David; Crawford, Paul; Hujsak, Richard; Kelso, Ted (2006). "Revisiting Spacetrack Report #3" (PDF). *American Institute of Aeronautics and Astronautics*.

[4] Lane, Max; Cranford, Kenneth (1969). "An improved analytical drag theory for the artificial satellite problem". *AIAA*.

[5] *ADCOM Form 2012* (PDF) (Technical report).

[6] Lane, Max; Hoots, Felix (December 1979). *General Pertubations Theories Derived from the 1965 Lane Drag Theory* (PDF) (Technical report). Project Space Track, Aerospace Defense Command.

[7] Hoots, Felix; Roehrich, Ronald (December 1980). *Models for Propagation of NORAD Element Sets* (PDF) (Technical report). Project Space Track, Aerospace Defense Command.

[8] Kelso, Ted (January 1992). "Two-Line Element Set Checksum Controversy". *CelesTrak*.

[9] "Space Track". Space-track.org. Retrieved 28 November 2014.

[10] "NASA, *Definition of Two-line Element Set Coordinate System*". Spaceflight.nasa.gov. Retrieved 28 November 2014.

[11] Kelso, T.S. (29 January 2007). "Validation of SGP4 and IS-GPS-200D Against GPS Precision Ephemerides". Celestrak.com. Retrieved 28 November 2014. AAS paper 07-127, presented at the 17th AAS/AIAA Space Flight Mechanics Conference, Sedona, Arizona.

[12] Carrico, Timothy; Carrico, John; Policastri, Lisa; Loucks, Mike (2008). "Investigating Orbital Debris Events using Numerical Methods with Full Force Model Orbit Propagation" (PDF). *American Institute of Aeronautics and Astronautics* (AAS 08-126).

Chapter 41

Proper orbital elements

The **proper orbital elements** of an orbit are constants of motion of an object in space that remain practically unchanged over an astronomically long timescale. The term is usually used to describe the three quantities:

- *proper semimajor axis* (a_p),

- *proper eccentricity* (e_p), and

- *proper inclination* (i_p).

The **proper elements** can be contrasted with the osculating Keplerian orbital elements observed at a particular time or epoch, such as the semi-major axis, eccentricity, and inclination. Those osculating elements change in a quasi-periodic and (in principle) predictable manner due to such effects as perturbations from planets or other bodies, and precession (e.g. perihelion precession). In the Solar System, such changes usually occur on timescales of thousands of years, while proper elements are meant to be practically constant over at least tens of millions of years.

For most bodies, the osculating elements are relatively close to the proper elements because precession and perturbation effects are relatively small (See diagram). For over 99% of asteroids in the asteroid belt, the differences are less than 0.02 AU (for semi-major axis a), 0.1 (for eccentricity e), and 2° (for inclination i).

Nevertheless, this difference is non-negligible for any purposes where precision is of importance. As an example, the asteroid Ceres has osculating orbital elements (at epoch November 26, 2005)

while its proper orbital elements (independent of epoch) are[1]

A notable exception to this small-difference rule are asteroids lying in the Kirkwood gaps, which are in strong orbital resonance with Jupiter.

To obtain proper elements for an object, one usually conducts a detailed simulation of its motion over timespans of several millions of years. Such a simulation must take into account many details of celestial mechanics including perturbations by the planets. Subsequently, one extracts quantities from the simulation which remain unchanged over this long timespan; for example, the mean inclination, eccentricity, and semi-major axis. These are the proper orbital elements.

Historically, various approximate analytic calculations were made, starting with those of Kiyotsugu Hirayama in the early 20th century. Later analytic methods often included thousands of perturbing corrections for each particular object. Presently, the method of choice is to use a computer to numerically integrate the equations of celestial dynamics, and extract constants of motion directly from a numerical analysis of the predicted positions.

At present the most prominent use of proper orbital elements is in the study of asteroid families, following in the footsteps of the pioneering work of Hirayama. As a Mars-crosser asteroid 132 Aethra is the lowest numbered asteroid to not have any proper orbital elements.

41.1 See also

- Perturbation (astronomy)

41.2 References

[1] "AstDyS-2 Ceres Synthetic Proper Orbital Elements". Department of Mathematics, University of Pisa, Italy. Retrieved 2011-09-19.

- Z. Knežević et al., *The Determination of Asteroid Proper Elements*, p. 603-612 in Asteroids III, University of Arizona Press (2002).

41.3 External links

- Latest calculations of proper elements for numbered minor planets at astDys.

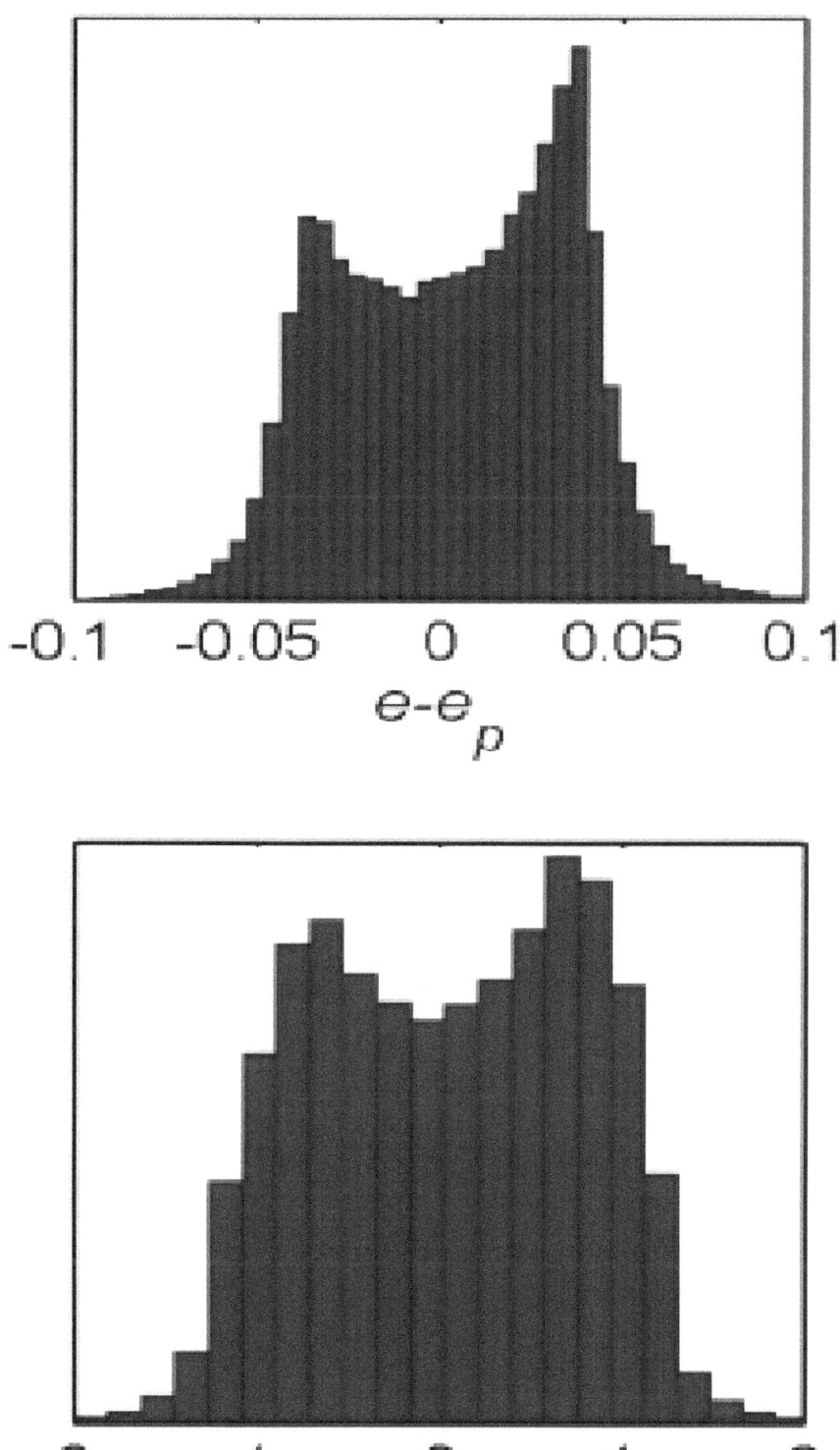

Distribution of the difference between proper and osculating orbital elements for asteroids with semi-major axes lying between 2 and 4 AU.

41.3. EXTERNAL LINKS

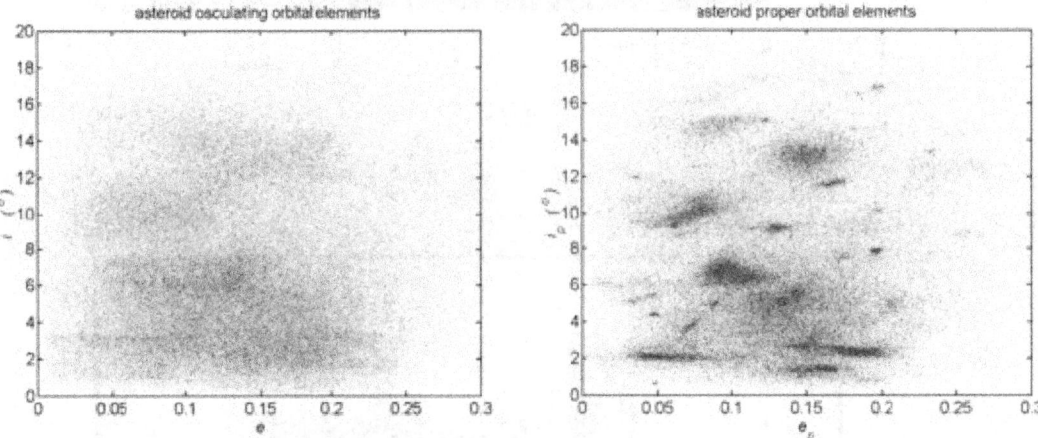

proper (right) and osculating (left) orbital elements for asteroids in the asteroid belt. Note how asteroid family clumps are not discernible on the left.

41.4 Text and image sources, contributors, and licenses

41.4.1 Text

- **Orbital mechanics** *Source:* https://en.wikipedia.org/wiki/Orbital_mechanics?oldid=682009021 *Contributors:* XJaM, SimonP, Stevertigo, Edward, Patrick, Michael Hardy, J-Wiki, Lancevortex, Doradus, Almak, Nickshanks, Stormie, Altenmann, Hadal, JesseW, Alexwcovington, Giftlite, DavidCary, Jyril, Wolfkeeper, Wwoods, Alison, Bobblewik, LiDaobing, Beland, WhiteDragon, Blazotron, Jkl, Discospinster, Cfailde, Vsmith, ArnoldReinhold, Ben Standeven, CanisRufus, Edward Z. Yang, Quinobi, Spearhead, Sietse Snel, Spoon!, 0.39, Shoujun, Michaf~enwiki, Teeks99, Infectbda, MrB, Mlessard, PAR, Allen McC.~enwiki, Gene Nygaard, Siafu, Pdn~enwiki, BD2412, Lasunncty, Lionelbrits, Antilived, YurikBot, Wavelength, Hairy Dude, Eleassar, Skiffer, CecilWard, Virogtheconq, Geoffrey.landis, HereToHelp, Algae, SmackBot, Gilliam, Saros136, Colonies Chris, WDGraham, Dragon695, DMacks, JorisvS, Cowbert, Novangelis, Jack Waugh, Happy-melon, JForget, Amalas, N2e, Thijs!bot, Marek69, Xeno, .anacondabot, TechPerson, Swpb, Ekotkie, R'n'B, Leyo, AstroHurricane001, Maurice Carbonaro, Ohms law, 28bytes, Sdsds, Rollo44, Rei-bot, SieBot, Gamesguru2, The way, the truth, and the light, EMBaero, Gits (Neo), Wwheaton, Franamax, Easphi, Sun Creator, Limitcycle, STPitcher, Avoided, Mitch Ames, MystBot, Addbot, سی‌مرغ, Lightbot, Luckas-bot, Yobot, Ptbotgourou, Skogs-Ola, Stamcose, AnakngAraw, Sjeannethomas, Citation bot, Man on the Mars, Obersachsebot, NOrbeck, FrescoBot, Trewal, Tom.Reding, Ivanvector, VernoWhitney, JustinTime55, Mmeijeri, JSquish, ZéroBot, Crua9, Singh18d-NJITWILL, Teapeat, Rememberway, Widr, MenteMagica, Helpful Pixie Bot, Գարիկ Ավագյան, PhilipTerryGraham, Crh23, Jayadevp13, Dexbot, Aymankamelwiki, Rfassbind, CsDix, Pdcalculus, YarLucebith, Lizia7, Tyler.jackoliver, Paradyne3g, Monkbot, Arnonspitzer and Anonymous: 78

- **Orbit** *Source:* https://en.wikipedia.org/wiki/Orbit?oldid=682299220 *Contributors:* Mav, Bryan Derksen, The Anome, AstroNomer~enwiki, -- April, Alex.tan, XJaM, Ray Van De Walker, Heron, Karl Palmen, Youandme, Nealmcb, Patrick, JohnOwens, Michael Hardy, Tim Starling, FrankH, Alan Peakall, GTBacchus, (, Minesweeper, Alfio, Looxix~enwiki, Ahoerstemeier, Glenn, Cherkash, Pizza Puzzle, Seth ze, Tom Peters, Dysprosia, Doradus, Topbanana, Nickshanks, Francs2000, Shantavira, Robbot, Hadal, Carnildo, Marc Venot, Tosha, Giftlite, Jyril, Harp, Mintleaf~enwiki, Haeleth, Wolfkeeper, Fosse8, Smurfix, Wwoods, Curps, Yekrats, Eequor, Mackerm, Gadfium, Andycjp, Beland, Karol Langner, The Land, Icairns, Karl-Henner, Sam Hocevar, Sam, Urhixidur, Jimius, Eisnel, Mike Rosoft, Venu62, Jiy, JTN, Discospinster, Zaheen, Rich Farmbrough, Vsmith, Saintswithin, Paul August, Bender235, Nabla, RJHall, Art LaPella, Spoon!, 0.39, Causa sui, Your vexation, Bobo192, Smalljim, Duk, Dvogel, Rje, Obradovic Goran, QuantumEleven, Daniel Arteaga~enwiki, Danski14, Alansohn, Sowelilitokiemu, Mlm42, Eddie Dealtry, Gene Nygaard, Shimeru, Siafu, Stemonitis, WilliamKF, Angr, Firsfron, Woohookitty, Pi@k~enwiki, Ch'marr, Plrk, SeventyThree, Alec Connors, Smartech~enwiki, Christopher Thomas, Reddwarf2956, Cgsguy2, Palica, Dysepsion, Mandarax, Graham87, Li-sung, Kbdank71, Rjwilmsi, Salix alba, Mike s, Nneonneo, Zambani, FlaBot, Mishuletz, Gurch, Chobot, Scoops, YurikBot, Wavelength, KamuiShirou, Zhatt, DavidMcKenzie, Test-tools~enwiki, Rjensen, Gbm, RL0919, Lomn, Davidmckenzie, Wknight94, Sandstein, 2over0, Nae'blis, HereToHelp, Emc2, Argo Navis, GrinBot~enwiki, FyzixFighter, That Guy, From That Show!, A bit iffy, SmackBot, Unyoyega, Blue520, KocjoBot~enwiki, Jrockley, Gilliam, Algont, Hmains, Andy M. Wang, Saros136, Metacomet, Baa, DHN-bot~enwiki, Dragice, Can't sleep, clown will eat me, Egg plant, Voyajer, TheKMan, Zirconscot, Flyguy649, Bowlhover, Nakon, Freakytim, Dreadstar, Marcus Brute, Richard0612, Kukini, EmirA~enwiki, SashatoBot, Dr. Sunglasses, Lakinekaki, JorisvS, Mgiganteus1, Bjankuloski06, Dicklyon, Meco, Ryulong, Eastfrisian, Levineps, Iridescent, Michaelbusch, Lvzon, Igoldste, CapitalR, Tau'olunga, Tawkerbot2, The Letter J, Emote, Kurtan~enwiki, Harold f, Fsswsb~enwiki, Hildenja, Drinibot, Harej bot, N2e, MarsRover, Neelix, Iokseng, Equendil, Logicus, Cydebot, Clappingsimon, Zginder, Rracecarr, Michael C Price, Landroo, Richhoncho, Saintrain, Epbr123, El pak, Martin Hogbin, Mojo Hand, Tapir Terrific, NickNak, Joymmart, Leon7, M0ffx, Einsidler, AntiVandalBot, A.M.962, Rjwoer, Leuqarte, JAnDbot, Dreamster, Propaniac, Magioladitis, Bongwarrior, VoABot II, EFletcher, Swpb, Catgut, Will ward, 28421u2232nfenfcenc, Taka2007, Jacobko, DerHexer, JaGa, TheRanger, DancingPenguin, Sly Marbro 03, Jerry teps, J.delanoy, Captain panda, Elva7025, Cpiral, Acalamari, Lunokhod, Jer10 95, Wuyanhuiyishi, Bobianite, Classicalhistory009, Cometstyles, Idioma-bot, Cole.christensen, VolkovBot, Daleband, Sdsds, TXiKiBoT, Rollo44, Qxz, Dendodge, Wassermann~enwiki, Raymondwinn, Entropy1963, FKmailliW, Leafnode, AlleborgoBot, PGWG, EmxBot, Ptitelise, SieBot, TJRC, Tresiden, Hmschallenger, Bastiche, Krawi, The way, the truth, and the light, Oda Mari, Paolo.dL, Doc Perel, Faradayplank, Goustien, Bagatelle, Hobartimus, Lrmcfarland, OKBot, CharlesGillingham, Anyeverybody, WikiLaurent, WikipedianMarlith, Martarius, ClueBot, The Thing That Should Not Be, Jan1nad, Wwheaton, Arakunem, Franamax, Fleem, Speed CG, Easphi, DanielDeibler, SuperHamster, Agge1000, Djr32, Excirial, MinnieRae~enwiki, PixelBot, Eeekster, BobKawanaka, Grey Matter, Pot, PhySusie, Aitias, SoxBot III, DumZiBoT, He6kd, InternetMeme, XLinkBot, Terry0051, Feyrauth, Addbot, Crazysane, Fgnievinski, Fluffernutter, Ederiel, Sam13024, FDT, Cst17, MrOllie, Glane23, Beleaf33, 84user, Tide rolls, Everyme, Luckas-bot, Tohd8BohaithuGh1, Fraggle81, Ccraccnam, Ultimatewikisurfer, Stamcose, Kissnmakeup, THEN WHO WAS PHONE?, KamikazeBot, Azcolvin429, Juliancolton Alternative, AnomieBOT, Ciphers, Galoubet, JackieBot, The High Fin Sperm Whale, Citation bot, Nifky?, Xqbot, Timir2, Unigfjkl, Invent2HelpAll, Turk oğlan, Srich32977, NOrbeck, J04n, Mario777Zelda, Chrismiceli, Doulos Christos, Creator editor, Erik9, FrescoBot, Parvons, Doctor sw27, Tom.Reding, Calmer Waters, A8UDI, Quantum4, שועל123, Dac04, Whalesmith, FoxBot, Vrenator, Duoduoduo, DARTH SIDIOUS 2, Jfmantis, Whisky drinker, Onel5969, TjBot, DASHBot, EmausBot, Twoe gappes, Smarthead88, Syncategoremata, Tommy2010, Netheril96, Mmeijeri, Wikipelli, ZéroBot, Empty Buffer, MajorVariola, GianniG46, Frigotoni, Aidarzver, Brandmeister, L Kensington, Zayzya, Donner60, ChiZeroOne, Teapeat, Dweymouth, ClueBot NG, Moneya, Cinnamon345, Rezabot, CasualVisitor, Theopolisme, Helpful Pixie Bot, Pine, Xonus, Vagobot, Mark Arsten, Ninney, Lozzabates, SodaAnt, Eah2498, Glacialfox, RudolfRed, Tfr000, Pratyya Ghosh, Hghyux, Khazar2, Ipodkylebacus1, Dark Silver Crow, Dexbot, Banfire, JeffAEdmonds, Fvfvdc, Tab8, The Herald, Crow, JaconaFrere, Rcast12, Kalebmcardle, Sascha Grusche, Loraof, TheKingBumi and Anonymous: 367

- **Kepler orbit** *Source:* https://en.wikipedia.org/wiki/Kepler_orbit?oldid=680463295 *Contributors:* Patrick, Michael Hardy, Giftlite, Beland, Bender235, Woohookitty, Hairy Dude, Bo Jacoby, SmackBot, W!B:, Chris the speller, A. Pichler, CRGreathouse, Myasuda, Bob Stein - VisiBone, Swpb, Hue White, Wassermann~enwiki, Paolo.dL, Superwj5, The Thing That Should Not Be, Pomona17, JMiscreant, UnCatBot, Nettings, Terry0051, Addbot, Fgnievinski, Yobot, Stamcose, AnomieBOT, JackieBot, Obersachsebot, NOrbeck, FrescoBot, EmausBot, Mmeijeri, ZéroBot, Iiar, Miroslav T, Teapeat, Ethg242, BG19bot, 2pem, ChrisGualtieri, JeffAEdmonds, Jaxcp3, MatthewJ00, Jeanhausser, H.utkuunlu and Anonymous: 27

- **Orbital state vectors** *Source:* https://en.wikipedia.org/wiki/Orbital_state_vectors?oldid=681224797 *Contributors:* Patrick, Michael Hardy, Ancheta Wis, Jyril, Karn, MJSS, 0.39, Simone, Gosgood, Lasunncty, Angusmclellan, Ryk, RussBot, AndrewBuck, Trekphiler, Serein (renamed because of SUL), JorisvS, Swpb, Spiral5800, Leafnode, Neparis, Zbvhs, Anchor Link Bot, ClueBot, Narasimhavarman10, Rror, Addbot, Fgnievinski, XZise, Erik9bot, DrilBot, Jfmantis, Zfeinst, Teapeat and Anonymous: 10

41.4. TEXT AND IMAGE SOURCES, CONTRIBUTORS, AND LICENSES

- **Semi-major axis** *Source:* https://en.wikipedia.org/wiki/Semi-major_axis?oldid=672759701 *Contributors:* Bryan Derksen, Patrick, Michael Hardy, Ahoerstemeier, Cherkash, Charles Matthews, Fairandbalanced, TravelingDude, Flockmeal, Aliekens, Zandperl, PedroPVZ, Seth Ilys, Matt Gies, Giftlite, Jyril, Wolfkeeper, Curps, Karol Langner, Zfr, Urhixidur, Mani1, 0.39, Tabletop, SeventyThree, Smartech~enwiki, Pfalstad, Ashmoo, Graham87, Salix alba, Chobot, YurikBot, E2mb0t~enwiki, TheMadBaron, GrinBot~enwiki, Finell, Metacomet, DHN-bot~enwiki, Trekphiler, Chlewbot, LuchoX, EVula, Mini-Geek, Chaotic Mind, JorisvS, Bjankuloski06en~enwiki, 041744, Hyperquantization, RobI, Thijs!bot, Sagaciousuk, AntiVandalBot, John.d.page, JAnDbot, Swpb, Isentropiclift, Ian.thomson, IdLoveOne, M-le-mot-dit, Barraki, Sdsds, TXiKiBoT, Hqb, SieBot, Ceroklis, Portalian, PlanetStar, Gerakibot, PixelBot, Razorflame, DumZiBoT, Darkicebot, Addbot, Roentgenium111, Narayansg, Kisbesbot, Tide rolls, Zorrobot, Luckas-bot, JackieBot, Xqbot, GrouchoBot, RibotBOT, SassoBot, Trafford09, Sae1962, Craig Pemberton, Velociostrich, RedBot, IVAN3MAN, Duoduoduo, Jfmantis, Уральский Кот, EmausBot, Paul Torry, Rami radwan, Norbeck, Teapeat, Henrybest3, 26 Ramadan, Jaime Vicente, Sndeep81 and Anonymous: 48

- **Orbital eccentricity** *Source:* https://en.wikipedia.org/wiki/Orbital_eccentricity?oldid=678135527 *Contributors:* Atlan, Patrick, JakeVortex, Robbot, Smb1001, Jyril, Pgan002, ELApro, GregBenson, Syp, RJHall, El C, 0.39, Harley peters, Shenme, SpeedyGonsales, Neitram, Keenan Pepper, Ricky81682, Simone, Alec Connors, Smartech~enwiki, Palica, RuM, Rjwilmsi, Erkcan, Nihiltres, ScottAlanHill, Wormholio, RobotE, Hairy Dude, RussBot, Pigman, Rsrikanth05, Test-tools~enwiki, RazorICE, Beanyk, Caerwine, Bartsas~enwiki, Geoffrey.landis, Erik J, Smack-Bot, Tennekis, Kmarinas86, OrphanBot, A5b, Lambiam, JorisvS, Ckatz, Douglas Spencer, Novangelis, W0lfie, Lvzon, King Hildebrand, Gogo Dodo, Sergei Schmalz, Thijs!bot, Konradek, AntiVandalBot, After Shock~enwiki, Tweesdad, JAnDbot, CosineKitty, WolfmanSF, Swpb, Ling.Nut, EstebanF, DerHexer, Kheider, ChrisfromHouston, Gurchzilla, Philip tao, Mustafa 03011, NatePhysics, Larryisgood, Technopat, Madhero88, Spinningspark, SieBot, Portalian, I Like Cheeseburgers, Callipides~enwiki, Pfvlloyd, Dawn Bard, Driftwood87, Pinkadelica, Vinay Jha, ClueBot, Renacat, Excirial, BobKawanaka, Warren oO, DumZiBoT, Hyunrosa91, Addbot, Wælgæst wæfre, HerculeBot, Yobot, Amirobot, Stamcose, PianoDan, ArthurBot, Dougofborg, Dave3457, Almuhammedi, Tom.Reding, عباد ديراني, FoxBot, Double sharp, Duoduoduo, Jfmantis, EmausBot, Padurar2009, Octaazacubane, JasonKnade, MrGachapon, ZéroBot, ChuispastonBot, Teapeat, Ebehn, Fjörgynn, ClueBot NG, CocuBot, Moneya, Helpful Pixie Bot, 26 Ramadan, HMSSolent, Snaevar-bot, Nimesh Mistry, Mogism, Jayy357, Lugia2453, Rfassbind, Deepanshu xkr and Anonymous: 102

- **Orbital inclination** *Source:* https://en.wikipedia.org/wiki/Orbital_inclination?oldid=680037939 *Contributors:* Bryan Derksen, Lir, Tom Peters, Lkesteloot, Robbot, PedroPVZ, Jyril, Harp, Zeimusu, Javier Carro, DragonflySixtyseven, Icairns, B.d.mills, Urhixidur, Oliver Jennrich, RJHall, Kwamikagami, 0.39, Alec Connors, Smartech~enwiki, Lasunncty, Mike s, Piet Delport, Rodasmith, DouglasHeld, Ageekgal, Argo Navis, Attilios, SmackBot, Bluebot, Trekphiler, Vladislav, Valich, Aldaron, Dreadstar, Hgilbert, Eliyak, JorisvS, Bjankuloski06en~enwiki, 16@r, KurtRaschke, Novangelis, JMK, Newone, Tawkerbot2, MiShogun, N2e, Martin Rizzo, AntiVandalBot, Spencer, JAnDbot, WolfmanSF, Avjoska, Swpb, Kheider, Hans Dunkelberg, Treisijs, Indubitably, AllanManangan, Spiral5800, Jantaro, Natox, SieBot, Portalian, PlanetStar, Enfenion, Nergaal, Martarius, ClueBot, Gits (Neo), Pomona17, Firth m, Niceguyedc, Gtstricky, BodhisattvaBot, Addbot, Roentgenium111, PaterMcFly, 102orion, Luckas-bot, Yobot, MacTire02, AnomieBOT, JackieBot, Theoprakt, RibotBOT, Sbooth, Tom.Reding, Zbayz, Double sharp, Andattaca2010, Teapeat, ClueBot NG, سردبیر, Helpful Pixie Bot, BattyBot, Jwoodward48wiki, Astredita, Loraof and Anonymous: 59

- **Argument of periapsis** *Source:* https://en.wikipedia.org/wiki/Argument_of_periapsis?oldid=660835118 *Contributors:* Michael Hardy, Rparle, Karl-Henner, Urhixidur, Edsanville, Oliver Jennrich, 0.39, Monk, Lasunncty, Barklund, BlueMoonlet, Bomble, Mirecki, BOT-Superzerocool, Poulpy, SmackBot, Winterheart, DHN-bot~enwiki, Vladislav, Radagast83, JorisvS, George100, Vaughan Pratt, Sax Russell, Nemnkim, Kevin Nelson, Swpb, SharkD, Venny85, AlleborgoBot, Portalian, Twinsday, Pomona17, Addbot, Some jerk on the Internet, Njaelkies Lea, Azcolvin429, VectorField, Nfr-Maat, RibotBOT, FrescoBot, HRoestBot, Teapeat, Helpful Pixie Bot, Kdt31415, Exoplanetaryscience, Alpe9942 and Anonymous: 13

- **Longitude of the ascending node** *Source:* https://en.wikipedia.org/wiki/Longitude_of_the_ascending_node?oldid=588538692 *Contributors:* Charles Matthews, Robbot, Alison, Karl-Henner, Urhixidur, Edsanville, Oliver Jennrich, RJHall, 0.39, Simone, Japanese Searobin, Siafu, Smartech~enwiki, Palica, Monk, Lasunncty, Nihiltres, Mirecki, Ronebofh, YurikBot, Spacepotato, Thiseye, Poulpy, Mejor Los Indios, Smack-Bot, Kemperb, Vladislav, Aldaron, Cybercobra, Sax Russell, WinBot, JAnDbot, DuncanHill, Swpb, SieBot, Portalian, Gerakibot, Mikeharris111, Pomona17, Easphi, Addbot, Luckas-bot, TaBOT-zerem, Amirobot, Sriram.aeropsn, LucienBOT, Spacerock1, EmausBot, ZéroBot, Teapeat, Rememberway, Vagobot and Anonymous: 10

- **Longitude of the periapsis** *Source:* https://en.wikipedia.org/wiki/Longitude_of_the_periapsis?oldid=681227039 *Contributors:* Jason Quinn, 0.39, R. S. Shaw, Chessphoon, Lasunncty, Marasama, Spacepotato, SmackBot, Aldaron, Jbergquist, JorisvS, Kevin Nelson, Incredio, STBotD, VolkovBot, Alexbot, Addbot, Zorrobot, Luckas-bot, LucienBOT, AsceticRose, Teapeat, Mayaz khattak and Anonymous: 6

- **True anomaly** *Source:* https://en.wikipedia.org/wiki/True_anomaly?oldid=649338387 *Contributors:* Michael Hardy, Jorge Stolfi, Beland, Urhixidur, Pt, Kwamikagami, 0.39, Simone, Andreas -horn- Hornig, Lasunncty, RE, YurikBot, Bo Jacoby, SmackBot, AndrewBuck, Darth Panda, Yeus, Vina-iwbot~enwiki, Albmont, Swpb, Tbonejoo, Neodymion, VolkovBot, Synthebot, Daigaku2051, Addbot, Njaelkies Lea, Stamcose, Phantom Hoover, Guyy333, FrescoBot, Simbotin, I dream of horses, Curiousranger, Sirkablaam, Jfmantis, Mmeijeri, Teapeat, Kkddkkdd, Poetsirrah! and Anonymous: 15

- **Mean anomaly** *Source:* https://en.wikipedia.org/wiki/Mean_anomaly?oldid=660657361 *Contributors:* XJaM, Michael Hardy, Kwantus, Nikitadanilov, Karn, Jorge Stolfi, Beland, Urhixidur, Liberatus, 0.39, PAR, Cmapm, Japanese Searobin, Chochopk, Smartech~enwiki, Palica, Lasunncty, Mike s, CiaPan, YurikBot, SmackBot, Ron E, AndrewBuck, Tamfang, Nviladkar, Sax Russell, Thijs!bot, JAnDbot, CosineKitty, Albmont, Swpb, AstroMark, FractalFusion, Addbot, Luckas-bot, Stamcose, JackieBot, NOrbeck, Louperibot, Jfmantis, WikitanvirBot, Iyer.arvind. sundaram, Mentibot, ChuispastonBot, Teapeat, Rezabot, سردبیر, Poetsirrah!, Shonzey and Anonymous17

- **Eccentric anomaly** *Source:* https://en.wikipedia.org/wiki/Eccentric_anomaly?oldid=683104294 *Contributors:* Patrick, Michael Hardy, Timwi, Jorge Stolfi, Ato, Beland, Wroscel, Urhixidur, Mdf, 0.39, Lasunncty, YurikBot, Nitefood, Bo Jacoby, SmackBot, AndrewBuck, Torrazzo, Lavateraguy, Thijs!bot, JamesBWatson, Albmont, Swpb, VolkovBot, TXiKiBoT, Adam37, Sound effx, AllHailZeppelin, PipepBot, Alexbot, Brews ohare, Addbot, Stamcose, Phantom Hoover, LilHelpa, TobeBot, Jfmantis, Mmeijeri, ChuispastonBot, Teapeat, Helpful Pixie Bot, Poetsirrah!, BG19bot, Franzl aus tirol, Pascuass and Anonymous: 20

- **Epoch (astronomy)** *Source:* https://en.wikipedia.org/wiki/Epoch_(astronomy)?oldid=668202209 *Contributors:* Heron, Alfio, Looxix~enwiki, Ellywa, Ahoerstemeier, Cherkash, Nickshanks, Robbot, Curps, Joe Kress, Karol Langner, Nike, Flex, Haruo, Smyth, El C, Art LaPella, Hashar-

- **Orbital node** *Source:* https://en.wikipedia.org/wiki/Orbital_node?oldid=677481669 *Contributors:* Charles Matthews, Rhys~enwiki, Gadfium, Urhixidur, Rich Farmbrough, Murtasa, 0.39, ㅁㅁㅁ, Monk, Lasunncty, Rjwilmsi, Dracontes, FlaBot, YurikBot, Eirik, Spacepotato, RussBot, PrologFan, Zwobot, Tcblack, SmackBot, C.Fred, W!B:, WinstonSmith, Flyguy649, Jack Waugh, JMK, JoeBot, Sax Russell, Thijs!bot, WinBot, DuncanHill, Idioma-bot, TXiKiBoT, AlleborgoBot, Astrologist, Easphi, Copyeditor42, Addbot, John Belushi, Luckas-bot, Oftopladb, Sriram.aeropsn, XZeroBot, Orion 8, عباد ديوراني ة, Teapeat, ClueBot NG, IluvatarBot, Murúg, Sh00ter87, Rfassbind and Anonymous: 12

- **Precession** *Source:* https://en.wikipedia.org/wiki/Precession?oldid=677387513 *Contributors:* AxelBoldt, Paul Drye, Vicki Rosenzweig, Mav, Bryan Derksen, The Anome, Wayne Hardman, Andre Engels, Karl E. V. Palmen, Eclecticology, Karl Palmen, Hephaestos, Lir, Patrick, Michael Hardy, Deadstar, Cyde, Shimmin, Stw, Looxix~enwiki, William M. Connolley, Caid Raspa, Mark Foskey, Glenn, Nikai, Kwekubo, Mulad, Tom Peters, Charles Matthews, Wetman, Jerzy, Robbot, Romanm, Henrygb, JesseW, Wereon, Enochlau, Leighxucl, Ancheta Wis, Giftlite, Harp, Tom Radulovich, Joe Kress, Vadmium, Gadfium, Gdr, Xmnemonic, LucasVB, Karol Langner, Lumidek, Dbernat32, Mormegil, Spiffy sperry, Guanabot, Cacycle, Pjacobi, HeikoEvermann, Ponder, Lemontea, GregBenson, Dfan, Livajo, Pt, Miraceti, Marx Gomes, Worldtraveller, Cwolfsheep, Aramael, SpeedyGonsales, La goutte de pluie, Kjkolb, Hesperian, Sam Korn, Ekhalom, Daniel Arteaga~enwiki, Jic, Ungtss, Logologist, Melaen, Tony Sidaway, TenOfAllTrades, ReubenGarrett, Mosesofmason, Natalya, ReelExterminator, Linas, Cleonis, Pdn~enwiki, SDC, Wdanwatts, Marudubshinki, Graham87, Kbdank71, AllanBz, Phoenix-forgotten, Canderson7, BlueMoonlet, Maurog, Vuong Ngan Ha, Ian Pitchford, Ewlyahoocom, Chobot, Whosasking, YurikBot, RobotE, Jimp, MattWright, Markus Schmaus, Big Brother 1984, Mysid, Ke6jjj, Paul Magnussen, StuRat, Reyk, Kungfuadam, KasugaHuang, SmackBot, Incnis Mrsi, InverseHypercube, KocjoBot~enwiki, Müslimix, Kmarinas86, Persian Poet Gal, B00P, CSWarren, Portnadler, Benet Allen, Drphilharmonic, Dmh~enwiki, JorisvS, Ocatecir, Akhen3sir, Odedee, Maestlin, Abel Cavaşi, U04617, Igoldste, Cydebot, W.F.Galway, Hkyriazi, WISo, Doug Weller, Juansempere, AlexDusty, AndrewDressel, Simeon H, Nemti, Ingolfson, JAnDbot, WolfmanSF, Fuchsias, Wabernat, LorenzoB, MarcusMaximus, Edward321, Kegrad, Ryangdotexe, Jorfer, Dhaluza, Joshua Issac, CardinalDan, VolkovBot, TXiKiBoT, Littlealien182, MDSL2005, Graymornings, Smg6512, Chenzw, Karajade, Doverbeach01, Drake144, SieBot, Portalian, Martin Kealey, The way, the truth, and the light, Adam37, Anchor Link Bot, P.Schellart, Nergaal, XDanielx, ClueBot, AcePilot101, XLinkBot, Lousedamouse, Gazouille, Crastinbic, The Rationalist, DavidGPeters, Addbot, Xp54321, Angelobear, Iceblock, MrOllie, Download, LaaknorBot, Tide rolls, Lightbot, Visionismind, Yobot, IW.HG, AnomieBOT, VanishedUser sdu9aya9fasdsopa, 9258fahsflkh917fas, Kingpin13, Materialscientist, Jmundo, NOrbeck, Ruy Pugliesi, GrouchoBot, TonyHagale, DrilBot, Zbayz, Series35, Peacedance, Hhawks12, WikitanvirBot, Mmeijeri, Mattkwyatt, Ὁ οἰστρος, Bilbo571, Andattaca2010, Samoojas, RockMagnetist, Albert Nestar, Stündle, Rocketrod1960, Gary Dee, Svetsveti, ClueBot NG, Enopet, LikeH2O, Akash pagla, Dan653, Mgermaine8, Anderson, Rfassbind, Espotom, MDHWYRM, Muzzi77, Monkbot, Griggyoldboy, Michael Fairchild, Lauriehurman, KasparBot and Anonymous: 202

- **Orbital station-keeping** *Source:* https://en.wikipedia.org/wiki/Orbital_station-keeping?oldid=678301653 *Contributors:* Patrick, Elembis, Chairboy, 0.39, Cmdrjameson, Mikeo, Mandarax, Changcho~enwiki, Gaius Cornelius, Aeusoes1, Smurrayinchester, SmackBot, Bazonka, Namangwari, Wyoskier, Javit, Rob Shanahan, EdC~enwiki, N2e, Swpb, Sm8900, T3rminatr, Andreas Carter, Iain99, CultureDrone, Binksternet, The Beta Queen, Alexrakia, Smeagol 17, Yobot, Stamcose, AnomieBOT, Prari, DrilBot, EmausBot, Helpful Pixie Bot and Anonymous: 28

- **Ground track** *Source:* https://en.wikipedia.org/wiki/Ground_track?oldid=662154855 *Contributors:* Master of Puppets, Vertium, Swpb, Xa Zvinx, Sdsds, Addbot, LaaknorBot, AnomieBOT, Savant-fou, Zbayz, ZéroBot, Zingzang2410, Rs wrangler, TheMPG and Anonymous: 6

- **Orbital maneuver** *Source:* https://en.wikipedia.org/wiki/Orbital_maneuver?oldid=654785301 *Contributors:* Patrick, Wolfkeeper, Micru, Solitude, Rich Farmbrough, Loren36, Livajo, 0.39, Stesmo, M3tainfo, Andreas -horn- Hornig, SchuminWeb, ViriiK, Krishnavedala, YurikBot, PhilipO, Reyk, Geoffrey.landis, Wbrameld, Mejor Los Indios, WDGraham, Can't sleep, clown will eat me, Adsllc, Ryulong, N2e, Thijs!bot, Swpb, Elva7025, JohnBlackburne, Sdsds, Fedos, Pdfpdf, Aalox, ImageRemovalBot, Enenn, Addbot, Luckas-bot, Stamcose, AnomieBOT, Dkassing, Erik9bot, Steve Quinn, WikitanvirBot, Mmeijeri, ZéroBot, ChiZeroOne, Teapeat, Rememberway, Incompetence, Helpful Pixie Bot, YFdyh-bot, Angelov1988, ChristianGaertner and Anonymous: 15

- **Hohmann transfer orbit** *Source:* https://en.wikipedia.org/wiki/Hohmann_transfer_orbit?oldid=683246863 *Contributors:* Bryan Derksen, Maury Markowitz, Patrick, Gmalivuk, Sannse, Karada, Looxix~enwiki, Rlandmann, Lancevortex, Doradus, Nickshanks, Giftlite, 0x0077BE, Wolfkeeper, Inkling, Wwoods, Urhixidur, O'Dea, Vsmith, Loren36, El C, PhilHibbs, 0.39, Hooperbloob, Snowolf, Roboshed, DonPMitchell, JabberWok, Stephenb, Voidxor, Ospalh, Light current, Mattg2k4, Darabuc~enwiki, Montestruc, Mejor Los Indios, AndrewWTaylor, KnightRider~enwiki, SmackBot, Midway, Chris the speller, Nsg, Vladislav, Chlewbot, Jmnbatista, Fitzhugh, DMacks, Dpu2002, JorisvS, Ft1~enwiki, Michael-busch, JForget, CmdrObot, N2e, Aeddub, Tewapack, Andyjsmith, Widefox, AstroLynx, JAnDbot, Deflective, Karlhahn, Albmont, SHCarter, Swpb, Misibacsi, Crazychris2704, Cole.christensen, Luciopaiva, Jasonc1201, Barneca, Sdsds, Thrawn562, Dbeachy1, Marcosaedro, Aajack-soniv, Leafnode, Mxlyons, Kinzele, Wwheaton, Eeekster, Wilsone9, Estirabot, SchreiberBike, DumZiBoT, Dradler, Wyatt915, Addbot, Light-bot, Luckas-bot, Yobot, AnomieBOT, Citation bot, ArthurBot, Invent2HelpAll, TechBot, BtM909, Fotaun, GliderMaven, Prari, FrescoBot, BenzolBot, Tom.Reding, IVAN3MAN, WikitanvirBot, Rami radwan, JustinTime55, Mmeijeri, Hhhippo, ZéroBot, Qniemiec, Chuispaston-Bot, ChiZeroOne, Teapeat, Rememberway, ClueBot NG, GeoffreyTFox, Helpful Pixie Bot, Ivzeivze, SiriusLH, BattyBot, Nimesh Mistry, Kekamohan, ChrisGualtieri, A R Buschert, Indronil Ghosh, Nydoc1, Astrozot, ChristianGaertner, Humabout, TECHnOhunk, Pulkitmidha, Ashishrs, Ioandbbsbsbsn, Rabbil02 and Anonymous: 72

- **Delta-v** *Source:* https://en.wikipedia.org/wiki/Delta-v?oldid=681474190 *Contributors:* Bryan Derksen, The Anome, Maury Markowitz, Stevertigo, Patrick, JohnOwens, Ixfd64, Cherkash, Charles Matthews, Doradus, Chris Roy, Wolfkeeper, Karn, Mboverload, Oliver Jennrich, Arnol-

41.4. TEXT AND IMAGE SOURCES, CONTRIBUTORS, AND LICENSES

dReinhold, Huntster, RoyBoy, 0.39, Foobaz, Stack, Mikeo, Gene Nygaard, Tonigonenstein, Mike s, FlaBot, Wavelength, Gaius Cornelius, Zwobot, Geoffrey.landis, NeilN, Mejor Los Indios, KnightRider~enwiki, SmackBot, Lotse, Kristod, Ged UK, Ourai, JorisvS, Andyops, Etatoby, Joseph Solis in Australia, Courcelles, N2e, SJ Zero, Headbomb, Dtgriscom, CairoTasogare, Slothman32, TechPerson, LorenzoB, Gary63, TempestCA, Ohms law, VolkovBot, Midoriko, Philip Trueman, TXiKiBoT, Someguy1221, PaulTanenbaum, Anchor Link Bot, LP-mn, Excirial, Alexbot, NuclearWarfare, Sanddune777, DumZiBoT, Feyrauth, MystBot, Addbot, Tarbos, Jasper Deng, Luckas-bot, Yobot, Fraggle81, TaBOT-zerem, Stamcose, JmCor, NOrbeck, J04n, GliderMaven, JustinTime55, Mmeijeri, Solomonfromfinland, ZéroBot, SporkBot, Bomazi, Teapeat, Rememberway, ClueBot NG, Incompetence, Kendall-K1, JMtB03, Theemathas, Frinthruit, Kitsunedawn, AKS.9955, Wavechaser and Anonymous: 48

- **Bi-elliptic transfer** *Source*: https://en.wikipedia.org/wiki/Bi-elliptic_transfer?oldid=674586585 *Contributors*: Wolfkeeper, Onco p53, Loren36, Roboshed, DonPMitchell, Trewornan, Nimur, PaulWay, Dauto, Chris the speller, AndrewBuck, Nsg, Qwerty 0, Spiel 496, Meithan, N2e, D4g0thur, Headbomb, Swpb, Waacstats, Saurc zlunag, Keermalec, Lucaswilkins, Sdsds, Brianonn, Venny 85, Peggymeng, Mumiemonstret, MystBot, Addbot, Yobot, AnomieBOT, Dkassing, Jonesey 95, JustinTime 55, ChiZeroOne, Teapeat, Helpful Pixie Bot, Duxwing, Stillyslalom, Cheerioswithmilk, EvergreenFir, Sanil gupta, Ca2james and Anonymous:12

- **Gravity assist** *Source*: https://en.wikipedia.org/wiki/Gravity_assist?oldid=682486183 *Contributors*: Bryan Derksen, Ap, Malcolm Farmer, Deb, Ray Van De Walker, Maury Markowitz, Patrick, JohnOwens, Michael Hardy, Collabi, Looxix~enwiki, CatherineMunro, Notheruser, Beck, Doradus, Morn, Robbot, SchmuckyTheCat, Carnildo, Fabiform, Mintleaf~enwiki, Wolfkeeper, Karn, Wwoods, Python eggs, Tagishsimon, TerokNor, Zeimusu, ShakataGaNai, Icairns, Urhixidur, CohenTheBavarian, Rich Farmbrough, ArnoldReinhold, Kbh3rd, JustinWick, El C, Bobo192, Kfogel, Foobaz, Timecop, Hooperbloob, Lysdexia, Alansohn, Wricardoh~enwiki, RobertStar20, A.T.M.Schipperijn, Dhartung, Simone, Cmapm, Gene Nygaard, Bignoter, Igny, Tckma, Mpatel, Someone42, Eyreland, Evilmoo, Kotukunui, Rjwilmsi, Jivecat, Gnarg~enwiki, MZMcBride, FlaBot, SchuminWeb, Ewlyahoocom, Chobot, Schwern, YurikBot, Wavelength, RobotE, Rylz, Gaius Cornelius, Schlockading, ErkDemon, Saulpwanson, Closedmouth, Geoffrey.landis, Phr en, Flumpaphone, Cmglee, SmackBot, RDBury, F, Dslayer202, Slashme, UrsaFoot, Saros136, Chris the speller, Qasrani, Stevage, SEIBasaurus, Beatgr, WikiPedant, Can't sleep, clown will eat me, LouScheffer, Elendil's Heir, Matt Whyndham, Daniel.Cardenas, Erimus, Gobonobo, Khono, JorisvS, Mgiganteus1, IronGargoyle, Etatoby, Tac2z, Sinistrum, Meco, Joseph Solis in Australia, Woodshed, Jonathan W, Ale jrb, Klossner, Phædrus, Necessary Evil, Scottymoze, Thijs!bot, Hazmat2, Dtgriscom, Seaphoto, Alphachimpbot, Pictor~enwiki, IanOsgood, Pigdome, QuizzicalBee, Albmont, Swpb, Email4mobile, Custardninja, Cecilkorik, KTo288, Valaggar, Philcha, Hans Dunkelberg, Maurice Carbonaro, Chemicalrubber, NewEnglandYankee, Bob, TWCarlson, Idioma-bot, Zumthie, Margacst, Sdsds, TXiKiBoT, Red Act, Tomladams, Someguy1221, Driesman, Broadbot, Jackfork, Spiral5800, Paulsmith99, OlkhichaAppa, Ergateesuk, Cwkmail, Jan Winnicki, Dolphin51, Nergaal, Eric.nickel, ImageRemovalBot, Jjefferp, MBK004, ClueBot, UrsusArctosL71, Fleem, CounterVandalismBot, Oxnard27, AssegaiAli, Alexbot, Jefflayman, Jshilliday, Certes, Addbot, Roentgenium111, Paintballerca, Download, LaaknorBot, Bendavis78, Numbo3-bot, Tide rolls, Lightbot, Anxietycello, A4arpan, Yobot, Amble, POcca, Kostan1, Sergei Kholmskiy, Archon 2488, The High Fin Sperm Whale, ArthurBot, Wcoole, Ruy Pugliesi, GrouchoBot, Microfilm, RibotBOT, Alan.A.Mick, GliderMaven, FrescoBot, Perlscrypt, Tom.Reding, Codwiki, Magmaiceman, ClovisCO, Spaluch1, Full-date unlinking bot, IVAN3MAN, Double sharp, Whisky drinker, Alph Bot, Androstachys, Helium4, Geo d, Solarra, JCRJohnson, Mmeijeri, Thrownshadows, AvicBot, ZéroBot, Ted Sweetser, Narendran95, AndyT.hk, Samoojas, Surajt88, Megacellist, Niels241, Slugogum, ChiZeroOne, Teapeat, Rememberway, ClueBot NG, Gilderien, Sleddog116, Rezabot, Wpmarc, Nullzero, MillingMachine, Ofml, BG19bot, Wooden1, Furkhaocean, Rijinatwiki, Eric M. Jones, Nuke1st, NotWith, Anbu121, BattyBot, Hiparick, GoShow, Mogism, Jaxcp3, Metsusna, Corinne, RicardAnufriev, Shogiru, St gonzo, SKoOctopus, RabidLaconic, Anythingcouldhappen, 汁麪, Monkbot and Anonymous: 161

- **Lagrangian point** *Source*: https://en.wikipedia.org/wiki/Lagrangian_point?oldid=682569011 *Contributors*: Damian Yerrick, Magnus Manske, Paul Drye, The Epopt, Ansible, Brion VIBBER, Mav, Bryan Derksen, Zundark, Tarquin, Wayne Hardman, XJaM, JeLuF, PierreAbbat, Roadrunner, Camembert, Bdesham, Patrick, JohnOwens, Michael Hardy, Frank Shearar, Mcarling, Skysmith, (, SebastianHelm, Minesweeper, Ahoerstemeier, Muriel Gottrop~enwiki, Caid Raspa, Julesd, Whkoh, Cherkash, Emperorbma, Jitse Niesen, Audin, Doradus, Wik, PaulL~enwiki, Nickshanks, Pakaran, Finlay McWalter, Robbot, Astronautics~enwiki, Pigsonthewing, Sverdrup, Rorro, Kielsky, Rursus, Bkell, Netjeff, HaeB, GreatWhiteNortherner, Jason in MN, Giftlite, JamesMLane, DocWatson42, DavidCary, Wolfkeeper, Art Carlson, MSGJ, Herbee, Angmering, Wwoods, Curps, Pascal666, Golbez, Cam, Zeimusu, ConradPino, Fpahl, Jodamiller, Beland, Mako098765, OrangUtanUK, WhiteDragon, Mzajac, OwenBlacker, CanSpice, Tomwalden, Rich Farmbrough, Guanabot, Sladen, Vsmith, Florian Blaschke, ArnoldReinhold, Bo Lindbergh, Bender235, Kbh3rd, Karmafist, Kwamikagami, Kross, Tverbeek, PhilHibbs, Shanes, Dennis Brown, Euyyn, Mike Schwartz, Matt McIrvin, Larry V, Jumbuck, Grutness, Gary, Eric Kvaalen, Arthena, Apoc2400, RoySmith, Mlm42, Hu, Gomi no sensei, Dschwen, Suruena, Evil Monkey, Staeiou, Kindalas, Cmprince, Pauli133, Axeman89, LukeSurl, Kitch, BerserkerBen, ReelExterminator, Agingjb, WilliamKF, Woohookitty, Natcase, Duncan.france, CharlesC, Fxer, Zooks527, Mandarax, Graham87, Protargol, Rjwilmsi, Zbxgscqf, Vary, Strait, Bubba73, Nguyen Thanh Quang, Dracontes, Cjpuffin, Mathbot, Gparker, Quuxplusone, Chobot, PointedEars, Neitherday, Hairy Dude, Jimp, 4C~enwiki, RussBot, Chuck Carroll, Hellbus, Hydrargyrum, Ergzay, Johnny Pez, The Merciful, Robertvan1, Aeusoes1, RazorICE, BlackAndy, Raven4x4x, Crasshopper, Cycleskinla, Deeday-UK, Wikicheng, Chase me ladies, I'm the Cavalry, Closedmouth, Reyk, Back ache, Geoffrey.landis, HereToHelp, Skittle, Diegorodriguez~enwiki, Cmglee, Ffangs, Hide&Reason, SmackBot, Unschool, Robotbeat, Incnis Mrsi, Herostratus, C.Fred, Declare, Brianski, BirdValiant, Grokmoo, GovanRear, Salvo46, Riaanvn, Dreg743, Jprg1966, Thumperward, Rpspeck, Scienz Guy, Hgrosser, Audriusa, WDGraham, Trekphiler, Tamfang, Writtenright, MattHanlon, Spacecaldwell, Allan McInnes, Subheight640, TheLimbicOne, Romanski, Vina-iwbot~enwiki, Lambiam, JorisvS, Terry Bollinger, 5telios, Rsquid, Loadmaster, TheHYPO, Dicklyon, Megane~enwiki, EdC~enwiki, MrDolomite, TJ Spyke, Craigboy, Tmangray, Benabik, Hamish.MacEwan, Memetics, Ruslik0, N2e, Goatchurch, Necessary Evil, Cydebot, Peripitus, Savre, Reywas92, VAXHeadroom, LarryMColeman, Clh288, Christian75, Thijs!bot, Skiltao, AndrewDressel, Undomiel, Uruiamme, Noclevername, 1-54-24, Hawk16zz, Saimhe, Vendettax, Yellowdesk, Tulio22, Ingolfson, JAnDbot, Deflective, CosineKitty, Medconn, Chromosome, ReK, Mwarren us, 100110100, Dream Focus, Rollred15, Seanette, WolfmanSF, QuizzicalBee, Mrld, SHCarter, LosD, Swpb, RypER, Thermal0xidizer, Originalname37, JMyrleFuller, Just James, JaGa, Onigame, Alfachino, Gwern, Atarr, Jim.henderson, Sm8900, CommonsDelinker, Kevin Carmody, AstroHurricane001, Hans Dunkelberg, ChrisfromHouston, Sidhekin, Tdadamemd, Andrew Moise, Dargaud, Rod57, JavaJawaUK, Tarotcards, DjScrawl, Ohms law, Atropos235, Black Walnut, Pborenstein, Speciate, Xenonice, VolkovBot, Larryisgood, Pleasantville, Mrh30, Otherdumbname, Seattle Skier, Grammarmonger, Larry R. Holmgren, Sdsds, KimiSan, Phoenix2602, Baileypalblue, OlavN, BeIsKr, Dependent Variable, AmouDaria, Taho s, Friscorose, Peregrinoerick, AlleborgoBot, Kktor, BillHicksRulez, Hughey, 1Shaggy1, Coronellian~enwiki, SieBot, WereSpielChequers, Alex Essilfie, Scimike, KGyST, Agesworth, K.h.w.m, Paolo.dL, Roger.ritenour, Beast of traal, Sailing rich, Roujo, RSStockdale, Anchor Link Bot, PlantTrees, Martarius, ArdClose,

Number774, CounterVandalismBot, Piledhigheranddeeper, Deselliers, Lfast321, Brews ohare, Kittylyst, Brilliant Pebble, JKeck, Mchaddock, Ophello, Nathan Johnson, Astrofreak92, Feyrauth, Subversive.sound, Addbot, Immortal Imp, Roentgenium111, Llesslie, Obsidianspider, AnnaFrance, 84user, Numbo3-bot, Bigzteve, Lightbot, Legobot, Luckas-bot, Yobot, MarioS, Amirobot, Aldebaran66, AnomieBOT, JackieBot, Flewis, Citation bot, DirlBot, LilHelpa, Xqbot, DrRevXyzzy, Deeptext, Invent2HelpAll, Stevenenglund, Metafax1, Ataleh, Lidmann, 11cookeaw1, GliderMaven, FrescoBot, Wtthompson, Liiiii, EmilTyf, Kvidre!, Neilksomething, AstaBOTh15, Dcshank, DanteEspinoza1989, Tom.Reding, LagrangeCalvert, TeigeRyan, Cnwilliams, Meier99, Topaxi, TobeBot, Dinamik-bot, RjwilmsiBot, TjBot, Mathematici6n, Androstachys, Tesseract2, EmausBot, Lucien504, Mmeijeri, FlyAkwa, Hhhippo, Listmeister, ZéroBot, Riommar, Crua9, Mattedia, Δ, Giulianone, SkywalkerPL, RockMagnetist, Dkoz314, Teapeat, Karthikrrd, Fjörgynn, ClueBot NG, Giggett, Matsaball, Macarenses, Gd199210018, Muon, MobyDick71, Bibcode Bot, Gauravjuvekar, Doyna Yar, BG19bot, Rasheeq1, Anwalt-frankfurt, Wc620, Trevayne08, Zedshort, Gianamar, Cliff12345, Ibk4173hf3p, BattyBot, Fraulein451, Dexbot, Olly314, Hms1103, Lugia2453, Tony Mach, Andyhowlett, Tommy.Hudec, Reatlas, Rfassbind, Tbrien88, Wikiuser314159, RedZay7, Itaspence, Samatict, AlexeiVronsky, DJHerls, Monkbot, Warqer, 0Archimedes0, Pabnau, TuxedoMonkey, Tetra quark, Appable, Kahntagious, KasparBot, Martin Coles, Eric1024 and Anonymous: 351

- **N-body problem** *Source:* https://en.wikipedia.org/wiki/N-body_problem?oldid=679146164 *Contributors:* AxelBoldt, Mav, Bryan Derksen, Zundark, The Anome, Gareth Owen, Khendon, PierreAbbat, Heron, Stevertigo, Nealmcb, Patrick, Michael Hardy, Geoffrey~enwiki, Eric119, Looxix~enwiki, Stevenj, Ideyal, Revolver, Dying, Charles Matthews, Reddi, Jitse Niesen, Doradus, Saltine, Itai, Finlay McWalter, Jaredwf, Fredrik, MathMartin, Hadal, Wikibot, Wereon, Robinh, Xanzzibar, Cyrius, Giftlite, Gene Ward Smith, Wolfkeeper, Jacob1207, Curps, CyborgTosser, Murison, Pascal666, Wmahan, Gugganij, Neilc, CryptoDerk, Abu badali, LucasVB, Fpahl, Beland, Urhixidur, Masudr, Pjacobi, Smyth, Paul August, Bender235, Tompw, RJHall, Pt, Wareh, Aaron D. Ball, Jpceayene, Apostrophe, Espoo, Alansohn, Jeltz, Avenue, Joris Gillis, RJFJR, Gene Nygaard, Oleg Alexandrov, Linas, Benhocking, GregorB, Dmitry Gerasimov~enwiki, Waldir, SeventyThree, Marudubshinki, Graham87, Rjwilmsi, MarSch, BlueMoonlet, Nneonneo, Bubba73, Mathbot, Gurch, Woodardj, Kri, Nsteinberg, Chobot, Bgwhite, YurikBot, Hairy Dude, 4C~enwiki, Bjf, Zwobot, Denis Constales, Ingling, ClaesWallin, Urger48400, Tsiaojian lee, Bo Jacoby, That Guy, From That Show!, SmackBot, Oub, InverseHypercube, K-UNIT, CrypticBacon, Septegram, Gaff, Kmarinas86, Jayanta Sen, Gutworth, Kostmo, Tamfang, Chlewbot, Mr Snrub, Somnlaut, Mistamagic28, Agradman, Ourai, EdC~enwiki, Novangelis, Freelance Intellectual, Olaf Davis, Mattbuck, Yaris678, Stebbins, A876, WillowW, Dimacq, Michael C Price, Tewapack, Jauricchio, Nisselua, Mattva01, Spartaz, Barek, Coolhandscot, MegX, Engelbaet, Swpb, Gwern, David J Wilson, CommonsDelinker, Vanderbei, Hu Totya, Hans Dunkelberg, ChrisfromHouston, Dispenser, Tarotcards, Michaelban, Sdommers~enwiki, TXiKiBoT, Jacob Lundberg, Red Act, MusicScience, PaulTanenbaum, Suriel1981, Timothy Cooper, Paolo.dL, Mangledorf, Anchor Link Bot, JL-Bot, Kjkatusc, ClueBot, PixelBot, Cr7i, Cuz183, Togaen, DumZiBoT, YouRang?, Terry0051, LuciferJ, Addbot, The Equilibrium, Canrosin, Anders Sandberg, Protonk, 84user, Lightbot, Wmrwiki, Yobot, Fraggle81, AnomieBOT, ThaddeusB, Piano non troppo, Csigabi, Constructive editor, GliderMaven, Keen Commander, Lookhigh, Parvons, Tom.Reding, Double sharp, Puzl bustr, Duoduoduo, Diannaa, Olawlor, Math+Wine, EmausBot, Tuankiet65, Dewritech, Kristian Larsen, Sk!d, Mmeijeri, Dcirovic, ZéroBot, BrokenAnchorBot, Maschen, Zfeinst, Teapeat, Llightex, Morgis, ClueBot NG, Anagogist, Chrisminter, Moneya, Ernest3.141, Bibcode Bot, BG19bot, Camrto, Negativecharge, Northamerica1000, 155blue, Solomon7968, Jobojobocat, Mogism, Rudrene, Mark viking, Jrmrjnck, Smirglis, Abitslow, KasparBot and Anonymous: 131

- **Kepler's laws of planetary motion** *Source:* https://en.wikipedia.org/wiki/Kepler'{}s_laws_of_planetary_motion?oldid=683730212 *Contributors:* AxelBoldt, WojPob, AstroNomer~enwiki, Andre Engels, Ben-Zin~enwiki, Heron, Lir, Patrick, Michael Hardy, Alan Peakall, Bcrowell, Lquilter, Eric119, SebastianHelm, Stw, Looxix~enwiki, Pbn~enwiki, AugPi, BenKovitz, Andres, Pizza Puzzle, Seth ze, Charles Matthews, Stan Lioubomoudrov, Jitse Niesen, Timc, Roachmeister, Xaven, JorgeGG, Robbot, Hankwang, Fredrik, Zandperl, Jredmond, Gandalf61, MathMartin, Stewartadcock, Sverdrup, Rtfisher, Hadal, Jleedev, Ancheta Wis, Giftlite, Tnewell, Harp, Tom harrison, MSGJ, Herbee, Karn, Peruvianllama, Wwoods, Zellerin~enwiki, Doshell, Telso, Sreyan, SURIV, LucasVB, Antandrus, Beland, Anythingyouwant, Bosmon, Icairns, Urhixidur, Fg2, Oliver Jennrich, ELApro, D6, Perey, Archer3, Ocon, Bornintheguz, Johan Elisson, Rich Farmbrough, Guanabot, Cfailde, Vsmith, Mat cross, ArnoldReinhold, Aonaran, Sam Derbyshire, Dbachmann, Paul August, SpookyMulder, Bender235, ESkog, Wfisher, Kwamikagami, Hayabusa future, Shanes, Kernco, Bobo192, Harley peters, BillCook, Nhandler, Tms, Jumbuck, Alansohn, Riana, Calton, Emvee~enwiki, Jon Cates, Mikeo, H2g2bob, Gearspring, Dan100, Oleg Alexandrov, Woohookitty, DonPMitchell, StradivariusTV, Kzollman, Cleonis, Drostie, Pdn~enwiki, Frungi, Zzyzx11, Smartech~enwiki, RuM, Graham87, Lasunncty, BorgHunter, Jake Wartenberg, Dr-Tall, Salix alba, Mike Peel, The wub, Drrngrvy, FlaBot, RexNL, David H Braun (1964), Physchim62, Imnotminkus, Chobot, Krishnavedala, DVdm, RashBold, Antiuser, YurikBot, Wavelength, Xihr, Zhatt, NawlinWiki, Bachrach44, Borbrav, ErkDemon, E2mb0t~enwiki, Syrthiss, Kkmurray, Cstaffa, Hirak 99, Modify, CharlesHBennett, Tevildo, GraemeL, Fram, Shyam, QmunkE, Argo Navis, Katieh5584, TLSuda, Sinan Taifour, GrinBot~enwiki, Bo Jacoby, Serendipodous, FyzixFighter, Mejor Los Indios, Sbyrnes321, Finell, Harthacnut, Attilios, SmackBot, Michaelliv, Ashill, Reedy, InverseHypercube, David Shear, Gregory j, Vilerage, W!B:, Munky2, Gilliam, Saros136, JCSantos, B00P, AndrewBuck, Metacomet, SEIBasaurus, Kungming2, Croquant, Jwillbur, Thisisbossi, SundarBot, Aldaron, Waprap, Korako, Wen D House, Radagast83, Decltype, Speedplane, Theodore7, KeithB, Bejnar, Wikiklaas, SashatoBot, Eliyak, Drieux, Kuru, Cronholm144, Loodog, Gobonobo, Shlomke, Hemmingsen, Accurizer, Morshem, Chrisch, Mets501, Hu12, MystRivenExile, GDallimore, Octane, Courcelles, Eluchil404, Gunslinger of Gilead, Xod, CmdrObot, RedRollerskate, Imamathwiz, Equendil, Kribbeh, Icek~enwiki, WillowW, JFreeman, UrbanLegend, Richhoncho, Phi*n!x, Epbr123, Jplvnv, Lupogun, Sry85, Martin Hogbin, Headbomb, Marek69, Iviney, Monkeyfett8, Diskid, Mentifisto, AntiVandalBot, Cyrilthemonkey, Orionus, JHFTC, QuiteUnusual, Leftynm, AlphaAquilae, LaQuilla, Alphachimpbot, Sluzzelin, JAnDbot, MER-C, IanOsgood, Yill577, Acroterion, Magioladitis, VoABot II, Swpb, Fabricebaro, Vssun, BMF203, Youkai no unmei, MartinBot, Ricardopn, David J Wilson, Schildwaechter, J.delanoy, Fatka, Jerry, Katalaveno, Johnbod, McSly, Hut 6.5, Nwbeeson, Srrizvi, Heyitspeter, Superdrew515, Sscruggs, KylieTastic, STBotD, Ricefountian, Idioma-bot, 28bytes, Pleasantville, JohnBlackburne, Thurth, Barneca, IamCanadianEh, Astronomyphile, TXiKiBoT, Oshwah, Vipinhari, Hqb, Qxz, Waleffe, LeaveSleaves, Wassermann~enwiki, Q Science, Millancad, Andy Dingley, SieBot, Ceroklis, Gerakibot, Cwkmail, Flyer22, Green6592, Hxhbot, Techman224, Schwabac, Anchor Link Bot, Zeyn1, Vreezkid, ClueBot, Ignacio Javier Igjav, Fyyer, The Thing That Should Not Be, CyrilThePig4, Polyamorph, Niceguyedc, Richerman, Agge1000, Djr32, Excirial, Simonmckenzie, Wumborti, GhuonU, Brews ohare, PhySusie, Kentgen1, The Red, C628, Aitias, BunnyFlying, Johnuniq, GabrielVelasquez, Forbes72, Terry0051, DCCougar, Gonfer, SilvonenBot, Mifter, Keegannmann, Addbot, Some jerk on the Internet, Jojhutton, Dgroseth, FDT, Proxima Centauri, Glane23, Sam lowry2002, Debresser, Favonian, TStein, HAHS 25, Bigzteve, Tide rolls, Charles Leedham-Green, Lightbot, Vasil', Luckas-bot, Fraggle81, Nradam, Stamcose, AnomieBOT, Ciphers, Jim1138, IRP, Prasenjit1988, Ulric1313, Materialscientist, Spirit469, RobertEves92, The High Fin Sperm Whale, Citation bot, Kotika98, ArthurBot, LilHelpa, Xqbot, Sionus, Cureden, TechBot, Grim23, NFD9001, NOrbeck, 11kravitzn, Omnipaedista, Frankie0607, Bloodstruck, Nedim Ardoğa, 78.26, Keldino,

41.4. TEXT AND IMAGE SOURCES, CONTRIBUTORS, AND LICENSES

TASDELEN, Acannas, Dave3457, Max.Casasco, FrescoBot, AllCluesKey, Sławomir Biały, Craig Pemberton, Citation bot 1, Relke, DrilBot, I dream of horses, Tom.Reding, Lithium cyanide, BRUTE, Jaoswald, Mjs1991, كاشف عقيل, Etincelles, Lotje, Vrenator, Yong, Duoduoduo, Reaper Eternal, Diannaa, Gegege13, Katovatzschyn, Jfmantis, RjwilmsiBot, WildBot, EmausBot, Schwartz paul, Super48paul, Laurifer, Dewritech, Syncategoremata, Tommy2010, Mmeijeri, Slawekb, Solomonfromfinland, JSquish, Crua9, StringTheory11, Thisibelieve, Wayne Slam, Samoojas, Donner60, Zfeinst, Fanyavizuri, Teapeat, MFJoergen, Brycehughes, Xanchester, ClueBot NG, Wcherowi, Smeagol25, Delusion23, Braincricket, Widr, Pjbussey, Helpful Pixie Bot, HMSSolent, Gob Lofa, Bibcode Bot, DBigXray, BG19bot, 1994bhaskar, MusikAnimal, Dan653, Mark Arsten, Cstalg, HTML2011, BattyBot, The Illusive Man, ChrisGualtieri, Tree2q, Lizard03, Dexbot, Mogism, Lugia2453, Greatuser, 77Mike77, Pforpickaxe, Faizan, CsDix, Eyesnore, JakeWi, Blackroseent98, NeapleBerlina, Hansmuller, Bingston, Bartekltg, Iojknm9090, Crow, Monkbot, Phdaerospace, Pikunsia, Jburdettelinn, Rolbit, Sorryhadtodothis, Hirumeshi, Alaura151695, Coolman468764, Llatosz, ImAwesomeSoDealWithIt, Shaelja, VexorAbVikipædia, KasparBot, Peaceful mind ap, Niecethe, AnkanDas5 and Anonymous: 613

- **Tsiolkovsky rocket equation** Source: https://en.wikipedia.org/wiki/Tsiolkovsky_rocket_equation?oldid=670489270 Contributors: The Anome, Heron, Patrick, Michael Hardy, Ixfd64, Julesd, Charles Matthews, Bemoeial, Doradus, Altenmann, Giftlite, Wolfkeeper, Karn, N328KF, Jkl, Discospinster, Rich Farmbrough, ArnoldReinhold, YUL89YYZ, Bender235, Jonas Olson, PAR, Suruena, Simone, Cmapm, Gene Nygaard, MiguelTremblay, Oleg Alexandrov, Woohookitty, DonPMitchell, Zzyzx11, RedBLACKandBURN, Skorkmaz, Mathbot, Ryan Hardy, Krishnavedala, YurikBot, Hairy Dude, MartinRudat, Enormousdude, Petri Krohn, Kungfuadam, Algae, KnightRider~enwiki, SmackBot, Prodego, Seifip, Kemperb, Autopilot, JorisvS, Physis, Hypnosifl, Mets501, Lakers, RSido, N2e, Gjsr~enwiki, Arb, JAnDbot, Olaf, Andrew Swallow, .anacondabot, Magioladitis, Jatkins, MarcusMaximus, J.delanoy, Salih, Sanchazo, Skrelk, DorganBot, JohnBlackburne, Sdsds, Rei-bot, Bphenry, Mikemoral, Triwbe, Nr4ps, Anchor Link Bot, Denisarona, ImageRemovalBot, ClueBot, Douglas Heriot, Niceguyedc, Gulmammad, Alexey Muranov, RP459, D.M. from Ukraine, Addbot, F Notebook, Yobot, TaBOT-zerem, KamikazeBot, AnomieBOT, JackieBot, Obersachsebot, Xqbot, RibotBOT, GliderMaven, Arthree, Tom.Reding, TheLongTone, Reach Out to the Truth, Jfmantis, EmausBot, John of Reading, JustinTime55, Mmeijeri, Solomonfromfinland, JSquish, MonoAV, Zueignung, Teapeat, DASHBotAV, Teaktl17, ClueBot NG, Incompetence, Zounds011, Muon, Helpful Pixie Bot, YodaWhat, Lugia2453, Jamesx12345, Faizan, CsDix, Challenger l, Ginsuloft, Marikanessa, Monkbot, OrangeOnyxDragon and Anonymous: 84

- **Vis-viva equation** Source: https://en.wikipedia.org/wiki/Vis-viva_equation?oldid=682746696 Contributors: Patrick, Stevan White, Doradus, Cutler, Wolfkeeper, Wwoods, Mihoshi, 0.39, GregorB, Plrk, Stoph, Chobot, EDG, JabberWok, Gaius Cornelius, SmackBot, Steve Pucci, JorisvS, Gjp23, Brianhicks, BobQQ, Albmont, Rich257, Salih, Student7, Nr4ps, The Thing That Should Not Be, Phidus, Addbot, Luckas-bot, Yobot, NOrbeck, Erik9bot, Fortdj33, RedBot, Teapeat, ClueBot NG, Latifahphysics, Indronil Ghosh, Nehpyhxin, CsDix and Anonymous: 20

- **Payload fraction** Source: https://en.wikipedia.org/wiki/Payload_fraction?oldid=653864244 Contributors: Maury Markowitz, Patrick, Nikai, Doradus, Hooperbloob, Vslashg, DonPMitchell, Arado, Dialectric, SmackBot, IanOsgood, Yobot, Hbquax, Sketchapancake and CsDix

- **Propellant mass fraction** Source: https://en.wikipedia.org/wiki/Propellant_mass_fraction?oldid=663467634 Contributors: The Anome, Tarquin, Maury Markowitz, Patrick, Doradus, Securiger, Rasmus Faber, Staz69uk, Wolfkeeper, Brockert, Vsmith, Night Gyr, Duk, Cmdrjameson, Gene Nygaard, Avraham, SmackBot, Slashme, Underpants, Thijs!bot, AntiVandalBot, IanOsgood, Dhaluza, Gouveia2, Sdsds, Shinyplastic, JackStonePGD, Tom.Reding, Jfmantis, K6ka, Solomonfromfinland, Artvill, MenteMagica, MerlIwBot, Jfittje, CsDix and Anonymous: 14

- **Mass ratio** Source: https://en.wikipedia.org/wiki/Mass_ratio?oldid=544735558 Contributors: Doradus, Wolfkeeper, D6, Mairi, A2Kafir, Wendell, Gene Nygaard, DonPMitchell, Jeff3000, Tsca.bot, Drdelorean, JAnDbot, Dhaluza, VolkovBot, Logan, Addbot, Yobot, Citation bot, HRoestBot, Unbitwise, Jfmantis, ClueBot NG, MenteMagica, 32alpha4tango, CsDix and Anonymous: 6

- **Apsis** Source: https://en.wikipedia.org/wiki/Apsis?oldid=681958388 Contributors: Mav, Bryan Derksen, XJaM, Karl Palmen, Patrick, Michael Hardy, DopefishJustin, Looxix~enwiki, Julesd, Evercat, Cherkash, Pizza Puzzle, Nikola Smolenski, Timwi, Scott Sanchez, RedWolf, Merovingian, Rursus, Ojigiri~enwiki, Wereon, Sj, Harp, Herbee, Wwoods, Curps, Alison, Pauldanon, 20040302, Bob Loblaw, ChicXulub, Knutux, Karol Langner, B.d.mills, Urhixidur, TJSwoboda, Jimius, Adashiel, Thorwald, ArnoldReinhold, Slipstream, TheOuthouseMouse, ESkog, RJHall, Kwamikagami, Worldtraveller, Art LaPella, 0.39, Bobo192, La goutte de pluie, BillCook, Devolution, Haham hanuka, Quaoar, Alansohn, CyberSkull, Garzo, Pwqn, Gene Nygaard, Carbenium, Jef-Infojef, Etacar11, Rocastelo, Nakos2208~enwiki, Tabletop, Pdn~enwiki, Waldir, Zzyzx11, Alec Connors, Smartech~enwiki, Nema Fakei, Bebenko, RuM, Mandarax, Dweekly, Monk, Ketiltrout, Ash211, Rjwilmsi, HappyCamper, Vuong Ngan Ha, Geimas5~enwiki, Chobot, Mordicai, DVdm, Bgwhite, Huw Powell, Jimp, Rsrikanth05, NawlinWiki, Anomie, Curtis Clark, Dysmorodrepanis~enwiki, Aeusoes1, Hymyly, Juanpdp, Cstaffa, Ms2ger, Square87~enwiki, Chesnok, Closedmouth, Tim Parenti, Jake Spooky, Geoffrey.landis, ArielGold, Argo Navis, Cmglee, Eog1916, Eigenlambda, SmackBot, The Angriest Man Alive, Declare, Kintetsubuffalo, CrypticBacon, Hmains, ERcheck, Saros136, Bluebot, Stubblyhead, Rick7425, Hibernian, Metacomet, Tamfang, Vladislav, Akmathes, Aldaron, Zachbenman, The PIPE, Chymicus, Rodri316, Microchip08, JorisvS, Novangelis, Newone, Epistemos, Beno1000, BBuchbinder, Poolkris, Chris55, Vaughan Pratt, Ispollock, Danial79, Bill K, Sax Russell, Neelix, Yaris678, BobQQ, RZ heretic, Zginder, SteveMcCluskey, Omicronpersei8, Saintrain, Caslider, Anshuk, Headbomb, Tham153, Cyclonenim, CodeWeasel, Orionus, Opelio, Fayenatic london, Rjwoer, Incredio, JAnDbot, Deflective, Struthious Bandersnatch, Naval Scene, KEKPΩΨ, WolfmanSF, Rafuki_33, Swpb, JoergenB, Dr. Morbius, MartinBot, Jim.henderson, Uriel8, Alro, Fconaway, Lilac Soul, Artaxiad, J.delanoy, All Is One, BobEnyart, Rominandreu, Perinigricon, Steel1943, VolkovBot, Midoriko, Tesscass, AlnoktaBOT, Philip Trueman, Raymondwinn, UnitedStatesian, Dutchsatellites.com, Fanatix, SieBot, Portalian, Gerakibot, Smsarmad, The way, the truth, and the light, BTH, KoshVorlon, Lightmouse, RW Marloe, Twinsday, Martarius, ClueBot, AstroMark, Narasimhavarman10, Easphi, Auntof6, Jotterbot, Unmerklich, SoxBot III, Forbes72, BodhisattvaBot, Feyrauth, Triptropic, SilvonenBot, Jbeans, Addbot, NjardarBot, Cypkerth, LaaknorBot, AndersBot, Alpinwolf, Mpfiz, Lightbot, م.نبيل, Zorrobot, Jarble, Michaello, Luckas-bot, TaBOT-zerem, Backslash Forwardslash, AnomieBOT, Xosema, Galoubet, Citation bot, DirlBot, Xqbot, TheAMmollusc, Gap9551, GrouchoBot, Omnipaedista, Kyng, CRea80, SouthernCritic111, Laterensis, Metalindustrien, Nagualdesign, FrescoBot, Paine Ellsworth, Part Time Security, Tkuvho, 10metreh, Tom.Reding, AmphBot, December21st2012Freak, Elmoro, Pezanos, Double sharp, Lotje, Duoduoduo, TjBot, DASHBot, RA0808, JustinTime55, Mmeijeri, QuentinUK, A2soup, Ruislick0, Wayne Slam, JoeSperrazza, Jguy, Tomásdearg92, SkywalkerPL, Teapeat, Ebehn, Petrb, ClueBot NG, Jack Greenmaven, CocuBot, Gyrfalcon23, Tinpisa, Widr, 760bbgun96, MerlIwBot, Gob Lofa, KLBot2, Kinaro, Periheliondi, Davidiad, Thóumas, GoShow, YFdyh-bot, MSUGRA, AutomaticStrikeout, Mogism, Angelislington, Jaxcp3, Kendram, Cezar teodosiu, Exoplanetaryscience, Khlopenkov, TheWhistleGag and Anonymous: 166

- **Eccentricity vector** Source: https://en.wikipedia.org/wiki/Eccentricity_vector?oldid=682916361 Contributors: Patrick, Michael Hardy, AugPi, Jyril, Bender235, 0.39, Simone, Xnk, Makaristos, Roboto de Ajvol, SmackBot, SJP, Thurth, Synthebot, Neparis, YonaBot, Paolo.dL, Addbot, Stamcose, Mm1671, Erik9bot, FrescoBot, Jfmantis, EmausBot, Teapeat, Serasuna, Jaxcp3, Sguerin2 and Anonymous: 3

- **Non-inclined orbit** *Source:* https://en.wikipedia.org/wiki/Non-inclined_orbit?oldid=661836015 *Contributors:* Cherkash, VeryVerily, RJHall, 0.39, Hooperbloob, Spacepotato, SmackBot, Tamfang, Tony Fox, Alaibot, Swpb, GroveGuy, Yobot, NOrbeck, Erik9bot, Sparky the Wolf, Teapeat, Kafishabbir and Anonymous: 1

- **Euler angles** *Source:* https://en.wikipedia.org/wiki/Euler_angles?oldid=681004679 *Contributors:* The Anome, XJaM, Patrick, Michael Hardy, Karada, Looxix~enwiki, AugPi, CarlKenner, Charles Matthews, Hyacinth, Robinh, Znode, Tosha, Giftlite, Jyril, Fropuff, Everyking, Dratman, Leonard G., Pgan002, Beland, B.d.mills, Urhixidur, Mike Rosoft, Wesha, CALR, BBUCommander, Ebelular, Mechgyver, Thuresson, Bobo192, Grue, Teeks99, PAR, Njahnke, Jheald, Cmprince, Oleg Alexandrov, Linas, GregorB, Waldir, BD2412, Rjwilmsi, Tadeu, Lionelbrits, Jonschlueter, RexNL, Chobot, Wavelength, KSmrq, Tony1, Bota47, David Biddulph, RG2, KnightRider~enwiki, Jsnx, SmackBot, Incnis Mrsi, Andres Agudelo, Kdliss, Thumperward, MalafayaBot, SEIBasaurus, Drewnoakes, Nbarth, Fmalan, Mhym, Ziggle, Jrvz, JorisvS, Wholmestu, Phancy Physicist, Brunetton~enwiki, Mets501, Yoderj, Abel Cavaşi, CRGreathouse, CBM, Marc W. Abel, Juansempere, Thijs!bot, Kborer, NocNokNeo, EdJohnston, JAnDbot, Asmeurer, Magioladitis, Jarekt, ELinguist, MarcusMaximus, Chessfan, Nono64, Terrek, Ahtih, VolkovBot, Error9312, JohnBlackburne, Thurth, Maxtremus, Sintaku, Saibod, Alangowitz, Dwbradway, YohanN7, SieBot, Paolo.dL, Rocket Laser Man, Novas0x2a, Xjwiki, ClueBot, Ideal gas equation, Rhubarb, Guntherwagner, XLinkBot, Dthomsen8, Insertesla, Claystein, Gjacquenot, Addbot, AkhtaBot, Niac2, MrOllie, Girolamous, Luckas-bot, Yobot, GoldAndBronze, Snaily, Luckas-bot, Yobot, Stamcose, Jim1138, Bill the Galactic Hero, Sbstn, McSush, Citation bot, Ksimek, Guillaume Ponchel, Shadowjams, Ckambler, Lani101, TobyNorris, Citation bot 1, DrilBot, Guentherwagner, Carlospascal, Joostdhw, Rausch, Ryandle, Savuor, StephenNewby, John of Reading, Netheril96, Mmeijeri, Phtempel, Bkocsis, SporkBot, Wmayner, Wayne Slam, Thelastcraftman, True bugman, Zephyrus Tavvier, Mikhail Ryazanov, ClueBot NG, Snotbot, Ryan Vesey, MerlIwBot, Secret of success, Helpful Pixie Bot, BG19bot, Papadim.G, Moradmandali, MusikAnimal, Dringend, James.ransley, JamesHaigh, Colonel angel, SteenthIWbot, Cthissen, Lekthor, Eilonb3, Mnrka, Esspe and Anonymous: 219

- **Drag (physics)** *Source:* https://en.wikipedia.org/wiki/Drag_(physics)?oldid=683281654 *Contributors:* Heron, Stevertigo, Michael Hardy, Kku, Arpingstone, CatherineMunro, Miernik, Mina86, Donreed, Moink, Giftlite, DocWatson42, Wolfkeeper, Bobblewik, Icairns, Canterbury Tail, Vsmith, JoeSmack, Chewie, Laurascudder, Bookofjude, Spoon!, Bobo192, Meggar, Notafish, Alansohn, Free Bear, Eric Kvaalen, Neonumbers, Ashley Pomeroy, Snowolf, BRW, RainbowOfLight, Mikeo, Allen McC.~enwiki, Gene Nygaard, Richwales, Rtdrury, GregorB, Mandarax, Qwertyus, Ohanian, Eyas, SouthernNights, Fresheneesz, Peregrine Fisher, Sceptre, RussBot, Arado, Bcebul, JabberWok, Polluxian, Hogghogg, NawlinWiki, Wiki alf, Aeusoes1, Raven4x4x, Gadget850, DeadEyeArrow, Enormousdude, WikiJon, Sean Whitton, Mustafarox, AGToth, Kungfuadam, Asterion, Malfita, SmackBot, Reedy, InverseHypercube, Thunderboltz, Dpwkbw, Gilliam, Skizzik, Frédérick Lacasse, Agateller, Jopsen, Persian Poet Gal, Baa, Mion, Ohconfucius, John, Loodog, JorisvS, Slakr, Optakeover, Mets501, Hu12, Iridescent, Joseph Solis in Australia, Courcelles, Tawkerbot2, Chetvorno, ChrisCork, Mmdoogie, Ale jrb, Fnfal, Nunquam Dormio, MiShogun, Lawler, Safalra, RGSchmitt, Gogo Dodo, JFreeman, Llort, Palmiped, Thijs!bot, Epbr123, Headbomb, F l a n k e r, AntiVandalBot, Jj137, Fayenatic london, Smartse, Gef756, Steelpillow, Golgofrinchian, JAnDbot, MER-C, Arch dude, Primarscources, PhilKnight, Gildos, VoABot II, Email4mobile, Rich257, Destynova, EagleFan, BatteryIncluded, Jodi.a.schneider, Joseph.slater, Robin S, MartinBot, Melamed katz, Eliz81, Arrgh406, Daytonduck, StevenWarren, DH85868993, ACBest, Ja 62, Inwind, SoCalSuperEagle, Mechanical Rose, 28bytes, Indubitably, Dqeswn, Philip Trueman, TXiKiBoT, Oshwah, Plenumchamber~enwiki, NPrice, Ann Stouter, Z.E.R.O., Darthjt, Matthewslf, Eatabullet, Suriel1981, Sammybat, Falcon8765, Insanity Incarnate, Dedede224, SieBot, Happysailor, Flyer22, Paolo.dL, Oxymoron83, Jenssss, Lightmouse, Dolphin51, Escape Orbit, YSSYguy, ClueBot, Phoenix-wiki, Bob1960evens, The Thing That Should Not Be, Drmies, Ikhmelin, DragonBot, Digitally-Crazy, Awickert, Quercus basaseachicensis, Alexbot, Pgj98r, Arjayay, Thingg, RedCure, Crowsnest, DumZiBoT, Aaron north, Jovianeye, LeheckaG, Avoided, NellieBly, Badgernet, Officially Mr X, Addbot, Basilicofresco, Innv, Ronhjones, CanadianLinuxUser, NjardarBot, Morning277, Michael Belisle, 5 albert square, Peti610botH, Luckas Blade, Yobot, Estudiarme, AmeliorationBot, AnomieBOT, Jim1138, Kingpin13, Materialscientist, Citation bot, Another Stickler, LouriePieterse, Capricorn42, Hammersbach, Fredwood, AbigailAbernathy, Omnipaedista, RibotBOT, Killerhhh, Kaybet, WaysToEscape, Constructive editor, GliderMaven, FrescoBot, VS6507, Steve Quinn, Airplaneherd, Citation bot 1, PigFlu Oink, DrilBot, HRoestBot, Skyraider1, RedBot, Chompa123, Keri, Tim1357, Dinamik-bot, Aerokiwi, Gfoley4, Montgolfière, JustinTime55, Edw400, Wikipelli, NicatronTg, Tolly4bolly, Donner60, Planetscared, Petrb, ClueBot NG, Wikiphysicsgr, Jack Greenmaven, CocuBot, नरेश, Frietjes, Widr, Bloomcounty, Sugajen, Helpful Pixie Bot, HMSSolent, Bibcode Bot, Doorknob747, Mongolengineer, Northamerica1000, MusikAnimal, Mark48torpedo, Robert the Devil, Sni56996, Real doing, BattyBot, Felamaslen, Mikehulslander, Kylebarthelson, Lugia2453, Rodney Aho, Reatlas, Jnims2, Melonkelon, RandoomJD, Wolfstuxs, Hoppeduppeanut, Sportsrockbd, Ugog Nizdast, JaconaFrere, A6tf3t, Tarun Thathvik, Lor, GoblinsDrool, Saisirajahmed, Juanm123, Arlo660, Oyshonib, Paulmag91 and Anonymous: 390

- **Theory of relativity** *Source:* https://en.wikipedia.org/wiki/Theory_of_relativity?oldid=683478283 *Contributors:* AxelBoldt, The Cunctator, Dan~enwiki, Koyaanis Qatsi, Css, Ed Poor, Andre Engels, Roadrunner, DrBob, Heron, FlorianMarquardt, Lir, Michael Hardy, Tim Starling, TimShell, Llywrch, Oliver Pereira, Liftarn, Gabbe, Chinju, IZAK, Delirium, Shimmin, Looxix~enwiki, Ahoerstemeier, William M. Connolley, Theresa knott, Snoyes, Notheruser, Andres, Mxn, Loren Rosen, Reddi, Kbk, Tpbradbury, Mir Harven, Jerzy, David.Monniaux, BenRG, Garo, Wilinckx~enwiki, Craig Stuntz, Kizor, Rfc1394, Dersonlwd, Intangir, Hadal, Wikibot, JackofOz, Anthony, SoLando, Carnildo, Tobias Bergemann, Nagelfar, Decumanus, Giftlite, Ævar Arnfjörð Bjarmason, Fropuff, Piquan, Wwoods, Everyking, Curps, Frencheigh, Ehabkost, Andris, Yekrats, Just Another Dan, Jackol, Tian2992, SoWhy, Knutux, Exigentsky, APH, DragonflySixtyseven, Jokestress, Kadambarid, Adashiel, Trevor MacInnis, Lacrimosus, Mike Rosoft, Mormegil, Jørgen Friis Bak, JimJast, Discospinster, Rich Farmbrough, Rhobite, ThomasK, Pjacobi, Vsmith, Metamatic, Jowr, Pavel Vozenilek, ESkog, MyNameIsNotBob, Aranel, Goto, El C, Gilgamesh he, Frankenschulz, RoyBoy, Bookofjude, Bobo192, Shaene~enwiki, Touriste, Marco Polo, Stesmo, Longhair, Smalljim, Tronno, GTubio, .:Ajvol:., Joe Jarvis, Jerryseinfeld, Matt McIrvin, Jojit fb, Nk, Helix84, Espoo, Jumbuck, Alansohn, Gary, Evaa, Mr Adequate, Riana, SlimVirgin, Cdc, Snowolf, Shinjiman, Wtmitchell, Velella, Yuckfoo, DonQuixote, Vuo, DV8 2XL, Netkinetic, SmthManly, Loxley~enwiki, Zntrip, Gmaxwell, Roland2~enwiki, MickWest, OwenX, Hello5959us, RHaworth, Etacar11, Camw, Nuggetboy, IHendry, Morganic, Robert K S, Pol098, WadeSimMiser, Jeff3000, GeorgeOrr, Mpatel, Nakos2208~enwiki, GalaazV, Alec Connors, Christopher Thomas, Mandarax, Graham87, A Train, Chun-hian, FreplySpang, Edison, Sjakkalle, Rjwilmsi, Koavf, Ems57fcva, The wub, Bhadani, THE KING, Yamamoto Ichiro, Dabuek, Doc glasgow, Truman Burbank, Alfred Centauri, Ewlyahoocom, Gurch, Pevernagie, DerGraph~enwiki, Tysto, Srleffler, King of Hearts, Chobot, Caleb Aaron Osment, Moocha, DVdm, Gwernol, Wavelength, Brandmeister (old), Petiatil, SpuriousQ, Yamara, Stephenb, Rsrikanth05, Bullzeye, EngineerScotty, NawlinWiki, EWS23, Wiki alf, Grafen, Voyevoda, DarthVader, Tailpig, Nick, Retired username, Jpbowen, Raven4x4x, DeadEyeArrow, Bota47, Szhaider, Nlu, Dna-webmaster, Pejhman, Closedmouth, Josh3580, Petri Krohn, Alias Flood, Tyrenius, RunOrDie, Poulpy, M.A.Dabbah, Sbyrnes321, DVD R W, Finell, Bibliomaniac15, Luk, SmackBot, Thaagenson, Dweller, Unschool, InverseHypercube, KnowledgeOfSelf, Martinp, Pgk, C.Fred, Bomac, Jagged 85, Chairman S., Althasil, Hardyplants, Hoipolloi, Swerdnaneb,

41.4. TEXT AND IMAGE SOURCES, CONTRIBUTORS, AND LICENSES

Macintosh User ,Gilliam ,Ohnoitsjamie ,Indium ,Hmains ,Skizzik ,Bluebot ,Keegan ,Persian Poet Gal ,Sirex 98,OrangeDog ,SchfiftyThree , Tianxiaozhang ~enwiki,RayAYang ,Jfsamper ,Tyr Anasazi ,DHN-bot~enwiki,Colonies Chris ,Darth Panda ,Scwlong ,AK7,Can't sleep,clown will eat me ,Scott3,Shalom Yechiel ,Onorem ,Domen~enwiki ,Rrburke ,VMS Mosaic ,Grover cleveland ,Meel,Tim Fellows ,666666 th user, Cybercobra ,Khukri ,Dylanrush ,Jiddisch~enwiki ,J.Wolfe@unsw.edu.au,BullRangifer ,SashatoBot ,ArglebargleIV ,Calidarien ,AThing ,John, General Ization, JI982, JohnCub, Youallknowme2000, Shadowlynk, Ph89~enwiki, IronGargoyle, Don't fear the reaper, Mets501, Sgore, Dr.K.,KJS 77,Qwilleranfan ,Quaeler ,Iridescent ,Shoeofdeath ,Wjejskenewr ,MOBle ,Jaksmata ,Igoldste ,Az1568 ,Tawkerbot 2,Fvasconcel-los,Mosaffa,Ewc 21,Nunquam Dormio ,Retrovirus ,THF,Zgod ,Smoove Z ,Tex,Gregbard ,Funnyfarmofdoom ,Tzaphkiel ,Cydebot ,Abebasal , Kanags ,Peterdjones ,Suttkus ,Gogo Dodo ,Was a bee ,Flowerpotman ,Nmajdan ,Odie 5533 ,Michael C Price ,Tawkerbot 4,Dynaflow,Colorprobe ,FastLizard 4,Optimist on the run ,Omicronpersei 8,Yogarine ,Adoto ,Tortillovsky ,Thijs!bot,Epbr123 ,Ante Aikio ,Daniel ,Martin Hogbin ,Andyjsmith ,Mojo Hand ,Headbomb ,Marek 69,John 254,E.Ripley ,Scottmsg ,The Hybrid ,D.H.Blathnaid ,I already forgot ,Mentifisto ,AntiVandalBot ,Seaphoto ,Der alte Hexenmeister ,Elmoosecapitan ,John1987,Scepia ,Tim Shuba ,Qwerty Binary ,Elfkin ,Gökhan , Res2216firestar, Deadbeef, JAnDbot, Krishvanth, Achalmeena, MER-C,CosineKitty, Tosayit, Instinct, Andonic, Roleplayer, Tstrobaugh, PhilKnight ,Clementvidal ,Boleslaw ,Acroterion ,Bongwarrior ,VoABot II ,JNW ,JamesBWatson ,Think outside the box ,Soulbot ,Dan Slurry , BrianGV ,Animum ,Kameejl ,Ahecht ,28421 u2232nfenfcenc ,Elliotb 2,Shijualex ,Dan Pelleg ,DerHexer ,Ejm634,Wikipedia Incarnate ,WLU, Gjd001, S3000, MartinBot, Ollycity, Cheifsguy, Hillinpa, NAHID,SuperMarioMan, Anaxial, Squogfloogle, Lilac Soul, Rirl4, J.delanoy, Lgelgand ,Trusilver ,R.Baley ,Bogey 97,Numbo 3,Hans Dunkelberg ,Uncle Dick ,Fleiger ,Vanished user 342562 ,SU Linguist ,Brashmadcap , Lantonov ,Wandering Ghost ,Katalaveno ,Smeira ,Tarotcards ,Gurchzilla ,Belovedfreak ,Richard D .LeCour ,NewEnglandYankee ,Profes 001, SJP,Tooncast ,Barraki ,DMCer ,Bonadea ,Halmstad ,Ale 2006 ,CardinalDan ,Abhishka ,Lights ,Deor ,Andrea moro ,Memattl,Izbitzer ,JeffG., JohnBlackburne ,Auggiehesch ,Optokinetics ,Station1,Pparazorback ,Philip Trueman ,DoorsAjar ,TXiKiBoT ,Vipinhari ,Charlesdrakew ,Qxz, Someguy 1221,Retiono Virguinian ,HansMair ,Oxfordwang ,09woodh ,Seraphim ,Superchucknorris ,Martin 451,Maharashtraexpress ,Amog, Mannafredo ,Cremepuff222,762080d,Raiug199999994444444 ,RadiantRay ,Mwilso 24,Hellodutil ,Eubulides ,Pea Jitngamplang ,BigDunc , RandomXYZb ,Andy Dingley ,Woawitchking ,Strangerer ,Bacchio ,Falcon 8765 ,Everwrest ,Kaori ,Ellocia ,Sue Rangell ,AlleborgoBot ,My favourite teddy bear ,RobthecoolestkideverJK ,Deconstructhis ,SaltyBoatr ,Ponyo ,M.V.E.i.,SieBot ,Tiddly Tom ,Hertz1888,Winchelsea ,Yintan,Orereta ,Bootha ,Keilana ,Toddst 1,Flyer 22,Paolo .dL,Redmarkviolinist ,Burpolon 2,Oxymoron 83,Al Gand ,AnonGuy ,Steven Crossin , Udirock ,Coldcreation ,StaticGull ,Hamiltondaniel ,Sacuar ,S3o3b3e31~enwiki ,Vfrg ,Dolphin 51,GDRamsey 8,ElectronicsEnthusiast ,RegentsPark ,Martarius ,Immutable 1,ClueBot ,CloneDeath ,The Thing That Should Not Be ,Quantumgeek ,Acorahrama ,SuperHamster ,Boing! said Zebedee ,LawrencRJ ,Viran ,Blanchardb ,Sairilian ,Oxnard 27,Flight Of The neo ,Neoshka ,Excirial ,Jusdafax ,Wikitumnus ,Vanisheduser12345 ,Leonard ^Bloom ,Lartoven ,NuclearWarfare ,Jotterbot ,Tnxman 307,Swirlface ,Snacks ,Razor flame ,Dekisugi ,Thehelpfulone , Aitias ,Galor 612,Shauharda ,Littleteddy ,Versus 22,Egmontaz ,Thinking Stone ,Vanished User 1004,XLinkBot ,Spitfire,Nathan Johnson , Emmjem ,Bass119922 ,Little Mountain 5,Ilikepie 2221,Skarebo ,WikHead ,NellieBly ,Mifter ,Blahdeblah 123456 ,Good Olfactory ,Jackdale , Regieg ,D.M.from Ukraine ,Addbot ,Cxz111,Some jerk on the Internet ,Jojhutton ,Tcncv ,Fyrael ,Kcoltra 1,TutterMouse ,Fieldday-sunday , Startstopl 23, Trustnextl, Shirtwaist, MrOllie, Download, CarsracBot, EconoPhysicist, Jennma-y06, Favonian, Geniusl 995, LinkFA-Bot, Phoxinus ,Dunnyman ,Numbo 3-bot,Diegohuyke ,VASANTH S .N.,Issyl0,Scottmsul ,Tide rolls ,Bigdanny 123,Golden 0103 ,Hand-ige Harrie , Duhtherangers ,MuZemike ,LuK3,Gameseeker ,Alfie66,Angrysockhop ,Luckas-bot ,TaBOT-zerem ,Ninjalemming ,THEN WHO WAS PHONE ?,Azcolvin429,Magog the Ogre ,AnomieBOT, "only those who have experienced it can understand it .",Jiml 138,Piano nontroppo,Keithbob, AdjustShift ,Ufim,EryZ,RandomAct ,Materialscientist ,90Auto,Glass 2008,Lkd85,Srinivas ,Ashword ,Garen Avenizian ,ArthurBot ,ChristianH , LilHelpa ,Ritvick ,Obersachsebot ,Xqbot ,Projecthome 2010,Capricorn 42,Wapondaponda ,Bihco ,Zerim ,Luckylea-fus, Jmundo, Tyrol5, AbigailAbernathy , Norlabs , АлександрBв, Shaymchone , Shirik , SassoBot , حامدمیرزاحسینی, N419 BH, Yulracso ,Nyuszikaa , Shadowjams ,Schekinov Alexey Victorovich ,Nolesfan 657,Aaron Kauppi ,Thehelpfulbot ,Celuici ,FrescoBot ,Kyleshirley 1980,Doug Bundy , Dogposter ,Goodbye Galaxy ,Chrissmith 22,HJ Mitchell ,Steve Quinn ,MrAnonymousUser ,Kwiki ,HamburgerRadio ,T3chl0v3r,Beerahndunn , Biker Biker ,Pinethicket ,I dream of horses ,Hert1234,Dujeu11,Tom Reding ,DanielGlazer ,Achaemenes ,Aaarrrhhh ,Jschnur ,SpaceFlight 89,Σ, Weekeepeer ,Henryg 52,TrapShooterPageCreator ,Barras ,Aknochel ,Robvanvee ,FoxBot ,Trappist the monk ,Reality Dis-tortion,Callanecc , Michael 9422, Dinamik-bot,Mrblobby 1928 ,Vrenator ,Xx3nvyxx ,Diablanco ,Christoph v .Mettenheim ,Hfitzhug ,Reaper Eternal , Diannaa , PowerPaul 86, Suffusion of Yellow , Deanmullen 09, DARTH SIDIOUS 2, Whisky drinker , Stows 99, Twisted Crowbar , NerdyScienceDude ,Shinespark ,Balph Eubank ,Slon02,Twastvedt,Buckeyetigre ,EmausBot ,JoeDG, Wilhelm-physiker,Leuma1234,Noob-slayer 88,Racerx 11,Puppy 389,RA 0808 ,Islamuslim , Tommy 2010,TuHan-Bot,Wikipelli ,Mz7,Lickling ,John Cline ,Bollyje ff,Shuipzv 3,Resitate , NicatronTg ,Empty Bu ffer,Shiny 1069 ,Herp Derp ,Lyubomir T ,Gruyitch ,DBSSURFER ,Monterey Bay ,Infazahmet ,Kilopi ,Frederick Van superness ,Ahmetbulut ,Wayne Slam ,Coolbob 2422,IGeMiNix ,Danmuz ,Mystieriawaterfalls ,GIAN PHIL ,Sailsbystars ,Orange Suede Sofa ,RockMagnetist ,Peter Karlsen ,DASHBotAV ,28bot,Joshfaia ,Daniel 55423 ,JdirksTheMessiah ,Lewisrooney 93,ClueBot NG , Joeblowlikescandy ,Augvillar ,This lousy T-shirt,I Hunta x ,Braincricket ,Rezabot ,Widr ,Dr std ,Diegorecht ,Names are hard to think of , Bmancain,Theopolisme ,Пуикаре,Helpful Pixie Bot ,Bibcode Bot ,DBigXray ,SteeledStriker ,Lowercase sigmabot ,M0rphzone ,MusikAn-imal, Fcdd,Mark Arsten ,Iurie Cojocari ,Jordatech ,Waitedavid 137,Spandaize ,Aranea Mortem ,Razbuznik ,Knaffster ,Ihaveasixinch ,SnowBlizzard, Somnibus ,Glacialfox ,Loriendrew ,Klilidiplomus ,Anbu121,Mrbreer ,BattyBot ,I'llShowYouDad ,Stigmatella aurantiaca ,Saurabh singh bazzad 9999856547 ,Dude5456,Drla8th!,Dexbot,Charging8642,CuriousMind 01,1Todd1,Lugia2453,Lettilucy,Jg6606,Ram2003,Reatlas,Smith33056 ,Epicgenius ,Hackceline ,PlanetEditor ,Bigdick 17,Wolfboy 6789 ,Capasl,DavidLeighEllis ,Soll,Jwratner 1,Dbknies ,George 8211,Quanta inc , Frinthruit, JaconaFrere ,TheFlash 1123 ,Anonymous black panther ,Gmu1993,Willjones 016,Dragonlord Jack ,Stu-dent227,Monkbot ,Juenni32, Fishcallederic ,Prof. Mc ,Rubhavan 10,Testing 625 ,Dwtechfanatic ,Kpcbrown ,Osa san ,Green fireinvermont ,Suhailkhan 7 7,Y-S.Ko, Theonlymrcat ,IGSHARD ,39 Debangshu ,Tetra quark ,PaulChan 24,Thorthugnasty ,Quwayne ,KasparBot ,Edpop ,Ujjwal 71761 , The guy who knows alot about stuff,ARUL ASHRI,Abhyudaya apoorva,Maxkh03and Anonymous:1211

- **Radiation pressure** *Source:* https://en.wikipedia.org/wiki/Radiation_pressure?oldid=681003679 *Contributors:* Bryan Derksen, XJaM, Fubar Obfusco, Peterlin~enwiki, Patrick, Michael Hardy, Jedimike, GTBacchus, Caid Raspa, Hike395, Chuunen Baka, Cutler, Karn, Chowbok, Beland, DragonflySixtyseven, Gene s, Urhixidur, Stepp-Wulf, Aperculum, RoyBoy, Gershwinrb, Wtshymanski, Stephan Leeds, Gene Nygaard, Japanese Searobin, Blaze Labs Research, DonPMitchell, BillC, Tabletop, Bob A, BlueMoonlet, TedTheHead, Kolbasz, Bgwhite, LucianSolaris, YurikBot, Wavelength, DragonHawk, Geoffrey.landis, SmackBot, Eskimbot, Hibernian, Colonies Chris, Georg-Johann, Wen D House, Vinaiwbot~enwiki, ZirbMonkey, Dr. Sunglasses, Chetvorno, Ollie, Vaughan Pratt, Icek~enwiki, Thijs!bot, Mathmoclaire, Markus Pössel, Dr. Phred Mbogo, D.H, IanOsgood, Leyo, GoatGuy, AntiSpamBot, C.lettingaAV, WarddrBOT, TXiKiBoT, SieBot, Thomasonline, ClueBot, Jerry Wright, Nnemo, Parkjunwung, Mild Bill Hiccup, Pgj98r, APh, Crowsnest, Fmitri, Addbot, PMarmottant~enwiki, Lightbot, Yinweichen, Luckas-bot, Easy n, Yobot, Castaflor, Nallimbot, AnomieBOT, Xqbot, ArielGenesis, AllCluesKey, Paine Ellsworth, OgreBot, Tom.Reding,

IVAN3MAN, Ripchip Bot, Upthetrail, Helpful Pixie Bot, Ymblanter, Vagobot, Northamerica1000, Rderdwien, Zedshort, Physicsch, Evanprs, ChrisGualtieri, Electricmuffin11, Mogism, Yuriy Af, Machosquirrel, Tpozar, Cuberoottheo, Reza.goodarzi.1989 and Anonymous: 72

- **Electromagnetism** *Source:* https://en.wikipedia.org/wiki/Electromagnetism?oldid=682255582 *Contributors:* AxelBoldt, Magnus Manske, Trelvis, Carey Evans, CYD, Mav, Bryan Derksen, Zundark, Szopen, The Anome, Malcolm Farmer, FreddyZ, Miguel~enwiki, William Avery, Maury Markowitz, Heron, Patrick, Tim Starling, Kku, Bcrowell, Delirium, Ahoerstemeier, William M. Connolley, Khorn, Babbo, Darkwind, Kevin Baas, Julesd, Glenn, Sray, Poor Yorick, Mm, Rawr, Mxn, Hike395, Emperorbma, Wikiborg, Reddi, Phys, Lumos3, Phil Boswell, Rob-bot, Hankwang, Pigsonthewing, ZimZalaBim, Arkuat, Hemanshu, Texture, Roscoe x, Sunray, Wikibot, Fuelbottle, Lupo, Diberri, Wile E.Heresiarch, Tobias Bergemann, Ancheta Wis, Decumanus, Giftlite, DocWatson42, Andries, Wolfkeeper, Lethe, Koyn~enwiki, Everyking, Snowdog, Dratman, Curps, Ssd, Tom-, Jason Quinn, Brockert, Sohanley, Karol Langner, APH, Maximaximax, Bodnotbod, Icairns, Lumidek, Iantresman, Tsemii, Slipstream (usurped), Adashiel, Mike Rosoft, EugeneZelenko, Rich Farmbrough, Bedel23, Pjacobi, Vsmith, Tbeiloth, MuDavid, Paul August, Bender235, El C, Shanes, Mkosmul, RoyBoy, Femto, Matt McIrvin, Bert Hickman, Physicistjedi, Sam Korn, SignorGiuseppe, Mareino, Ranveig, Jumbuck, Red Winged Duck, Alansohn, Gary, Pinar, ChristopherWillis, Arthena, Atlant, Andrewpmk, Lectonar, Snowolf, Melaen, BRW, Wtshymanski, Evil Monkey, Bob1817, Mikeo, Gene Nygaard, Ttownfeen, Oleg Alexandrov, Cimex, TheNightFly, MONGO, Nakos2208~enwiki, Macaddct1984, Eras-mus, Gimboid13, Rnt20, Graham87, Qwertyus, Ando228, Rjwilmsi, Collins.mc, Tan-gotango, Ligulem, SeanMack, Bhadani, Yamamoto Ichiro, Cjpuffin, FlaBot, Nihiltres, Nivix, RexNL, Ewlyahoocom, Gurch, Otets, Fresh-eneesz, TeaDrinker, Srleffler, Physchim62, Chobot, Roboto de Ajvol, YurikBot, Wavelength, Spacepotato, JabberWok, CambridgeBay-Weather, Salsb, Wimt, David R. Ingham, NawlinWiki, Bachrach44, Buster79, Tearlach, Anetode, Scottfisher, Figaro, Bota47, Nick123, FF2010, Light current, Orioane, Enormousdude, 21655, 2over 0, JoanneB, Phil Holmes, Willtron, Sizarieldor, AGToth, Katieh5584, RG2, GrinBot~enwiki, Sbyrnes321, DVD R W, Luk, SmackBot, Manu 0x0~enwiki, PEHowland, Prodego, KnowledgeOfSelf, McGeddon, Jagged85, Jrockley, Swerdnaneb, Rpmorrow, Skizzik, JAn Dudík, LinguistAtLarge, Master of Puppets, Complexica, CMacMillan, TheGerm, Frap, Ioscius, Avoidance, SundarBot, Stevenmitchell, Cybercobra, MichaelBillington, Blake-, Akriasas, Illnab1024, Daniel.Cardenas, LeoNomis, Sadi Carnot, Apoorvchebolu, Skinnyweed, TTE, WayKurat, Sarfa, DJIndica, Nmnogueira, Ozhiker, Wvbailey, Finejon, UberCryxic, Philoso-phus, Cronholm144, Ckatz, El Dahveed, Grapetonix, Alatius, Sinistrum, Dicklyon, Jon186, Waggers, Ryulong, Describer, KJS77, Newone, Adambiswanger1, Courcelles, Ziusudra, Tawkerbot2, Yanah, Xcentaur, Mosaffa, JForget, KyraVixen, Baiji, Vyznev Xnebara, Fjomeli, MarsRover, Musicalantonio, Marly88, Peripitus, Fifo, Ssilvers, UberScienceNerd, Thijs!bot, VoABot, Jb.schneider-electric, Headbomb, Marek69, GerryAshton, Leon7, D.H, MichaelMaggs, AntiVandalBot, Majorly, Seaphoto, Quintote, Gnixon, Jnyanydts, Tyco.skinner, Eleos, Steelpillow, JAnD-bot, Matthew Fennell, Mkch, Bongwarrior, VoABot II, Tails4, SHCarter, TxAlien, GBYork, WhatamIdoing, Vanished user ty12kl89jq10, Adrian J. Hunter, 28421u2232nfenfcenc, Coldwarrior, User A1, Khalid Mahmood, Ruhihumphries, PrattTA1, InvertRect, MartinBot, Jar-gon777, LedgendGamer, J.delanoy, Sasajid, Abecedare, Bogey97, JohnPritchard, Maurice Carbonaro, Extransit, Tarotcards, NewEnglandYan-kee, Wesino, Shoessss, Cometstyles, ACBest, DorganBot, Treisijs, Bently34, Lights, 28bytes, Part Deux, Thedjatclubrock, Constant314, Philip Trueman, TXiKiBoT, The Original Wildbear, Guillaume2303, Anonymous Dissident, Kevin Steinhardt, Monkey Bounce, Captin-John, Imasleepviking, Mbarrieau, DoktorDec, Atomicswoosh, TongueSpeaker, Andy Dingley, Dirkbb, Lova Falk, @pple, Sylent, Doc James, PGWG, NHRHS2010, SieBot, Coffee, Tresiden, Graham Beards, Work permit, Hertz1888, Avargasm, Winchelsea, Dawn Bard, Caltas, Yin-tan, Zoragotcha, Keilana, Flyer22, Qst, Csblack, Klmeze~enwiki, Joseph Banks, Oxymoron83, Fbarw, Maelgwnbot, StaticGull, Dolphin51, TreeSmiler, Kanonkas, Ainlina, ElectronicsEnthusiast, Llywelyn2000, WikipedianMarlith, Bschaeffer~enwiki, Atif.t2, Martarius, ClueBot, Stevekirst7, The Thing That Should Not Be, Jan1nad, Uncle Milty, Blanchardb, Twicemost, LizardJr8, Jackey0105, Electromagnetic, OttoTanaka, Excirial, Kocher2006, BlueLikeYou, Jusdafax, Abrech, Plastic Fish, MacedonianBoy, PhySusie, Friedlibend und tapfer, Thingg, Je-sus, murphy55567, Versus22, InternetMeme, XLinkBot, Emmette Hernandez Coleman, Jovianeye, Rror, Boratlike, HarlandQPitt, Rogimoto, Deineka, Addbot, Dustbin123, Allfor12008, CanadianLinuxUser, Download, EconoPhysicist, Redheylin, WikiDegausser, K Eliza Coyne, LinkFA-Bot, Kisbesbot, AgadaUrbanit, VASANTH S.N., Tide rolls, Lightbot, Gail, Kurtis, Will.M.Thompson, Luckas-bot, Yobot, Ptbot-gourou, Lichen from Hell, ScienceMind, Tempodivalse, Orion11M87, AnomieBOT, Jim1138, Piano non troppo, AdjustShift, Penguinatortoo, Materialscientist, Citation bot, Vuerqex, Xqbot, Konor org, Lolman33, Plumpurple, Melmann, DSisyphBot, GrouchoBot, Nayvik, Omni-paedista, RibotBOT, Jsleaby, Maplestory101, Slowart, A. di M., GliderMaven, Tobby72, Wikipe-tan, Sum33, Ian88800, Hippycaller, SteveQuinn, HamburgerRadio, Citation bot 1, Чаховіч Уладзіслаў, Pinethicket, Elockid, Hard Sin, LittleWink, PvsKllKsVp, Tinton5, Yahia.barie, Jschnur, Ezhuttukari, Corinne68, FoxBot, Рыцарь поля, Retired user 0001, SchreyP, Hickorybark, ItsZippy, Lotje, Cjlim, LilyKitty, An-tipastor, Reaper Eternal, DARTH SIDIOUS 2, Triden, RjwilmsiBot, Alph Bot, Agent Smith (The Matrix), WildBot, DASHBot, Emaus-Bot, John of Reading, WikitanvirBot, Mnkyman, Never give in, ITshnik, IncognitoErgoSum, RA0808, Tommy2010, Amrator, Wikipelli, Thecheesykid, JSquish, ZéroBot, Harddk, Leptonoggin, Fæ, Maypigeon of Liberty, WFarver, H3llBot, Quondum, Git2010, Makecat, Sonygal, Coasterlover1994, L Kensington, Rr2wiki, Maschen, Donner60, Rjowsey, RockMagnetist, Peter Karlsen, Sir sachin, Teapeat, Planetscared, Cgt, Xanchester, ClueBot NG, Jack Greenmaven, MelbourneStar, Hiperfelix, Hallaman3, Ant.acke, Enopet, Frietjes, Milikguay, Widr, HelpfulPixie Bot, Eemcginnis, Vagobot, Bolatbek, Manuelfeliz, MusikAnimal, Stephenwanjau, Rm1271, Thekillerpenguin, Cadiomals, Mariano Blasi, Whyamiadampandalol, Franz99, Physicsch, Brad7777, Glacialfox, Neumannjo, Acul132, ChrisGualtieri, Dexbot, Marlowfrontier, Lugia2453, Frosty, Josophie, Hamerbro, SubratamindPal, Trillig, Abhaikumar10, CsDix, Ihatedirac2k13, Hell to earth88, CROY123, Zenibus, Jwrat-ner1, YiFeiBot, JosephSpiral, SpecialPiggy, Kdmeaney, Yashshroff97, Csutric, Mahusha, Venomous Cobra, Trackteur, Gronk Oz, Meski33, Hkeyser, Lolbob12345, Nanophysisct12345, Simonessnygg, Noobmagnet, Mediavalia, Slayeredwarrior, Tetra quark, Isambard Kingdom, Ser-gioCruz2015, Aenfinger, KasparBot, Zeke Essiestudy, Kafishabbir, JJMC89, Cesarnajera56 and Anonymous: 702

- **Two-line element set** *Source:* https://en.wikipedia.org/wiki/Two-line_element_set?oldid=677363064 *Contributors:* Maury Markowitz, Michael Hardy, Karn, Beland, JTN, FranksValli, Hadlock, Siafu, RadioFan, SmackBot, WDGraham, Derek R Bullamore, Gahs, N2e, Thijs!bot, Arch dude, Magioladitis, Swpb, Ramireja, Koesper, Alexbot, Addbot, Dgroseth, Yobot, Stamcose, KamikazeBot, AnomieBOT, Xqbot, Bataleur57, Tom.Reding, Mts7665, Ted Sweetser, SporkBot, Donner60, Screenmutt, Cqdx, EuroCarGT, Dexbot, Quirian, Bvv2306, Jordanstephens and Anonymous: 26

- **Proper orbital elements** *Source:* https://en.wikipedia.org/wiki/Proper_orbital_elements?oldid=683354442 *Contributors:* Beland, Phe, Rich Farmbrough, Spacepotato, Black Falcon, Deville, Deuar, SmackBot, Saros136, Bluebot, JorisvS, Richhoncho, Kheider, Addbot, Fgnievinski, JackieBot, Gap9551, Full-date unlinking bot, Zbayz, Jfmantis, Teapeat and Anonymous: 4

41.4. TEXT AND IMAGE SOURCES, CONTRIBUTORS, AND LICENSES

41.4.2 Images

- **File:194144main_022_drag.ogg** *Source:* https://upload.wikimedia.org/wikipedia/commons/1/12/194144main_022_drag.ogg *License:* Public domain *Contributors:* http://www.nasa.gov/audience/foreducators/topnav/materials/listbytype/Drag_Lesson_4.html http://www.nasa.gov/mov/194144main_022_drag.mov *Original artist:* NASA
- **File:ACE_at_L1.png** *Source:* https://upload.wikimedia.org/wikipedia/commons/7/71/ACE_at_L1.png *License:* Public domain *Contributors:* http://helios.gsfc.nasa.gov/ace/gallery.html *Original artist:* Unknown
- **File:Albert_Einstein_1979_USSR_Stamp.jpg** *Source:* https://upload.wikimedia.org/wikipedia/commons/8/80/Albert_Einstein_1979_Stamp.jpg *License:* Public domain *Contributors:* own scan of stamp from my collection *Original artist:* П. Бендель
- **File:Ambox_important.svg** *Source:* https://upload.wikimedia.org/wikipedia/commons/b/b4/Ambox_important.svg *License:* Public domain *Contributors:* Own work, based off of Image:Ambox scales.svg *Original artist:* Dsmurat (talk · contribs)
- **File:An_image_describing_the_semi-major_and_semi-minor_axis_of_ellipse.svg** *Source:* https://upload.wikimedia.org/wikipedia/commons/7/76/An_image_describing_the_semi-major_and_semi-minor_axis_of_ellipse.svg *License:* CC BY-SA 4.0 *Contributors:* Own work *Original artist:* Sae1962
- **File:Andre-marie-ampere2.jpg** *Source:* https://upload.wikimedia.org/wikipedia/commons/7/74/Andre-marie-ampere2.jpg *License:* Public domain *Contributors:* ? *Original artist:* ?
- **File:Angular_Parameters_of_Elliptical_Orbit.png** *Source:* https://upload.wikimedia.org/wikipedia/commons/1/1d/Angular_Parameters_of_Elliptical_Orbit.png *License:* CC-BY-SA-3.0 *Contributors:* ? *Original artist:* ?
- **File:Anomalies.PNG** *Source:* https://upload.wikimedia.org/wikipedia/commons/7/7a/Anomalies.PNG *License:* CC BY-SA 3.0 *Contributors:* Own work *Original artist:* Brews ohare
- **File:Apogee_(PSF).png** *Source:* https://upload.wikimedia.org/wikipedia/commons/a/a4/Apogee_%28PSF%29.png *License:* Public domain *Contributors:* Archives of Pearson Scott Foresman, donated to the Wikimedia Foundation *Original artist:* Pearson Scott Foresman
- **File:Asteroid_osculating_vs_proper_elements.png** *Source:* https://upload.wikimedia.org/wikipedia/commons/9/9d/Asteroid_osculating_vs_proper_elements.png *License:* CC-BY-SA-3.0 *Contributors:* Originally from zh.wikipedia; description page is/was here. *Original artist:* Original uploader was 屋顶 at zh.wikipedia
- **File:Automation_of_foundry_with_robot.jpg** *Source:* https://upload.wikimedia.org/wikipedia/commons/8/8a/Automation_of_foundry_with_robot.jpg *License:* Public domain *Contributors:* KUKA Roboter GmbH, Zugspitzstraße 140, D-86165 Augsburg, Germany, Dep. Marketing, Mr. Andreas Bauer, http://www.kuka-robotics.com *Original artist:* KUKA Roboter GmbH, Bachmann
- **File:Bar_magnet.jpg** *Source:* https://upload.wikimedia.org/wikipedia/commons/d/d8/Bar_magnet.jpg *License:* CC-BY-SA-3.0 *Contributors:* ? *Original artist:* ?
- **File:Bi-elliptic_transfer.svg** *Source:* https://upload.wikimedia.org/wikipedia/commons/8/83/Bi-elliptic_transfer.svg *License:* GFDL *Contributors:* Own work *Original artist:* AndrewBuck
- **File:Cassini'()s_speed_related_to_Sun.png** *Source:* https://upload.wikimedia.org/wikipedia/commons/8/80/Cassini%27s_speed_related_to_Sun.png *License:* CC-BY-SA-3.0 *Contributors:* Transferred from en.wikipedia to Commons by Pline using CommonsHelper. *Original artist:* Python eggs at English Wikipedia
- **File:Cassini_interplanet_trajectory.svg** *Source:* https://upload.wikimedia.org/wikipedia/commons/b/bd/Cassini_interplanet_trajectory.svg *License:* Public domain *Contributors:* Likely http://saturn.jpl.nasa.gov/photos/imagedetails/index.cfm?imageId=776 . Transferred from en.wikipedia to Commons by Pline using CommonsHelper. *Original artist:* ?
- **File:Circular.Polarization.Circularly.Polarized.Light_Right.Handed.Animation.305x190.255Colors.gif** *Source:* https://upload.wikimedia.org/wikipedia/commons/8/81/Circular.Polarization.Circularly.Polarized.Light_Right.Handed.Animation.305x190.255Colors.gif *License:* Public domain *Contributors:* Own work *Original artist:* Dave3457
- **File:Comet_Hale-Bopp_1995O1.jpg** *Source:* https://upload.wikimedia.org/wikipedia/commons/b/ba/Comet_Hale-Bopp_1995O1.jpg *License:* CC BY-SA 3.0 *Contributors:* Own work *Original artist:* E. Kolmhofer, H. Raab; Johannes-Kepler-Observatory, Linz, Austria (http://www.sternwarte.at)
- **File:Commons-logo.svg** *Source:* https://upload.wikimedia.org/wikipedia/en/4/4a/Commons-logo.svg *License:* ? *Contributors:* ? *Original artist:* ?
- **File:Conic_sections,_orbits,_and_gravitational_potential.jpg** *Source:* https://upload.wikimedia.org/wikipedia/commons/8/89/Conic_2C_orbits%2C_and_gravitational_potential.jpg *License:* CC BY-SA 3.0 *Contributors:* Own work *Original artist:* Sascha Grusche
- **File:Conic_sections_with_plane.svg** *Source:* https://upload.wikimedia.org/wikipedia/commons/d/d3/Conic_sections_with_plane.svg *License:* CC BY 3.0 *Contributors:* Own work *Original artist:* Pbroks13
- **File:Cosine_squared_graph,_or_half_of_one_plus_the_cosine_of_twice_x.svg** *Source:* https://upload.wikimedia.org/wikipedia/en/8/8b/Cosine_squared_graph%2C_or_half_of_one_plus_the_cosine_of_twice_x.svg *License:* CC-BY-SA-3.0 *Contributors:*
  fooplot.com
  Original artist:
  Cuberoottheo
- **File:Crab_Nebula.jpg** *Source:* https://upload.wikimedia.org/wikipedia/commons/0/00/Crab_Nebula.jpg *License:* Public domain *Contributors:* HubbleSite: gallery, release. *Original artist:* NASA, ESA, J. Hester and A. Loll (Arizona State University)
- **File:David_A._Aguilar'()s_Red_Dwarf_Stars.jpg** *Source:* https://upload.wikimedia.org/wikipedia/commons/7/72/David_A._Aguilar%27s_Red_Dwarf_Stars.jpg *License:* Public domain *Contributors:* http://www.nasa.gov/images/content/126852main_image_feature_401_ys_full.jpg from http://www.nasa.gov/multimedia/imagegallery/image_feature_401.html *Original artist:* David A. Aguilar (CfA)

- File:Delta-Vs_for_inner_Solar_System.svg *Source:* https://upload.wikimedia.org/wikipedia/commons/7/74/Delta-Vs_for_inner_Solar_.svg *License:* Public domain *Contributors:* Originally from en:Image:Deltavs.svg by en:User:Wolfkeeper *Original artist:* en:User:Wolfkeeper
- File:Drag_Curve_2.jpg *Source:* https://upload.wikimedia.org/wikipedia/commons/0/04/Drag_Curve_2.jpg *License:* CC-BY-SA-3.0 *Contributors:* Transferred from en.wikipedia; transferred to Commons by User:Sfan00_IMG using CommonsHelper. *Original artist:* Original uploader was GRAHAMUK at en.wikipedia
- File:Drag_sphere_nasa.svg *Source:* https://upload.wikimedia.org/wikipedia/commons/c/c6/Drag_sphere_nasa.svg *License:* Public domain *Contributors:* http://www.grc.nasa.gov/WWW/k-12/airplane/dragsphere.html *Original artist:* NASA
- File:Earth-moon.jpg *Source:* https://upload.wikimedia.org/wikipedia/commons/5/5c/Earth-moon.jpg *License:* Public domain *Contributors:* NASA [1] *Original artist:* Apollo 8 crewmember Bill Anders
- File:Earth_precession.svg *Source:* https://upload.wikimedia.org/wikipedia/commons/4/43/Earth_precession.svg *License:* Public domain *Contributors:* Vectorized by Mysid in Inkscape after a NASA Earth Observatory image in Milutin Milankovitch Precession. *Original artist:* NASA, Mysid
- File:Eccentric_and_true_anomaly.PNG *Source:* https://upload.wikimedia.org/wikipedia/commons/e/e1/Eccentric_and_true_anomaly.PNG *License:* CC BY-SA 3.0 *Contributors:* Own work *Original artist:* Brews ohare
- File:Eccentricity_rocky_planets.jpg *Source:* https://upload.wikimedia.org/wikipedia/commons/9/98/Eccentricity_rocky_planets.jpg *License:* GPL *Contributors:* Data generated with Gravity Simulator written by Tony Dunn.
  Source JPG on server *Original artist:* frankuitaalst from the Gravity Simulator message board.
- File:Edit-clear.svg *Source:* https://upload.wikimedia.org/wikipedia/en/f/f2/Edit-clear.svg *License:* Public domain *Contributors:* The *Tango! Desktop Project. Original artist:*

  The people from the Tango! project. And according to the meta-data in the file, specifically: "Andreas Nilsson, and Jakub Steiner (although minimally)."
- File:Ellipse_latus_rectum.svg *Source:* https://upload.wikimedia.org/wikipedia/commons/4/47/Ellipse_latus_rectum.svg *License:* Public domain *Contributors:* Own work *Original artist:* Krishnavedala
- File:Equinox_path.png *Source:* https://upload.wikimedia.org/wikipedia/commons/b/bd/Equinox_path.png *License:* CC-BY-SA-3.0 *Contributors:* Own work *Original artist:* Dbachmann
- File:Euler2a.gif *Source:* https://upload.wikimedia.org/wikipedia/commons/8/85/Euler2a.gif *License:* CC BY-SA 3.0 *Contributors:* This file was derived from Euler2.gif: <a href='//commons.wikimedia.org/wiki/File:Euler2.gif' class='image'><img alt='Euler2.gif' src='https://upload.wikimedia.org/wikipedia/commons/thumb/c/c5/Euler2.gif/50px-Euler2.gif' width='50' height='47' srcset='https://upload.wikimedia.org/wikipedia/commons/thumb/c/c5/Euler2.gif/75px-Euler2.gif 1.5x, https://upload.wikimedia.org/wikipedia/commons/thumb/c/c5/Euler2.gif/100px-Euler2.gif 2x' data-file-width='368' data-file-height='348' /></a>
  Original artist: Euler2.gif: Juansempere
- File:EulerG.png *Source:* https://upload.wikimedia.org/wikipedia/commons/7/73/EulerG.png *License:* Public domain *Contributors:* Own work *Original artist:* DF Malan
- File:EulerProjections.svg *Source:* https://upload.wikimedia.org/wikipedia/commons/e/e5/EulerProjections.svg *License:* CC BY 3.0 *Contributors:* I Juan Sempere created this work entirely by myself, based on elements with Creative Commons license. *Original artist:* Juan Sempere
- File:EulerProjections2.svg *Source:* https://upload.wikimedia.org/wikipedia/commons/9/9e/EulerProjections2.svg *License:* CC BY-SA 3.0 *Contributors:* Own work (Original caption: "*I created this work entirely by myself based on previous work under Creative Commons*") *Original artist:* Juansempere at en.wikipedia
- File:EulerX.png *Source:* https://upload.wikimedia.org/wikipedia/commons/3/38/EulerX.png *License:* Public domain *Contributors:* Own work *Original artist:* DF Malan
- File:Eulerangles.svg *Source:* https://upload.wikimedia.org/wikipedia/commons/a/a1/Eulerangles.svg *License:* CC BY 3.0 *Contributors:* Hand drawn in Inkscape by me *Original artist:* Lionel Brits
- File:Flow_foil.svg *Source:* https://upload.wikimedia.org/wikipedia/commons/1/14/Flow_foil.svg *License:* CC BY-SA 3.0 *Contributors:* Own work *Original artist:* BoH
- File:Flow_plate.svg *Source:* https://upload.wikimedia.org/wikipedia/commons/8/8d/Flow_plate.svg *License:* CC BY-SA 3.0 *Contributors:* Own work *Original artist:* BoH
- File:Flow_plate_perpendicular.svg *Source:* https://upload.wikimedia.org/wikipedia/commons/c/ce/Flow_plate_perpendicular.svg *License:* CC BY-SA 3.0 *Contributors:* Own work *Original artist:* BoH
- File:Flow_sphere.svg *Source:* https://upload.wikimedia.org/wikipedia/commons/c/c8/Flow_sphere.svg *License:* CC BY-SA 3.0 *Contributors:* Own work *Original artist:* BoH
- File:Folder_Hexagonal_Icon.svg *Source:* https://upload.wikimedia.org/wikipedia/en/4/48/Folder_Hexagonal_Icon.svg *License:* Cc-by-sa-3.0 *Contributors:* ? *Original artist:* ?
- File:Gemini_7_in_orbit_-_GPN-2006-000035.jpg *Source:* https://upload.wikimedia.org/wikipedia/commons/d/de/Gemini_7_in_orbit_-_GPN-2006-000035.jpg *License:* Public domain *Contributors:* Great Images in NASA Description *Original artist:* NASA
- File:Geostationary_orbit-animation.gif *Source:* https://upload.wikimedia.org/wikipedia/commons/0/02/Geostationary_orbit-animation.gif *License:* Public domain *Contributors:* Transferred from pl.wikipedia; transferred to Commons by User:Masur using CommonsHelper. *Original artist:* Original uploader was Pixel at pl.wikipedia
- File:Gimbaleuler.svg *Source:* https://upload.wikimedia.org/wikipedia/commons/7/72/Gimbaleuler.svg *License:* Public domain *Contributors:*

41.4. TEXT AND IMAGE SOURCES, CONTRIBUTORS, AND LICENSES

- Gimbaleuler.gif *Original artist:*
- derivative work: McSush (<a href='//commons.wikimedia.org/wiki/User_talk:McSush' title='User talk:McSush'>talk</a>)
- File:Gimbaleuler2.svg *Source:* https://upload.wikimedia.org/wikipedia/commons/3/34/Gimbaleuler2.svg *License:* CC BY-SA 3.0 *Contributors:*
- Transferred from en.wikipedia by Ronhjones *Original artist:* Juan Sempere / Juansempere at en.wikipedia
- File:Grav_slingshot_diag.svg *Source:* https://upload.wikimedia.org/wikipedia/commons/6/62/Grav_slingshot_diag.svg *License:* CC BY-SA 4.0 *Contributors:* Own work *Original artist:* Paulsmith99
- File:Gravitational_slingshot.svg *Source:* https://upload.wikimedia.org/wikipedia/commons/3/3a/Gravitational_slingshot.svg *License:* CC BY-SA 3.0 *Contributors:* Own work *Original artist:* Leafnode
- File:Gravity_Wells_Potential_Plus_Kinetic_Energy_-_Circle-Ellipse-Parabola-Hyperbola.png *Source:* https://upload.wikimedia.org/wikipedia/commons/9/94/Gravity_Wells_Potential_Plus_Kinetic_Energy_-_Circle-Ellipse-Parabola-Hyperbola.png *License:* CC0 *Contributors:* http://preview.tinyurl\protect\char"007B\relax\dot\protect\char"007D\relax com/Thesis-EnergyPotentialAnalysis - Figure 3.2 on pdf pg34of64. *Original artist:* Invent2HelpAll
- File:Gyroscope_operation.gif *Source:* https://upload.wikimedia.org/wikipedia/commons/d/d5/Gyroscope_operation.gif *License:* Public domain *Contributors:* ? *Original artist:* ?
- File:Gyroscope_precession.gif *Source:* https://upload.wikimedia.org/wikipedia/commons/8/82/Gyroscope_precession.gif *License:* Public domain *Contributors:* Own work *Original artist:* LucasVB
- File:Gyroscopic_precession_256x256.png *Source:* https://upload.wikimedia.org/wikipedia/commons/6/68/Gyroscopic_precession_256x256.png *License:* CC BY-SA 2.5 *Contributors:* Image created with the help of POV-ray ray-tracing software. *Original artist:* Cleonis at en.wikipedia
- File:He1523a.jpg *Source:* https://upload.wikimedia.org/wikipedia/commons/5/5f/He1523a.jpg *License:* CC BY 4.0 *Contributors:* http://www.solstation.com/x-objects/he1523.htm *Original artist:* ESO, European Southern Observatory
- File:Hohmann_transfer_orbit.svg *Source:* https://upload.wikimedia.org/wikipedia/commons/d/df/Hohmann_transfer_orbit.svg *License:* CC BY-SA 2.5 *Contributors:* Own work based on image by Hubert Bartkowiak *Original artist:* Leafnode
- File:Ilc_9yr_moll4096.png *Source:* https://upload.wikimedia.org/wikipedia/commons/3/3c/Ilc_9yr_moll4096.png *License:* Public domain *Contributors:* http://map.gsfc.nasa.gov/media/121238/ilc_9yr_moll4096.png *Original artist:* NASA / WMAP Science Team
- File:Impulsive_maneuver.svg *Source:* https://upload.wikimedia.org/wikipedia/commons/0/04/Impulsive_maneuver.svg *License:* GFDL *Contributors:* Own work *Original artist:* Stamcose
- File:Inclinedthrow.gif *Source:* https://upload.wikimedia.org/wikipedia/commons/6/63/Inclinedthrow.gif *License:* CC BY-SA 3.0 *Contributors:* Own work *Original artist:* AllenMcC.
- File:Induced_drag_r.svg *Source:* https://upload.wikimedia.org/wikipedia/commons/6/61/Induced_drag_r.svg *License:* CC-BY-SA-3.0 *Contributors:* Own work *Original artist:* BillC
- File:Intermediateframes.svg *Source:* https://upload.wikimedia.org/wikipedia/commons/4/4a/Intermediateframes.svg *License:* GFDL *Contributors:* Own work *Original artist:* Juansempere
- File:Iss_ground_track.jpg *Source:* https://upload.wikimedia.org/wikipedia/en/9/95/Iss_ground_track.jpg *License:* PD *Contributors:* ? *Original artist:* ?
- File:James-clerk-maxwell3.jpg *Source:* https://upload.wikimedia.org/wikipedia/commons/6/6f/James-clerk-maxwell3.jpg *License:* Public domain *Contributors:* ? *Original artist:* ?
- File:Kennedy-Thorndike_experiment_DE.svg *Source:* https://upload.wikimedia.org/wikipedia/commons/d/d4/Kennedy-Thorndike_expe DE.svg *License:* CC BY-SA 3.0 *Contributors:* This vector image was created with Inkscape. *Original artist:* User:Stigmatella aurantiaca
- File:Kepler-first-law.svg *Source:* https://upload.wikimedia.org/wikipedia/commons/1/1a/Kepler-first-law.svg *License:* CC-BY-SA-3.0 *Contributors:* The original PNG version: <a href='//commons.wikimedia.org/wiki/File:Kepler-first-law.png' class='image'><img alt='Kepler-first-law.png' src='https://upload.wikimedia.org/wikipedia/commons/thumb/8/87/Kepler-first-law.png/24px-Kepler-first-law.png' width='24' height='18'srcset='https://upload.wikimedia.org/wikipedia/commons/thumb/8/87/Kepler-first-law.png/36px-Kepler-first-law.png 1.5x, https://upload.wikimedia.org/wikipedia/commons/thumb/8/87/Kepler-first-law.png/48px-Kepler-first-law.png 2x' data-file-width='250' data-file-height='186'/></a>Kepler-first-law.png *Original artist:* Original by Arpad Horvath
- File:Kepler-second-law.gif *Source:* https://upload.wikimedia.org/wikipedia/commons/6/69/Kepler-second-law.gif *License:* CC BY-SA 3.0 *Contributors:* Gonfer *Original artist:* Gonfer (talk)
- File:Kepler_laws_diagram.svg *Source:* https://upload.wikimedia.org/wikipedia/commons/9/98/Kepler_laws_diagram.svg *License:* CC BY 2.5 *Contributors:* Own work *Original artist:* Hankwang
- File:Kepler_orbits.svg *Source:* https://upload.wikimedia.org/wikipedia/commons/b/b7/Kepler_orbits.svg *License:* GFDL *Contributors:* Own work *Original artist:* Stamcose
- File:Kipler'{}s_Error.jpg *Source:* https://upload.wikimedia.org/wikipedia/commons/1/13/Kipler%27s_Error.jpg *License:* CC BY-SA 3.0 *Contributors:* Own work *Original artist:* Rudrene
- File:L4_diagram.svg *Source:* https://upload.wikimedia.org/wikipedia/commons/7/78/L4_diagram.svg *License:* CC-BY-SA-3.0 *Contributors:* ? *Original artist:* ?
- File:Lagrange_points2.svg *Source:* https://upload.wikimedia.org/wikipedia/commons/e/ee/Lagrange_points2.svg *License:* CC BY 3.0 *Contributors:*

- Lagrange_points.jpg *Original artist:* Lagrange_points.jpg: created by
- File:Lagrange_points_simple.svg *Source:* https://upload.wikimedia.org/wikipedia/commons/a/a5/Lagrange_points_simple.svg *License:* BY 3.0 *Contributors:* File:Lagrange_points2.svg *Original artist:* Xander89
- File:Lagrangian_points_equipotential.jpg *Source:* https://upload.wikimedia.org/wikipedia/commons/5/5f/Lagrangian_points_equipote.jpg *License:* CC BY-SA 3.0 *Contributors:* Own work *Original artist:* User:cmglee
- File:Lightning.0257.jpg *Source:* https://upload.wikimedia.org/wikipedia/commons/3/33/Lightning.0257.jpg *License:* Public domain *Contributors:* Lightning pics *Original artist:* Timothy Kirkpatrick
- File:MAUD-MTEX-TiAl-hasylab-2003-Liss.png *Source:* https://upload.wikimedia.org/wikipedia/commons/0/0d/MAUD-MTEX-TiAl-.png *License:* CC BY-SA 4.0 *Contributors:* Own work *Original artist:* Klaus-Dieter Liss
- File:MAVENnMars.jpg *Source:* https://upload.wikimedia.org/wikipedia/en/7/7c/MAVENnMars.jpg *License:* PD *Contributors:* http://photojournal.jpl.nasa.gov/jpeg/PIA14761.jpg Parent: http://www.jpl.nasa.gov/spaceimages/details.php?id=PIA14761 *Original artist:* NASA
- File:Mdis_depart_anot.ogv *Source:* https://upload.wikimedia.org/wikipedia/commons/e/e0/Mdis_depart_anot.ogv *License:* Public domain *Contributors:* http://messenger.jhuapl.edu/gallery/sciencePhotos/image.php?page=30&gallery_id=2&image_id=159 *Original artist:* NASA / JHU/APL
- File:Michelson-Morley_experiment_(en).svg *Source:* https://upload.wikimedia.org/wikipedia/commons/0/06/Michelson-Morley_%28en%29.svg *License:* Public domain *Contributors:* Created by bdesham in Inkscape. *Original artist:* Benjamin D.Esham(bdesham)
- File:Molniya.jpg *Source:* https://upload.wikimedia.org/wikipedia/commons/b/bf/Molniya.jpg *License:* Public domain *Contributors:* en.wikipedia.org/wiki/Image:Molniya.jpg *Original artist:* Hartzell
- File:MontreGousset001.jpg *Source:* https://upload.wikimedia.org/wikipedia/commons/4/45/MontreGousset001.jpg *License:* CC-BY-3.0 *Contributors:* Self-published work by ZA *Original artist:* Isabelle Grosjean ZA
- File:N-body_problem_(3).gif *Source:* https://upload.wikimedia.org/wikipedia/commons/f/f9/N-body_problem_%283%29.gif *License:* lic domain *Contributors:* ? *Original artist:* ?
- File:Newton_Cannon.svg *Source:* https://upload.wikimedia.org/wikipedia/commons/7/73/Newton_Cannon.svg *License:* CC-BY-SA-3 *tributors:* Own work *Original artist:* user:Brian Brondel
- File:NewtonsLawOfUniversalGravitation.svg *Source:* https://upload.wikimedia.org/wikipedia/commons/0/0e/NewtonsLawOf.svg *License:* CC BY 3.0 *Contributors:* Self-made by User:Dna-Dennis *Original artist:* User:Dna-Dennis
- File:Newtons_proof_of_Keplers_second_law.gif *Source:* https://upload.wikimedia.org/wikipedia/commons/6/60/Newtons_proof_of_second_law.gif *License:* Public domain *Contributors:* Own work *Original artist:* Lucas V. Barbosa
- File:Northnode-symbol.png *Source:* https://upload.wikimedia.org/wikipedia/commons/b/b5/Northnode-symbol.png *License:* CC-BY-3.0 *Contributors:*
- Transferred from en.wikipedia to Commons by User:Vinhtantran using CommonsHelper. *Original artist:* Samuella at en.wikipedia
- File:Nuvola_apps_edu_mathematics_blue-p.svg *Source:* https://upload.wikimedia.org/wikipedia/commons/3/3e/Nuvola_apps_edu_math blue-p.svg *License:* GPL *Contributors:* Derivative work from Image:Nuvola apps edu mathematics.png and Image:Nuvola apps edu mathematics-p.svg *Original artist:* David Vignoni (original icon); Flamurai (SVG convertion); bayo (color)
- File:Nuvola_apps_kalzium.svg *Source:* https://upload.wikimedia.org/wikipedia/commons/8/8b/Nuvola_apps_kalzium.svg *License:* LGPL *Contributors:* Own work *Original artist:* David Vignoni, SVG version by Bobarino
- File:Optimal_Transfer_Orbit_using_Electric_Propulsion.png *Source:* https://upload.wikimedia.org/wikipedia/commons/b/b2/Optimal_Transfer_Orbit_using_Electric_Propulsion.png *License:* CC BY-SA 4.0 *Contributors:* Own work *Original artist:* Arnonspitzer
- File:Orbit1.svg *Source:* https://upload.wikimedia.org/wikipedia/commons/e/eb/Orbit1.svg *License:* CC-BY-SA-3.0 *Contributors:* Lasunncty (talk) *Original artist:* Lasunncty (talk).
- File:Orbit2.gif *Source:* https://upload.wikimedia.org/wikipedia/commons/f/f2/Orbit2.gif *License:* Public domain *Contributors:* Own work *Original artist:* User:Zhatt
- File:OrbitalEccentricityDemo.svg *Source:* https://upload.wikimedia.org/wikipedia/commons/8/89/OrbitalEccentricityDemo.svg *License:* CC-BY-SA-3.0 *Contributors:* ScottAlanHill 600×480 (24,403 bytes) (Examples of orbital trajectories with various eccentricities. Created by submitter.) English Wikipedia *Original artist:* ScottAlanHill
- File:Orbital_General_Transfer.svg *Source:* https://upload.wikimedia.org/wikipedia/commons/1/12/Orbital_General_Transfer.svg *License:* CC BY-SA 3.0 *Contributors:* Own work *Original artist:* MenteMagica
- File:Orbital_Hohmann_Transfer.svg *Source:* https://upload.wikimedia.org/wikipedia/commons/7/70/Orbital_Hohmann_Transfer.svg *License:* CC BY-SA 3.0 *Contributors:* Own work *Original artist:* MenteMagica
- File:Orbital_Planes.svg *Source:* https://upload.wikimedia.org/wikipedia/commons/2/23/Orbital_Planes.svg *License:* Public domain *Contributors:* Own work *Original artist:* T3rminatr 07:29, 17 February 2008 (UTC)
- File:Orbital_Two-Impulse_Transfer.svg *Source:* https://upload.wikimedia.org/wikipedia/commons/b/bf/Orbital_Two-Impulse_Transfer.svg *License:* CC BY-SA 3.0 *Contributors:* Own work *Original artist:* MenteMagica
- File:Orbital_inclination_from_momentum_vector.gif *Source:* https://upload.wikimedia.org/wikipedia/commons/2/28/Orbital_inclinati from_momentum_vector.gif *License:* Public domain *Contributors:* Own work *Original artist:* Theoprakt
- File:Orbital_motion.gif *Source:* https://upload.wikimedia.org/wikipedia/commons/4/4e/Orbital_motion.gif *License:* GFDL *Contributors:*

41.4 TEXT AND IMAGE SOURCES, CONTRIBUTORS, AND LICENSES

- Earth derived from this image (public domain) *Original artist:* Own work
- File:Orbital_state_vectors.png *Source:* https://upload.wikimedia.org/wikipedia/commons/2/24/Orbital_state_vectors.png *License:* CC-BY-SA-3.0 *Contributors:* Own work of 0.39 *Original artist:* 0.39 at en.wikipedia
- File:Orbitalaltitudes.jpg *Source:* https://upload.wikimedia.org/wikipedia/commons/8/82/Orbitalaltitudes.jpg *License:* GFDL *Contributors:* Own work *Original artist:* Rrakanishu
- File:People_icon.svg *Source:* https://upload.wikimedia.org/wikipedia/commons/3/37/People_icon.svg *License:* CC0 *Contributors:* OpenClipart *Original artist:* OpenClipart
- File:Pillars_of_Creation.jpeg *Source:* https://upload.wikimedia.org/wikipedia/commons/6/6a/Pillars_of_Creation.jpeg *License:* Public domain *Contributors:* http://www.sun.org/images/pillars-of-creation *Original artist:* NASA, Jeff Hester, and Paul Scowen (Arizona State University)
- File:Plane.svg *Source:* https://upload.wikimedia.org/wikipedia/commons/6/67/Plane.svg *License:* CC BY 3.0 *Contributors:* Own work. Transferred from en.wikipedia to Commons by User:WaldirusingCommonsHelper. *Original artist:* Original uploader was Juansempereat en.wikipe.
- File:Planet_orbit_nodes_2_animation.gif *Source:* https://upload.wikimedia.org/wikipedia/commons/b/b9/Planet_orbit_nodes_2_.gif *License:* CC BY-SA 3.0 *Contributors:* Own work, created with Google SketchUp v7.1 *Original artist:* Orion 8
- File:Planetary_Orbits.jpg *Source:* https://upload.wikimedia.org/wikipedia/commons/1/12/Planetary_Orbits.jpg *License:* CC BY-SA 3.0 *Contributors:* Own work *Original artist:* Kalebmcardle
- File:Portal-puzzle.svg *Source:* https://upload.wikimedia.org/wikipedia/en/f/fd/Portal-puzzle.svg *License:* Public domain *Contributors:* ? *Original artist:* ?
- File:Praezession.svg *Source:* https://upload.wikimedia.org/wikipedia/commons/b/bb/Praezession.svg *License:* CC-BY-SA-3.0 *Contributors:*

Original artist: User Herbye (German Wikipedia). Designed by Dr. H. Sulzer

- File:Precessing_Kepler_orbit_280frames_e0.6_smaller.gif *Source:* https://upload.wikimedia.org/wikipedia/commons/8/89/Precessing_Kepler_orbit_280frames_e0.6_smaller.gif *License:* CC BY 3.0 *Contributors:* Own work *Original artist:* WillowW
- File:PrecessionOfATop.svg *Source:* https://upload.wikimedia.org/wikipedia/commons/2/2b/PrecessionOfATop.svg *License:* CC BY-SA 2.5 *Contributors:* Drawn in Inkscape. *Original artist:* Xavier Snelgrove
- File:Precession_N.gif *Source:* https://upload.wikimedia.org/wikipedia/commons/1/16/Precession_N.gif *License:* CC BY-SA 2.5 *Contributors:* self, 4 bit GIF *Original artist:* Tau'olunga
- File:Proper_osculating_element_difference.png *Source:* https://upload.wikimedia.org/wikipedia/en/f/f5/Proper_osculating_element_.png *License:* Cc-by-sa-3.0 *Contributors:* ? *Original artist:* ?
- File:Qualitive_variation_of_cd_with_mach_number.png *Source:* https://upload.wikimedia.org/wikipedia/commons/0/0e/Qualitive_of_cd_with_mach_number.png *License:* Public domain *Contributors:* Own work *Original artist:* Wolfkeeper
- File:Question_book-new.svg *Source:* https://upload.wikimedia.org/wikipedia/en/9/99/Question_book-new.svg *License:* Cc-by-sa-3.0 *Contributors:*

Created from scratch in Adobe Illustrator. Based on Image:Question book.png created by User:Equazcion *Original artist:* Tkgd2007

- File:Restricted_3-Body_1.jpg *Source:* https://upload.wikimedia.org/wikipedia/commons/9/97/Restricted_3-Body_1.jpg *License:* CC SA 3.0 *Contributors:* Own work *Original artist:* Rudrene
- File:RocketSunIcon.svg *Source:* https://upload.wikimedia.org/wikipedia/commons/d/d6/RocketSunIcon.svg *License:* Copyrighted free *Contributors:* Self made, based on File:Spaceship and the Sun.jpg *Original artist:* Me
- File:SS-faraday.jpg *Source:* https://upload.wikimedia.org/wikipedia/commons/1/12/SS-faraday.jpg *License:* Public domain *Contr* Arthur Shuster & Arthur E. *Original artist:* Based on a painting by A. Blakely.
- File:STS-130_Endeavour_flyaround_5.jpg *Source:* https://upload.wikimedia.org/wikipedia/commons/5/5b/STS-130_Endeavour_flyaround_5.jpg *License:* Public domain *Contributors:* http://spaceflight.nasa.gov/gallery/images/shuttle/sts-130/html/s130e012142.html *Original artist:* NASA
- File:Sail-Force1.gif *Source:* https://upload.wikimedia.org/wikipedia/commons/8/8f/Sail-Force1.gif *License:* CC BY-SA 3.0 *Cont* Own work *Original artist:* Jerry Wright
- File:Southnode-symbol.png *Source:* https://upload.wikimedia.org/wikipedia/commons/e/e2/Southnode-symbol.png *License:* CC-3.0 *Contributors:*
- Transferred from en.wikipedia to Commons by User:Vinhtantran using CommonsHelper. *Original artist:* Samuella at en.wikipedia
- File:Soyuz_TMA-7_spacecraft_2edit1.jpg *Source:* https://upload.wikimedia.org/wikipedia/commons/b/bc/Soyuz_TMA-7_spacecraft_2edit1.jpg *License:* Public domain *Contributors:* Transferred from en.wikipedia to Commons. *Original artist:* The original uploader was Thegreenj at English Wikipedia
- File:Spacetime_curvature.png *Source:* https://upload.wikimedia.org/wikipedia/commons/2/22/Spacetime_curvature.png *License:* SA-3.0 *Contributors:* ? *Original artist:* ?
- File:Speed_vs_time_for_objects_with_drag.png *Source:* https://upload.wikimedia.org/wikipedia/en/3/3a/Speed_vs_time_for_drag.png *License:* CC-BY-SA-3.0 *Contributors:*

I used Google to graph the equations, then edited in MS Paint

Original artist:

Loodog

- File:Stylised_Lithium_Atom.svg *Source:* https://upload.wikimedia.org/wikipedia/commons/e/e1/Stylised_Lithium_Atom.svg *License:* CC-BY-SA-3.0 *Contributors:* based off of Image:Stylised Lithium Atom.png by Halfdan. *Original artist:* SVG by Indolences. Recoloring and ironing out some glitches done by Rainer Klute.
- File:Taitbrianangles.svg *Source:* https://upload.wikimedia.org/wikipedia/commons/e/ea/Taitbrianangles.svg *License:* CC BY 3.0 *Contributors:*
- Eulerangles.svg *Original artist:* Eulerangles.svg: Lionel Brits
- File:Taitbrianzyx.svg *Source:* https://upload.wikimedia.org/wikipedia/commons/5/53/Taitbrianzyx.svg *License:* CC BY 3.0 *Contributors:* Own work *Original artist:* Juansempere
- File:Text_document_with_red_question_mark.svg *Source:* https://upload.wikimedia.org/wikipedia/commons/a/a4/Text_document_with_red_question_mark.svg *License:* Public domain *Contributors:* Created by bdesham with Inkscape; based upon Text-x-generic.svg from the Tango project. *Original artist:* Benjamin D. Esham (bdesham)
- File:Tle_first_row.jpg *Source:* https://upload.wikimedia.org/wikipedia/commons/4/4b/Tle_first_row.jpg *License:* CC BY-SA 4.0 *Contributors:* Own work *Original artist:* Bvv2306
- File:Tle_second_row.jpg *Source:* https://upload.wikimedia.org/wikipedia/commons/4/43/Tle_second_row.jpg *License:* CC BY-SA 4.0 *Contributors:* Own work *Original artist:* Bvv2306
- File:Tle_title.jpg *Source:* https://upload.wikimedia.org/wikipedia/commons/f/fe/Tle_title.jpg *License:* CC BY-SA 4.0 *Contributors:* Own work *Original artist:* Bvv2306
- File:Total_energy_during_Hohmann_transfer.png *Source:* https://upload.wikimedia.org/wikipedia/commons/5/59/Total_energy_during_Hohmann_transfer.png *License:* CC BY-SA 3.0 *Contributors:* Own work *Original artist:* Qniemiec
- File:Tsiolkovsky_rocket_equation.svg *Source:* https://upload.wikimedia.org/wikipedia/commons/f/f6/Tsiolkovsky_rocket_equation.svg *License:* CC0 *Contributors:* Own work *Original artist:* Krishnavedala
- File:Var_mass_system.PNG *Source:* https://upload.wikimedia.org/wikipedia/commons/f/fc/Var_mass_system.PNG *License:* Public domain *Contributors:* Transferred from en.wikipedia to Commons by Logan using CommonsHelper. *Original artist:* Skorkmaz at English Wikipedia
- File:Voyager.jpg *Source:* https://upload.wikimedia.org/wikipedia/commons/d/d2/Voyager.jpg *License:* Public domain *Contributors:* NASA website *Original artist:* NASA
- File:Voyager_2_path.svg *Source:* https://upload.wikimedia.org/wikipedia/commons/d/d2/Voyager_2_path.svg *License:* Public domain *Contributors:* This file was derived from Voyager 2 path.png: <a href='//commons.wikimedia.org/wiki/File:Voyager_2_path.png' class='image'><img alt='Voyager 2 path.png' src='https://upload.wikimedia.org/wikipedia/commons/thumb/7/78/Voyager_2_path.png/50px-Voyager_2_path.png' width='50' height='42' srcset='https://upload.wikimedia.org/wikipedia/commons/thumb/7/78/Voyager_2_path.png/75px-Voyager_2_path.png 1.5x, https://upload.wikimedia.org/wikipedia/commons/thumb/7/78/Voyager_2_path.png/100px-Voyager_2_path.png 2x' data-file-width='805' data-file-height='675' /></a>
Original artist: Voyager_2_path.png: *Voyager_Path.jpg: created by
- File:Voyager_2_velocity_vs_distance_from_sun.svg *Source:* https://upload.wikimedia.org/wikipedia/commons/2/2c/Voyager_2_velocity_vs_distance_from_sun.svg *License:* CC BY-SA 3.0 *Contributors:* Own work *Original artist:* Cmglee
- File:Voyager_Path.svg *Source:* https://upload.wikimedia.org/wikipedia/commons/5/53/Voyager_Path.svg *License:* Public domain *Contributors:*
- Original from http://solarsystem.nasa.gov/multimedia/display.cfm?IM_ID=2143 *Original artist:* Voyager_Path.jpg: created by
- File:Wiki_letter_w.svg *Source:* https://upload.wikimedia.org/wikipedia/en/6/6c/Wiki_letter_w.svg *License:* Cc-by-sa-3.0 *Contributors:* ? *Original artist:* ?
- File:Wiki_letter_w_cropped.svg *Source:* https://upload.wikimedia.org/wikipedia/commons/1/1c/Wiki_letter_w_cropped.svg *License:* CC-BY-SA-3.0 *Contributors:*
- Wiki_letter_w.svg *Original artist:* Wiki_letter_w.svg: Jarkko Piiroinen
- File:Wikibooks-logo-en-noslogan.svg *Source:* https://upload.wikimedia.org/wikipedia/commons/d/df/Wikibooks-logo-en-noslogan.svg *License:* CC BY-SA 3.0 *Contributors:* Own work *Original artist:* User:Bastique, User:Ramac et al.
- File:Wikiquote-logo.svg *Source:* https://upload.wikimedia.org/wikipedia/commons/f/fa/Wikiquote-logo.svg *License:* Public domain *Contributors:* ? *Original artist:* ?
- File:Wikisource-logo.svg *Source:* https://upload.wikimedia.org/wikipedia/commons/4/4c/Wikisource-logo.svg *License:* CC BY-SA 3.0 *Contributors:* Rei-artur *Original artist:* Nicholas Moreau
- File:Wikiversity-logo.svg *Source:* https://upload.wikimedia.org/wikipedia/commons/9/91/Wikiversity-logo.svg *License:* CC BY-SA 3.0 *Contributors:* Snorky (optimized and cleaned up by verdy_p) *Original artist:* Snorky (optimized and cleaned up by verdy_p)
- File:Wiktionary-logo-en.svg *Source:* https://upload.wikimedia.org/wikipedia/commons/f/f8/Wiktionary-logo-en.svg *License:* Public domain *Contributors:* Vector version of Image:Wiktionary-logo-en.png. *Original artist:* Vectorized by Fvasconcellos (talk · contribs), based on original logo tossed together by Brion Vibber
- File:Wooden_hourglass_3.jpg *Source:* https://upload.wikimedia.org/wikipedia/commons/7/70/Wooden_hourglass_3.jpg *License:* CC-BY-SA-3.0 *Contributors:* Own work *Original artist:* User:S Sepp
- File:Yaw_Axis_Corrected.svg *Source:* https://upload.wikimedia.org/wikipedia/commons/c/c1/Yaw_Axis_Corrected.svg *License:* CC BY-SA 3.0 *Contributors:*
- Yaw_Axis.svg *Original artist:* Yaw_Axis.svg: Auawise
- File:Orsted.jpg *Source:* https://upload.wikimedia.org/wikipedia/commons/7/79/%C3%98rsted.jpg *License:* Public domain *Contributors:* *Original artist:* Christoffer Wilhelm Eckersberg

41.4.3 Content license

- Creative Commons Attribution-Share Alike 3.0

www.ingramcontent.com/pod-product-compliance
Lightning Source LLC
Chambersburg PA
CBHW080650190526
45169CB00006B/2054